Electrochemistry in Ionic Liquids

Angel A.J. Torriero
Editor

Electrochemistry in Ionic Liquids

Volume 2: Applications

Springer

Editor
Angel A.J. Torriero
Centre for Chemistry and Biotechnology
School of Life and Environmental Sciences
Faculty of Science, Engineering and Built Environment
Deakin University
Melbourne Burwood Campus
Burwood, VIC, Australia

ISBN 978-3-319-15131-1 ISBN 978-3-319-15132-8 (eBook)
DOI 10.1007/978-3-319-15132-8

Library of Congress Control Number: 2015937618

Springer Cham Heidelberg New York Dordrecht London

Printed on acid-free paper

Springer International Publishing AG Switzerland is part of Springer Science+Business Media (www.springer.com)

Preface for First Volume

Electrochemistry is dominated by the study of species dissolved in solution. The use of a solvent as the reaction medium helps electrochemists to control important reaction conditions such as pH, rate of mass transfer, concentration of reactant, solubility, solvation, etc. Water and organic solvents are the most popular media. However, by using appropriate ionic liquids, reactants and products that are unstable in those media remain stable, and redox reactions that are impossible in water and organic solvents become possible. The reaction environments are markedly wider in some ionic liquids than in other solvent systems. In spite of this, some fundamental electrochemical concepts generally used in conventional solvent systems are not always valid in ionic liquids.

The aim of this book is to present the insights of experts on emerging experimental techniques and theoretical concepts that are or will be in the vanguard of the field of electrochemistry in ionic liquids. The two volumes of this book provides the reader with a broad and self-contained account of electrochemical techniques available to work in ionic liquids. It also gathers and critically discusses the important properties of protic ionic liquids, deep-eutectic solvents, task-specific ionic liquids, polymeric ion gels, and lithium-ion solvation, relevant to electrochemistry.

The two volumes of this book will be of interest for researchers in relevant industries as well as for academics, for scientists, and for the general chemical and electrochemical audience. It also can be given to graduate students and postdoctoral fellows looking for guidance to start working in this research field.

Burwood, VIC, Australia Angel A.J. Torriero

Preface for Second Volume

The field of electrochemistry in ionic liquids has blossomed in the last decade, since some of these ionic liquids tend to be good electrochemical solvents, exhibit interesting conductive properties, and offer the unique possibility to tune the physicochemical properties. These exclusive properties make them very useful as both the electrolyte and the solvent in electrochemical applications. There are several comprehensive reviews of this field, but none of them have covered the electrochemical applications of ionic liquids to the widest extent. Thanks to the contribution of scientists from different countries with strong expertise in the field, this new book collates the formerly dispersed knowledge into two handy volumes. Readers will find ten chapters gathered in three sections. The book commences with the latest understanding of electrodeposition, and electroless deposition, followed by a critical discussion of voltammetry of adhered microparticles, electrochemistry of organic and organometallic compounds, electrocatalytic reactions, oxygen reduction reaction, and ionic liquids in surface protection and lubrication. The book finishes with an overview on the current industrial application of ionic liquids and the challenges, issues, and recycling methods used in industrial developments.

The two volumes of this book will be of interest to researchers in relevant industries as well as academics, scientists, and anyone interested in general chemistry and electrochemistry. The book can also be used as a guide for graduate students and postdoctoral fellows who wish to start working in this field.

Burwood, VIC, Australia Angel A.J. Torriero

Contents

Volume I

Volume II

Contributors

Volume I

Mérièm Anouti Laboratoire PCM2E (EA 6296), Université François Rabelais de Tours, Parc de Grandmont, Tours, France

Cameron L. Bentley School of Chemistry, Monash University, Clayton South, VIC, Australia

Energy Flagship, Commonwealth Scientific and Industrial Research Organisation (CSIRO), Clayton South, VIC, Australia

Alan M. Bond School of Chemistry, Monash University, Clayton South, VIC, Australia

Hiroyuki Doi Graduate School of Science and Technology, Niigata University, Nishi-ku, Niigata City, Japan

Kenta Fujii Graduate School of Science and Engineering, Yamaguchi University, Ube, Yamaguchi, Japan

Joshua A. Hammons X-ray Science Division, Advanced Photon Source, Argonne National Laboratory, Argonne, IL, USA

Anthony F. Hollenkamp Energy Flagship, Commonwealth Scientific and Industrial Research Organisation (CSIRO), Clayton South, VIC, Australia

Jan Ilavsky X-ray Science Division, Advanced Photon Source, Argonne National Laboratory, Argonne, IL, USA

Peter J. Mahon Faculty of Science, Engineering and Technology, Swinburne University of Technology, Hawthorn, VIC, Australia

David Mecerreyes POLYMAT, University of the Basque Country UPV/EHU, San Sebastian, Spain

Abdul Rehman Department of Chemistry, Oakland University, Rochester, MI, USA

Laurent Rubatat EPCP, UMR IPREM 5256, Université de Pau et des Pays de l'Adour, Pau, France

Maitane Salsamendi POLYMAT, University of the Basque Country UPV/EHU, San Sebastian, Spain

Shiro Seki Materials Science Research Laboratory, Central Research Institute of Electric Power Industry, Komae, Tokyo, Japan

Angel A.J. Torriero Centre for Chemistry and Biotechnology, School of Life and Environmental Sciences, Faculty of Science, Engineering and Built Environment, Deakin University, Melbourne Burwood Campus, Burwood, VIC, Australia

Yasuhiro Umebayashi Graduate School of Science and Technology, Niigata University, Nishi-ku, Niigata City, Japan

Darren A. Walsh School of Chemistry, University of Nottingham, Nottingham, UK

Zhe Wang Department of Chemistry, Oakland University, Rochester, MI, USA

Xiangqun Zeng Department of Chemistry, Oakland University, Rochester, MI, USA

Fan Zhang Materials Measurement Science Division, Material Measurement Laboratory, National Institute of Standards and Technology, Gaithersburg, MD, USA

Jie Zhang School of Chemistry, Monash University, Clayton South, VIC, Australia

Hua Zhao Department of Chemistry and Forensic Science, Savannah State University, Savannah, GA, USA

Volume II

Joaquín Arias-Pardilla Grupo de Ciencia de Materiales e Ingeniería Metalúrgica, Departamento de Ingeniería de Materiales y Fabricación, Universidad Politécnica de Cartagena, Cartagena, Spain

Cameron L. Bentley School of Chemistry, Monash University, Clayton, VIC, Australia

María Dolores Bermúdez Grupo de Ciencia de Materiales e Ingeniería Metalúrgica, Departamento de Ingeniería de Materiales y Fabricación, Universidad Politécnica de Cartagena, Campus de la Muralla del Mar., Cartagena, Spain

Alan M. Bond School of Chemistry, Monash University, Clayton, VIC, Australia

Natalia Borisenko Institute of Electrochemistry, Clausthal University of Technology, Arnold-Sommerfeld-Strasse, Clausthal-Zellerfeld, Germany

Andinet Ejigu School of Chemistry, The University of Nottingham, Nottingham, UK

Tulia Espinosa Grupo de Ciencia de Materiales e Ingeniería Metalúrgica, Departamento de Ingeniería de Materiales y Fabricación, Universidad Politécnica de Cartagena, Campus de la Muralla del Mar., Cartagena, Spain

Michael D. Horne Mineral Resources Flagship, CSIRO, Clayton South, VIC, Australia

Yasushi Katayama Department of Applied Chemistry, Faculty of Science and Technology, Keio University, Kanagawa, Japan

Douglas MacFarlane School of Chemistry, Monash University, Clayton, VIC, Australia

Takeo Ohsaka Department of Electronic Chemistry, Interdisciplinary Graduate School of Science and Engineering, Tokyo Institute of Technology, Yokohama, Japan

Cristina Pozo-Gonzalo ARC Centre of Excellence for Electromaterials Science, IFM-Institute for Frontier Materials, Deakin University, Burwood, VIC, Australia

Amal I. Siriwardana TECNALIA, Parque Tecnológico de San Sebastián, Mikeletegi Pasealekua, Donostia-San Sebastián, Gipuzkoa, Spain

Angel A.J. Torriero Centre for Chemistry and Biotechnology, School of Life and Environmental Sciences, Faculty of Science, Engineering and Built Environment, Deakin University, Melbourne Burwood Campus, Burwood, VIC, Australia

Darren A. Walsh School of Chemistry, University of Nottingham, Nottingham, UK

Da Zhang School of Environmental and Materials Engineering, Shanghai Second Polytechnic University, Shanghai, China

Jie Zhang School of Chemistry, Monash University, Clayton, VIC, Australia

Shu-Feng Zhao School of Chemistry, Monash University, Clayton, VIC, Australia

About the Editor

Angel A.J. Torriero is a Lecturer of Chemistry and Electrochemistry at Deakin University, Melbourne, Australia. He has published more than 50-refereed papers (h-index = 18; Scopus, November 2014), six book chapters, several patents, and one book, *Electrochemical Properties and Applications of Ionic Liquids* in 2011.

Dr. Torriero has a broad interest in both fundamental and applied electrochemistry and has made significant contributions in a number of fields, including analytical electrochemistry, biosensor, bioelectrochemistry, organic and organometallic electrochemistry, and most recently internal reference systems for ionic liquids.

Chapter 11
Introduction

Angel A.J. Torriero

Electrochemistry is dominated by the study of species dissolved in solution. In principle, any liquid may be used as a solvent, being water and organic solvents such as dichloromethane and acetonitrile, for instance, commonly employed in this role. Ionic liquids (ILs), which consist of combinations of cations and anions, may provide alternative solvents, but they are far from new. In many cases, they are the typical oils that have surrounded synthetic chemists for centuries when the required product is an organic salt. Nonetheless, in the last 15 years or so they have attracted significant attention, and at the present are being sought as highly valuable solvents.

Although there is no agreement upon definition, an IL may be conveniently described as a compound composed entirely of ions that is a liquid at temperature less than 100 °C. However, this is an arbitrary definition employed to differentiate ILs from classically well-known molten salts, which generally are thought to refer to high-melting and very corrosive solvent media such as molten KCl or NaCl. The arbitrary line drawn between molten salts and ILs at a melting temperature of 100 °C has been justified by the sudden widening range of applications for liquid salts below this temperature [1].

It is a common practice from electrochemistry advocates of ionic liquid research to claim that ILs are (1) nonvolatile, (2) nonflammable, (3) comprised of simple ionic species,(4) highly thermally stable, (5) highly electrochemically stable, and sometimes (6) intrinsically green. However, in the last few years, partly as a result of maturity in the field derived from an expanded knowledge of their physical and chemical properties and also as a consequence of the rapidly widening range of ILs available, it has become clear that essentially none of these generic properties are ubiquitous to ILs. It is important to ask then: which are the ubiquitous properties of

A.A.J. Torriero (✉)
Centre for Chemistry and Biotechnology, School of Life and Environmental Sciences, Faculty of Science, Engineering and Built Environment, Deakin University, Melbourne Burwood Campus, 221 Burwood Highway, Burwood, VIC 3125, Australia
e-mail: angel.torriero@deakin.edu.au

A.A.J. Torriero (ed.), *Electrochemistry in Ionic Liquids*,
DOI 10.1007/978-3-319-15132-8_11

ionic liquids? MacFarlane has responded to this question by saying [2]: "If we accept the definition of an ionic liquid as being an ionic compound (a salt) which is liquid below 100 °C, then in fact the only defining properties that one can expect to observe, a priori, are that, at some temperature below 100 °C:

(a) The substance is liquid (its glass transition temperature and/or melting point is below 100 °C).
(b) It contains ions and therefore exhibits some level of ionic conductivity.

The glass transition aspect of this definition allows the exclusion of substances that are so viscous as to be glassy at the temperature of interest, so it simply adds a level of clarification to the word *liquid*." Consequently, some level of ionic conductivity is the only property that we can *always* expect to be generic in ionic liquids.

Over the past decade, we have witnessed the application of ILs to a large variety of fields such as electrodeposition, energy storage, sensor development, synthesis, catalysis, surface protection, etc. Following progress in fundamental and applied research in ILs could require extensive multidisciplinary efforts for meeting emerging challenges in these areas. This book seeks to cover a broad perspective of applied electrochemical research in ILs and its rapid expanding areas. Chapter 12 discusses recent progress in the electrodeposition of semiconductors from ionic liquids. Furthermore, Chap. 13 provides details for the electroless deposition of metal ions in ILs as a consequence of disproportionation reactions. The techniques used for this purpose as well as the factors affecting the reaction are discussed in detail.

Chapter 14 presents the attributes of voltammetry of adhered microparticles in contact with ILs for determining the thermodynamics and kinetic properties of electroactive species. This technique provides an alternative and powerful electroanalytical tool, which has great versatility in ILs under conditions where conventional voltammetry with dissolved analyte is impractical.

The application of ILs as solvent systems for the electrochemical study and electrosynthesis of organic molecules is discussed in Chap. 15, highlighting some particular examples, with the aim of demonstrating any similarities and differences observed in ILs respect to conventional aqueous or organic solvent systems. Chapter 14 also discusses some electrochemical reactions of organic compounds in ILs. Meanwhile, the electrode kinetics of organometallic complexes of some transition metal ions with tris(2,2′-bipyridine) have been presented and discussed in detail in Chap. 16.

Reactions which are sensitive to the electrode material, such as the oxygen reduction reaction, hydrogen oxidation reaction, and hydrogen evolution reaction, have been studied for decades in conventional solvents. An overview of these electrocatalytic reactions in ILs is presented in Chap. 17. The insights provided in this chapter could impact not only on the fundamental understanding of electrocatalysis, but also on the next-generation energy conversion technologies. The previously mentioned oxygen reduction reaction is probably one of the best studied electrochemical reactions in conventional solvent systems. However, the knowledge of the oxygen reduction mechanism in ILs is still scarce. Chapter 18 provides a comprehensive discussion of the information available for this reaction in ionic liquids, which complement those already mentioned in Chap. 17.

The protection of metals and alloys surfaces is highly important for our everyday life. Chapter 19 discusses the state-of-the-art in the use of ILs for corrosion protection coatings formation, and lubrication. The physicochemical interaction of ILs with the different material surfaces is discussed in detail.

Although the study of different electrochemical processes in ILs or at the metal/ionic liquid interface is currently one of the most active areas of research, the industrial application of ILs or processes involving ILs is still limited. Chapter 20 provides the first comprehensive survey on the pilot plants and industrial processes of ionic liquids. Moreover, challenges, issues, and recycling methods of ILs in industrial developments are also discussed.

This book aims to provide the reader with an extensive perspective on trends in the applied field of ionic liquid electrochemistry research. The broad spectrum of currently active topics part of this book provides a good basis for the continuing development of each of these areas.

References

1. Torriero AAJ, Muhammad S (eds) (2011) Electrochemical properties and application of ionic liquids. Nova Science, New York
2. MacFarlane DR, Seddon KR (2007) Ionic liquids—progress on the fundamental issues. Aust J Chem 60(1):3–5

Part III
Electrodeposition in Ionic Liquids

Chapter 12
Electrodeposition of Semiconductors in Ionic Liquids

Natalia Borisenko

12.1 Introduction

Today semiconductor technology is present in everyday life. Many modern day electronic devices (e.g., calculators, mobile phones, mp3 players, netbooks, laptops, personal and tablet computers, light detectors, solar cells, radars, aircraft guidance-control assemblies, etc.) would not exist without semiconductors. Since the beginning of electronics industry, the commercial-scale manufacturing of semiconductors to provide smaller, faster, and cheaper electronic devices has attracted a growing research interest. The production of semiconductor materials with the required properties is technologically complicated and usually involves four basic processing steps: deposition of a required material onto a wafer, removal of unwanted materials, patterning or lithography to create the desired connection, and modification of electrical properties. The ongoing development of the manufacturing process is a main reason that the price of electronic devices continuously decreases. Currently only five deposition processes are available: Physical Vapor Deposition (PVD), Molecular Beam Epitaxy (MBE), Chemical Vapor Deposition (CVD), Atomic Layer Deposition (ALD), and electrochemical deposition. In industrial processes, CVP or PVD methods are preferred to produce semiconductor thin films. In fundamental studies MBE under ultrahigh vacuum (UHV) conditions has been applied. PVD includes several vacuum deposition methods where the desired material is deposited by the condensation of a vaporized form onto substrate (wafer), while in CVP the volatile precursors react on the substrate surface to produce the desired deposit. In MBE the solid source materials are heated in UHV and form a molecular or atomic beam. The atoms of the beam interact on a surface of a heated crystalline substrate to form a desired material.

N. Borisenko (✉)
Institute of Electrochemistry, Clausthal University of Technology,
Arnold-Sommerfeld-Strasse 6, Clausthal-Zellerfeld 38678, Germany
e-mail: natalia.borissenko@tu-clausthal.de

A.A.J. Torriero (ed.), *Electrochemistry in Ionic Liquids*,
DOI 10.1007/978-3-319-15132-8_12

The growth process takes place close to thermodynamic equilibrium and can be well controlled. MBE is usually applied to produce single crystals. The deposits obtained by these methods are of a high quality, but such processes are cost-intensive and need vacuum conditions for preparation thus making the semiconductors quite expensive. An alternative process to the above mentioned techniques is electrochemical deposition, which allows an easy control of the process parameters and which is comparably cheap. In general, the electrochemically deposited films do not possess crystalline perfection. However the structure and the size of the deposit can be adjusted by variation of the parameters (e.g., composition of the electrolyte, electrode potential, current density, and temperature) and thus the semiconductors with the required properties can be fabricated.

Over 30 years the electrodeposition of semiconductors has been investigated in different solvents such as aqueous solutions, organic media and molten salts and a good overview can be found in refs. [1–6]. Some semiconductors (e.g., Si, Ge, GaAs) can be well electrodeposited from high-temperature molten salts but the operating conditions (>800 °C) make these electrolytes at least demanding. Attempts have been made to deposit silicon and germanium from organic solvents, but they are usually volatile that limits the temperature window. Furthermore, the obtained deposits were usually amorphous. Aqueous solutions have been applied to electrodeposit binary compound semiconductors (e.g., InSb, InAs, GaAs, CdTe, CdS, etc.). Furthermore, besides classical deposition, Electrochemical Atomic Layer Epitaxy (E-ALE) and Electrochemical Atomic Layer Deposition (E-ALD) methods have been applied to deposit various semiconductors in aqueous solutions [7–11]. In general, the electrodeposition of semiconductors from aqueous electrolytes is strongly disturbed by hydrogen evolution that limits the electrochemical window.

In direct comparison, Ionic Liquids (ILs) have much wider electrochemical windows (up to 6 V) and extremely low-vapor pressures, which allows deposition at temperatures above 100 °C. Furthermore, water can be easily removed from them ensuring water-free conditions. The technological and scientific importance of ILs is that the morphology and crystal size of deposits can be modified by varying cation/anion's chemical structure [12–15]. Due to a huge number of possible cations and anions a large number of possible ILs with different properties can be generated.

In the past several decades ionic liquids have attracted a fast-growing research interest as potential electrolytes for electrodeposition of semiconductor materials [16–19]. Processes that are impossible in common aqueous and organic solutions become viable if ILs are used. However, although the possibility of electrodeposition from ILs has been demonstrated for a variety of semiconductor materials, there are some fundamental issues that still have to be clarified. In this chapter attention to semiconductors deposition shall be attracted as such studies are still rare in ionic liquids.

12.2 II–VI Compound Semiconductors

In the past several decades wide band gap II–VI semiconductors have attracted a significant research interest as promising candidates for high-performance opto-electronic devices such as light-emitting diodes (LEDs) and laser diodes operating in the blue or ultraviolet spectral range. To date ionic liquids have been employed as a solvent media for electrodeposition of CdTe, CdS, ZnTe, ZnS, CdSe, and ZnO II–VI semiconductors.

12.2.1 CdTe

Cadmium telluride (CdTe) is one of the most extensively deposited compound semiconductors. This material exhibits a direct band gap of 1.44 eV at room temperature and can be both n- and p-semiconducting. This is one of the most promising photovoltaic materials for thin-film solar cells. High-quality CdTe-based photovoltaics are electrodeposited commercially. The electrodeposition of CdTe was investigated by Hsiu et al. [20] in the Lewis basic 1-ethyl-3-methylimidazolium chloride/tetrafluoroborate ($[C_2mim][Cl]/[BF_4]$) containing $CdCl_2$ and $TeCl_4$ and by Chauhan et al. [21] in 1-butyl-3-methylimidazolium chloride ($[C_4mim][Cl]$) containing $CdCl_2$ and TeO_2 at elevated temperatures. In both cases CdTe films were obtained by the electrochemical reduction of Te(IV) to Te followed by the underpotential deposition (UPD) of Cd on the deposited Te. Furthermore, an excess amount of Cd(II) is required to make CdTe deposits with a Cd:Te composition close to 1:1. It was shown that the semiconductor thin film can be deposited electrochemically on tungsten and titanium substrates in $[C_2mim][Cl]/[BF_4]$ ionic liquid at temperatures above 120 °C [20] and on fluorine-doped tin oxide (FTO) coated glass substrate in $[C_4mim][Cl]$ ionic liquid at 80 °C [21]. The crystallinity of the deposits improves by increasing the deposition temperature, which—again—demonstrates the high potential of the wide thermal windows of ionic liquids for compound semiconductor electrodeposition. The band gap of the CdTe film electrodeposited in $[C_4mim][Cl]$ was determined to be 1.44 eV by UV–Vis absorption spectroscopy [21], which is in good agreement with the reported bulk CdTe band gap value. Recently, choline chloride/urea (ChCl/urea) ionic liquid was applied as a solvent media for the electrodeposition of CdTe [22]. However, a characterization of the obtained deposits was not performed.

12.2.2 ZnTe

Zinc telluride (ZnTe) has a direct band gap of 2.25 eV at room temperature. It is usually applied in switching devices and in solar cells. The electrodeposition of ZnTe in ionic liquids was investigated by Lin et al. [23]. ZnTe was obtained on Ni in the Lewis basic $ZnCl_2$/1-ethyl-3-methylimidazolium chloride (40/60 mol %) ionic liquid containing propylene carbonate as a cosolvent at 40 °C. Furthermore, 8-quinolinol (oxine) was added to the solution to shift the reduction of Te(IV) to more negative potentials thus facilitating the codeposition. Similar to CdTe, the codeposits were obtained via UPD of Zn on tellurium. The composition of the ZnTe deposits is dependent on the deposition potential and on the concentration of Te(IV) in the solution. A lower Te(IV) concentration in the bath results in a higher zinc atomic fraction in the deposit. After thermal annealing the band gap was determined to be 2.3 eV by UV–Vis absorption spectroscopy, which is in good agreement with ZnTe films made by other methods.

12.2.3 CdS

Cadmium sulfide (CdS) is an n-type semiconductor with a direct band gap of 2.42 eV at room temperature. This material has been widely used as window layers for fabrication of heterojunction solar cells, such as CdS/CdTe and CdS/$CdInSe_2$. The electrodeposition of CdS from ionic liquids was reported first in 2007 by Dale et al. [24]. The semiconducting CdS was made electrochemically on FTO coated glass in ChCl/urea ionic liquid containing $CdCl_2$ and 0.1 M Na_2S (aq.) at 100 °C. It was shown that as-deposited yellow CdS film gives n-type photoresponse under illumination. However the quality of deposit was rather low. In 2009, the MacFarlane group showed for the first time that the methyltributylphosphonium tosylate ([P_{1444}]tos) ionic liquid is well suited for the electrodeposition of CdS [25]. Thin, adherent, homogeneous, and crack-free films of CdS were electrodeposited on FTO-coated glass by reduction of thiosulfate in the presence of Cd^{2+} ions in [P_{1444}]tos at 130–150 °C. $CdCl_2$ and sodium thiosulfate pentahydrate ($Na_2S_2O_3{\cdot}5H_2O$) were used as a source of Cd and S, respectively. Furthermore, deionized water was added into IL to assist with dissolution of precursors. The authors proposed the direct reduction of $S_2O_3^{2-}$ to S^{2-}, which then reacts with Cd^{2+} to form CdS. The observed films exhibit a band gap of 2.59 eV, which reduces to 2.52 eV after annealing.

12.2.4 ZnS

Zinc sulfide (ZnS) has attracted an interest as an appropriate material to replace a CdS buffer layer in photovoltaic devices. ZnS in a cubic form exhibits a band gap of 3.54 eV at room temperature, while the hexagonal form has a band gap of 3.91 eV. Furthermore, it can be doped as either an n-type or p-type semiconductor. There is only one study on the electrodepostition of ZnS in ionic liquids. Dale et al. [24] reported that thin film of ZnS with a band gap of 3.6 eV can be deposited on FTO-coated glass in ChCl/urea ionic liquid containing $ZnCl_2$ and

0.1 M Na_2SO_3 (aq.). In this paper the author showed only a photocurrent spectrum of the electrochemically made ZnS film. A detailed characterization of ZnS deposits obtained by this method is still missing.

12.2.5 CdSe

Cadmium selenide (CdSe) is an n-type semiconductor with a band gap of 1.74 eV at room temperature. The semiconductor nanoparticles with sizes below 100 nm are technologically interesting due to a quantum confinement effect. It has been reported that thin film of CdSe with a band gap of 1.8 eV can be made electrochemically on FTO-coated glass in ChCl/urea ionic liquid containing $CdCl_2$ and 0.1 M Na_2SO_3 (aq.) [24]. Similar to ZnS, the authors presented only a photocurrent spectrum of electrochemically made CdSe. A detailed characterization of CdSe deposits was not shown.

12.2.6 ZnO

Zinc oxide (ZnO) is a semiconductor with a direct band gap of 3.37 eV. In recent years ZnO has attracted a great interest as promising material for several technological applications (e.g., solar cells, UV light-emitting diodes, catalysis, photocatalysis, field emission, piezoelectric, gas sensing, etc.). In 2009, Azaceta et al. [26] reported that nanocrystalline ZnO can be electrodeposited on FTO-coated glass in 1-butyl-1-methylpyrrolidinium bis(trifluoromethylsulfonyl)imide ([C_4mPyr][$N(Tf)_2$]) ionic liquid saturated with O_2 and containing Zn[$N(Tf)_2$]$_2$. The authors reported that ZnO is deposited at temperatures above 100 °C due to the chemical reaction between Zn^{2+} and O_2^- generated from the O_2 reduction. The deposited films exhibit semiconducting behavior and room temperature photoluminescence emission [27]. It was shown that ZnO thin films can be obtained on Mo electrochemically in ChCl/urea ionic liquid saturated with O_2 and containing zinc perchlorate at 90 °C [28]. Recently, the electrodeposition of ZnO was investigated from Zn[$N(Tf)_2$]$_2$ in 1-ethyl-3-methylimidazolium bis(trifluoromethylsulfonyl)imide ([C_2mim][$N(Tf)_2$]), [C_4mPyr][$N(Tf)_2$], and 1-butyl-3-methylimidazolium trifluoromethanesulfonate ([C_4mim][CF_3SO_3]) ionic liquids saturated with O_2 [29]. The authors studied the influence of temperature, substrate, the nature of the ionic liquid, and electrochemical deposition conditions on the morphology and the crystal structure of electrochemically made ZnO film. It was shown that the deposition potential and temperature influence significantly morphology and crystal structure of the deposit. Furthermore, it was reported that the film growth orientation can be manipulated by addition of different cosolvents into ILs as a solvent dielectric constant seems to be a key parameter to tailor the deposit growth direction [29].

12.3 III–V Compound Semiconductors

The III–V compound semiconductors, in particular antimony-based semiconductors (AlSb, GaSb, and InSb), are important materials for electronic and optoelectronic devices. However, surprisingly, little work has been done on the electrodeposition of III–V semiconductors in ionic liquids.

12.3.1 GaAs

At room temperature gallium arsenide (GaAs) has a direct band gap of 1.43 eV and a high mobility of charge carriers that allows GaAs-based electronic devices to operate at higher frequencies than equivalent Si devices, resulting in faster electronics that makes this semiconductor interesting for many optoelectronic applications including semiconductor lasers, LEDs, and solar cells. In 1986, Wicelinski and Gale [30] showed that GaAs can—in principle—be electrodeposited from $GaCl_3$ and $AsCl_3$ in chloroaluminate ionic liquid. The deposition was performed at 40 °C in the Lewis acid chloroaluminate ionic liquid composed of $AlCl_3$ and 1-butylpyridinium chloride ($[C_4py][Cl]$). However it was reported that aluminum codeposition occurs in the UPD regime on the Ga surface. To minimize Al contamination, Carpenter and Verbrugge used a chlorogallate ionic liquid [31]. GaAs thin films were electrodeposited on glassy carbon at room temperature in the Lewis basic $GaCl_3$/1-methyl-3-ethylimidazolium chloride (40/60 mol%) ionic liquid containing $AsCl_3$. This method is promising—in principle—for the electrodeposition of GaAs-based compound semiconductors as the quality of the deposits can be improved by thermal annealing. However, $GaCl_3$ ionic liquids are extremely aggressive and $AsCl_3$ is extremely poisonous so that such liquids would involve considerable safety issues.

12.3.2 InSb

Indium antimonide (InSb) has the highest mobility of charge carries at room temperature and a direct band gap of 0.17 eV. The principle of InSb electrodeposition is the same as for GaAs. It was reported that InSb can be directly electrodeposited on Pt and glassy carbon at 45 °C in the Lewis basic chloroindate ionic liquid $InCl_3$/1-methyl-3-ethylimidazolium chloride containing $SbCl_3$ [32]. The In/Sb ratio in the deposit is strongly dependent on the applied electrode potential. Consequently elemental Sb and In can also be found in the films. In 2003, the Sun group [33] reported that InSb can be electrodeposited on nickel in the Lewis basic 1-ethyl-3-methylimidazolium chloride/tetrafluoroborate ($[C_2mim]$ $[Cl]/[BF_4]$) ionic liquid containing $InCl_3$ and $SbCl_3$. Polycrystalline InSb with a band gap of 0.2 eV can be directly electrodeposited at 120 °C without additional annealing. The composition of the InSb films depends on the deposition potential and the concentration of In(III) and Sb(III) in the solution. Furthermore, the crystallinity of the codeposits is strongly improved by increasing the deposition temperature.

12.3.3 AlSb

Aluminum antimonide (AlSb) exhibits a direct band gap of 2.5 eV and an indirect band gap of 1.2 eV at room temperature. This semiconductor attracts interest as a highly efficient solar cell material, p–n junction diode and an anode material for Li-ion batteries. The electrodeposition of AlSb in ionic liquids was investigated by several groups. It was reported that AlSb can be electrodeposited on Au(111) at room temperature in the Lewis neutral ionic liquid $AlCl_3$/1-butyl-3-methylimidazolium chloride ($AlCl_3$/[C_4mim][Cl]) containing $SbCl_3$ [34, 35]. In situ probe techniques were used to study the electrodeposition of the semiconductor. It was shown that the codeposits exhibit a band gap of about 2.0 eV. As in the case of GaAs and InSb, the codeposition of the elements occurs. Furthermore, strong doping effects by the elements are obtained if the deposition is performed at electrode potentials away from the compound deposition potential. It was reported that stoichiometric AlSb can also be obtained at room temperature in the Lewis acid $AlCl_3$/1-methyl-3-ethylimidazolium chloride ($AlCl_3$/[C_2mim][Cl]) to which $SbCl_3$ was added [36]. The as-deposited AlSb is amorphous, however thermal annealing improves the crystallinity of the deposits. In 2008, Hussey and Tsuda [37] employed a Lewis acid $AlCl_3$/[C_2mim][Cl] to study the electrodeposition of AlInSb ternary compound semiconductor. Solutions of In^+ and Sb^{3+} in the IL were prepared by electrochemical etching of In wire and Sb rod, respectively. It was reported that AlInSb deposits exhibit photocatalytic behavior and can catalyze the photodecomposition of water under illumination with visible light.

12.4 Other Compound Semiconductors

12.4.1 SnS

Tin monosulfide (SnS) is a promising material for thin-film photovoltaic applications. The semiconductor has an indirect band gap of 1.1 eV and a direct band gap of 1.3 eV. Furthermore, it shows intrinsic p-type conductivity. Recently, Steichen et al. [38] reported a direct electrodeposition of p-type SnS on molybdenum in 1-butyl-3-methylimidazolium dicyanamide ([C_4mim][$N(CN)_2$]) ionic liquid containing elemental sulfur and $SnCl_2$ above 100 °C. The authors proposed that the SnS formation proceeds via the electrochemical reduction of S_8 to S_4^{2-} followed by the thermally activated dissociation of S^{2-} anions, which subsequently react with Sn^{2+} cations at the interface. The as-deposited SnS exhibits an indirect band gap of 1.17 eV and high-optical adsorption at photon energy above 1.4 eV [38].

12.4.2 ZnSb

Zinc antimonide (ZnSb) is an indirect semiconductor with a bang gap of 0.53 eV. It is a promising candidate for efficient thermoelectric materials. The electrodeposition of ZnSb was investigated on Au(111) by in situ scanning probe methods at 50 °C in $ZnCl_2$/1-butyl-3-methylimidazolium chloride ([C_4mim][Cl]) ionic liquid containing $SbCl_3$ [39]. Near the semiconductor deposition potential an inhomogeneous distribution of ZnSb clusters on the top of Sb surface was obtained. The ZnSb clusters show semiconducting behavior with a band gap of about 0.6 eV. In the case of Zn excess a doping effect was reported.

12.5 Elemental Semiconductors

The conductivity of Si and Ge can be modified by doping with group III or V elements that lead to p- or n-type materials. Junctions of n- and p-doped Si are interesting for photovoltaic applications. Quantum confinement effects, such as the size-dependent photoluminescence of semiconductor nanoparticles, are especially interesting for electronic and optoelectronic devices.

12.5.1 Si

Silicon is an elemental semiconductor which exhibits an indirect band gap of 1.1 eV at room temperature in the microcrystalline phase. Silicon nanoparticles show a size-dependent photoluminescence. It is one of the most important semiconductors and widely used in electronic devices such as computers and integrated circuits, as well as in photovoltaic devices. In 2001, Katayama et al. [40] reported that a thin Si layer can be electrodeposited in 1-ethyl-3-methylimidazolium hexafluorosilicate ([C_2mim]$_2$[SiF_6]) at 90 °C. However, the deposit was completely oxidized upon exposure to air that made it difficult to decide whether the deposit was semiconducting or not. Nanoscale semiconducting Si was deposited at room temperature in 2004 [41]. By means of in situ current/voltage (I/U) tunneling spectroscopy it was proved that the layer with a thickness of 100 nm obtained in [C_4mPyr][N(Tf)$_2$] saturated with $SiCl_4$ exhibits a band gap of about 1 eV, indicating that semiconducting silicon had been electrodeposited. A first detailed in situ Scanning Tunneling Microscopy (STM) study of Si electrodeposition on Au(111) in [C_4mPyr][N(Tf)$_2$] containing $SiCl_4$ is presented in ref. [42]. It was shown that by reducing the electrode potential first, small Si clusters appear on the gold surface and then a continuous deposit forms by their merging. In situ I/U tunneling spectroscopy revealed that a layer of up to 5 nm in thickness has a band gap of 1.0 ± 0.2 eV, typical for silicon in the bulk phase. Furthermore, it was shown that an about 1 μm thick Si film has a spherical morphology: the larger grains of up to 300 nm in size consist of smaller particles in the 10–20 nm regime [43]. The energy dispersive X-ray (EDX) analysis of the deposit gives Si and a small amount of oxygen from ex situ oxidation attack. X-ray Photoelectron Spectroscopy (XPS) analysis of a micrometer thick electrodeposit proves that indeed elemental Si was obtained [44].

Recently, Komadina et al. [45] also studied the electrodeposition of Si from $SiCl_4$ in trimethyl-*n*-hexylammonium bis(trifluoromethylsulfonyl)imide ([N_{1116}][N(Tf)$_2$]) ionic liquid using Electrochemical Quartz Crystal Microbalance (EQCM). The measurements were performed at 40 °C. Current efficiencies indicated by EQCM were much higher than 100 % for four-electron reduction step, while XPS analysis revealed that only of about 70 wt% Si is present in the deposits prior to ex situ oxidation. The authors suggested that either Si is not completely reduced at the potentials investigated or a chemical reaction step occurs [45]. Furthermore, a large damping due to a high viscosity of the ionic liquid may introduce a significant error in the mass estimated by EQCM. Therefore the reaction mechanism is still unclear.

Very recently it was reported on the influence of temperature and the anion of the ionic liquid on the morphology and crystal size of electrochemically made Si [46]. For this purpose three different ionic liquids with the cation 1-butyl-1-methylpyrrolidinium ([C_4mPyr]) and three different anions, namely, trifluoromethylsufonate ([CF_3SO_3]), bis(trifluoromethylsulfonyl)imide ([N(Tf)$_2$]), and tris(pentafluoroethyl)trifluorophosphate ([FAP]) were employed as solvents. Si was electrodeposited from $SiCl_4$ dissolved in these ionic liquids on gold and on copper at room temperature and at 100 °C. Interestingly, the general features of the cyclic voltammetry behavior of the employed electrolyte are almost similar, apart from some differences in the surface process before the bulk deposition of silicon, and in the deposition potential. However, in situ STM on Au(111) revealed a significant difference in the interfacial behavior of $SiCl_4$/[C_4mPyr][N(Tf)$_2$] and $SiCl_4$/[C_4mPyr][FAP] [46]. In the case of [C_4mPyr][N(Tf)$_2$], small Si islands appear at the electrode potential of 1.5 V more negative than the open circuit potential (OCP). Their lateral and vertical growth leads to the formation of a rough layer. If [C_4mPyr][FAP] is employed, a number of randomly distributed two-dimensional (2DOM) islands are formed at the electrode potential of 0.5 V more negative than OCP. These islands grow in a layer-by-layer process. The difference for Si islands deposition between [C_4mPyr][N(Tf)$_2$] and [C_4mPyr][FAP] is roughly 1 V. The layers deposited in both ionic liquids exhibit a band gap of about 1.1 eV indicating the deposition of semiconducting Si. An IR spectroscopy analysis showed that $SiCl_4$ interacts differently with the employed ILs that can lead to different silicon

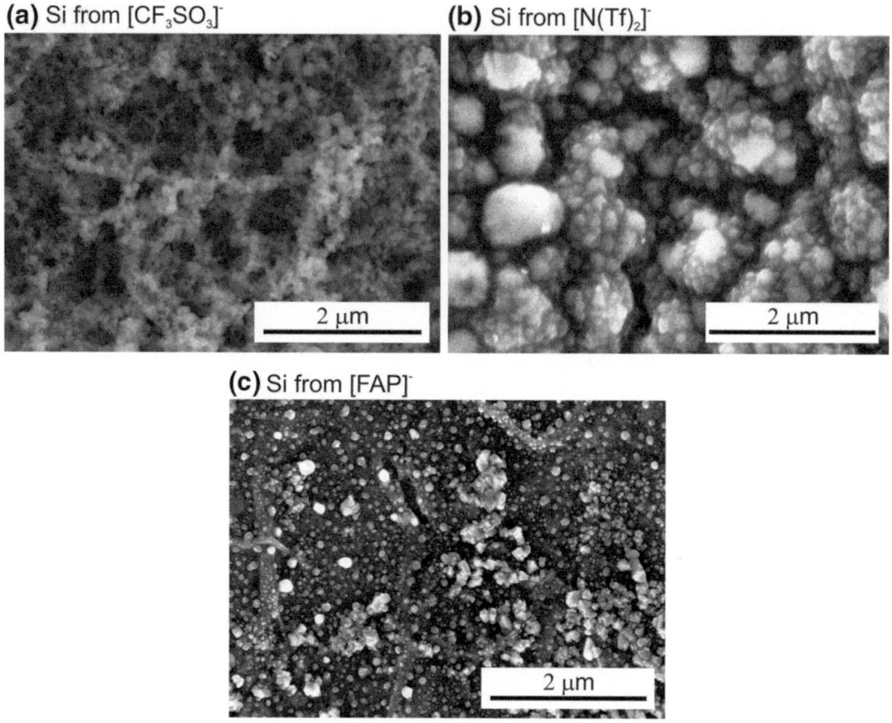

Fig. 12.1 SEM micrographs of silicon electrodeposits obtained from 0.15 M $SiCl_4$ in (**a**) $[C_4mPyr][CF_3SO_3]$, (**b**) $[C_4mPyr][N(Tf)_2]$, and (**c**) $[C_4mPyr][FAP]$ at 100 °C

complexes [46]. Furthermore, deposits with different morphologies were obtained in these ILs. The quality of the deposits significantly improves at 100 °C.

Silicon was made potentiostatically at 100 °C on mild steel at −1.8, −1.9, and −2.4 V vs. Pt-quasi ref. from $SiCl_4/[C_4mPyr][CF_3SO_3]$, $SiCl_4/[C_4mPyr][N(Tf)_2]$, and $SiCl_4/[C_4mPyr][FAP]$, respectively, for 1 h. The SEM images of the obtained deposits are presented in Fig. 12.1. The electrodeposited Si has a spherical morphology. Fairly thick and uniform deposits were obtained from $[C_4mPyr][CF_3SO_3]$ and $[C_4mPyr][N(Tf)_2]$ (Fig. 12.1a, b), while the deposit obtained from $[C_4mPyr][FAP]$ is thin and not uniform (Fig. 12.1c). In all cases, the average particle size is 50–70 nm. The electrodeposited silicon films on mild steel from the three ILs are amorphous as no peaks for crystalline silicon were recorded. In order to improve the crystallinity of the deposits, the samples were annealed at 700 °C under argon. The XRD patterns of the annealed silicon deposits on mild steel from $SiCl_4/[C_4mPyr][CF_3SO_3]$ and $SiCl_4/[C_4mPyr][N(Tf)_2]$ are shown in Fig. 12.2. Two sharp diffraction peaks at 33.2° and 51.7° corresponds to crystalline silicon (JCPDS No. 17-0901). Furthermore, there are some reflections originated from silica (JCPDS No. 14-0260) due to ex situ oxidation of the deposited Si. The average crystal size of Si was found to be ~50 nm using Scherrer equation [47]. These

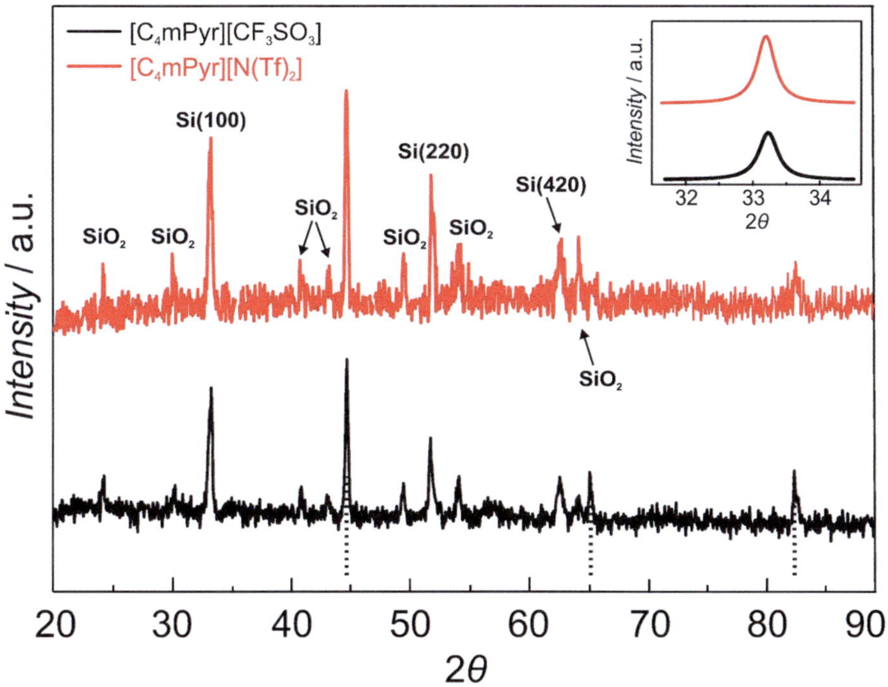

Fig. 12.2 XRD patterns of the silicon deposited on mild steel from [C_4mPyr][CF_3SO_3] (*black*) and [C_4mPyr][N(Tf)$_2$] (*red*) after annealing at 700 °C. The *dashed vertical lines* represent the peaks from the substrate

results reveal that the anion of the ionic liquid has an influence on the morphology of the electrodeposits. Furthermore, the crystallinity of the deposited Si can be improved by thermal annealing.

Recently, the mechanism of the electrodeposition of Si from its chloride in [C_4mPyr][N(Tf)$_2$] was investigated using UV–Vis spectroelectrochemistry [48]. It was found that the deposition phenomenon is complex in the initial stages. Furthermore, the solvation layers of the IL have a strong influence on the deposition process. A red-shift at higher wavelength indicated the possibility of quantum confinement effects at initial stages of Si deposition.

12.5.2 Ge

Germanium is an important elemental semiconductor which exhibits an indirect band gap of 0.67 eV at room temperature in the microcrystalline phase. In contrast to the microcrystalline element, nanocrystalline germanium is a direct semiconductor and a promising material in the optoelectronic industry. The electrodeposition of Ge in ionic liquids was primarily investigated by the Endres group [49–52]. It was shown that Ge can be electrodeposited on Au(111) from GeX_4 (where X = I, Br, Cl) in 1-butyl-3-methylimidazolium hexafluorophospate ($[C_4mim][PF_6]$) ionic liquid. In situ STM was used to investigate the initial stages of the electrodeposition at room temperature. It was found that even under potentiostatic control in the negative regime a chemical attack of Ge deposits by GeI_4 or maybe by decomposition products at the counter electrode occurs that leads to a chemical dissolution of the deposited films [52]. It was reported that micrometer thick Ge layers can be electrodeposited from saturated solutions of $GeBr_4$ or $GeCl_4$ at room temperature [53]. In situ I/U tunneling spectroscopy showed that a 300 pm thick UPD Ge layer has a metallic behavior, while about 100 nm thick Ge films clearly show a semiconducting behavior with a band gap of 0.7 eV, which is in good agreement with the band gap of intrinsic microcrystalline bulk Ge. A layer of about 20 nm in thickness exhibits both sites with semiconducting behavior and sites with rather metallic behavior [50]. XPS measurement of a micrometer thick Ge deposit made from GeX_4 proves that indeed elemental Ge was obtained [53]. These results show that ionic liquids based on $[BF_4]^-$ and $[PF_6]^-$ anions can be used for the nanoscale electrodeposition of Ge, however their chemical and physical properties can change upon exposure to moisture for a long time due to a slow hydrolysis of $[BF_4]^-$ and $[PF_6]^-$ liberating HF. Therefore, the use of ionic liquids based on more hydrophobic anions such as bis(trifluoromethylsulfonyl)imide ($[N(Tf)_2]$) is preferable. In 2008, it was reported that Ge can be electrodeposited in 1-butyl-1-methylpyrrolidinium bis(trifluoromethylsulfonyl)imide ($[C_4mPyr][N(Tf)_2]$) containing $GeCl_4$ [43]. It was found that the deposited Ge has a spherical morphology with smallest grains of around 10–20 nm. In order to avoid the chemical attack of Ge(IV) on the deposited Ge, $GeCl_2$ was electrochemically prepared to be used as Ge precursor instead of $GeCl_4$. However, the deposition of Ge was strongly inhibited in the Ge(II)-containing ionic liquid [43].

The room temperature electrodeposition of Ge and GeS_x has also been reported in *n*-methyl-*n*-propylpiperidinium bis(trifluoromethylsulfonyl)imide ($[C_3mpip][N(Tf)_2]$) ionic liquid containing $GeCl_4$ and 1,4-butanedithiol [54]. The deposited films of GeS_x were smooth, porous, and had an amorphous glassy character. Wu et al. [55] presented an EQCM study of Ge electrodeposition in 1-butyl-1-methylpyrrolidinium dicyanamide ($[C_4mPyr][N(CN)_2]$) and a mixture of $[C_4mPyr][N(CN)_2]$ and 1-butyl-1-methylpyrrolidinium chloride ($[C_4mPyr][Cl]$). $[GeCl_4(C_4im)_2]$ (C_4im = *n*-butylimidazol) was used as a source of Ge. In both cases, smooth, porous germanium films were electrodeposited on both copper and silicon substrates at elevated temperatures.

UV–Vis spectroelectrochemistry was applied to study the electrodeposition process of Ge in $[C_4mPyr][N(Tf)_2]$ containing $GeCl_4$ [48]. The measurements revealed the UPD of Ge on gold at the Ge(IV) to Ge(II) reduction process. During the deposition of bulk germanium a red-shift at higher wavelength was found which is related to increase in particle size and agglomeration. Furthermore, quantum confinement effects in Ge nanoparticles were observed.

The interesting observation is that UV light influences the reduction potential of germanium species and the morphology of the electrodeposits [56]. Ge was electrodeposited in the presence or absence of UV irradiation on indium tin oxide (ITO) and on Au from $[C_4mPyr][N(Tf)_2]$ containing $GeCl_4$. The results revealed that with decreasing the wavelength of the UV light, a shift to less negative values in the reduction potential takes place. Furthermore, the presence of UV irradiation significantly decreased the particle size of deposited Ge. In the presence of UV light a green luminescence was also observed in the electrolyte during cyclic voltammetry experiments. The electrochemical behavior of 0.1 M $GeCl_4$ in $[C_4mPyr][N(Tf)_2]$ on ITO in the presence and absence of UV light is shown in Fig. 12.3. Without UV illumination there are two cathodic peaks with maxima at −1.74 and −2.17 V prior to irreversible reduction of the organic cation at −3.0 V (Fig. 12.3a). The first process is correlated with the reduction of Ge(IV) to Ge(II). At the second process a black deposit clearly forms on the electrode surface due to the reduction of Ge(II) to Ge(0). A broad anodic process at $E > -1.0$ V is attributed to some incomplete dissolution of deposited Ge. At electrode potentials above +0.4 V the black Ge deposit disappears completely. Therefore a rise in anodic current at +0.4 V is attributed to the oxidation of deposited Ge. If the measurement is performed in the presence of 254 nm UV light, a shift in the reduction potential occurs (Fig. 12.3b). The first peak is obtained at −1.6 V, while the second peak appears at −1.93 V. Furthermore, a green luminescence appears during the reduction of Ge(IV) to Ge(II) in the presence of UV light, which is attributed to the presence of $GeCl_2$ in the electrolyte [56]. The microstructure of Ge films deposited for 30 min at −2.17 V without UV light (inset in Fig. 12.3a) and at −1.93 V in the presence of UV light (inset in Fig. 12.3b) shows spherical agglomerates built up of smaller clusters. In the absence of UV light the cluster size is 400–500 nm, while under UV irradiation clusters of about 100 nm in diameter are obtained. These results reveal that UV irradiation significantly decreases the particle size of electrodeposited Ge.

12.5.3 Si_xGe_{1-x}

Si_xGe_{1-x} is an interesting semiconductor material because its fundamental semiconducting properties can be modified by replacing some Si atoms with Ge atoms. In 2008, it was reported that Si_xGe_{1-x} can be made electrochemically at room temperature in $[C_4mPyr][N(Tf)_2]$ containing $SiCl_4$ and $GeCl_4$ as precursors [43]. It was found that during the cyclic voltammetry scan, the evolution of colors from red to orange and green occurs, which was qualitatively explained with the quantum confinement effect. In situ UV–Vis spectroscopy studies confirmed the quantum confinement effect, however the reaction mechanism was found to be complex [48]. The nature of the ionic liquid has an influence on the composition and the particle size of electrodeposit. The Si_xGe_{1-x} deposits obtained from an equimolar solution (0.1 M) of $SiCl_4$ and $GeCl_4$ in $[C_2mim][N(Tf)_2]$ have a lower Si content than those obtained from $[C_4mPyr][N(Tf)_2]$ at the same conditions (an overall Si:Ge ratio about 1–1.5:1 from $[C_4mPyr][N(Tf)_2]$ vs. about 1:3 from $[C_2mim][N(Tf)_2]$) [57].

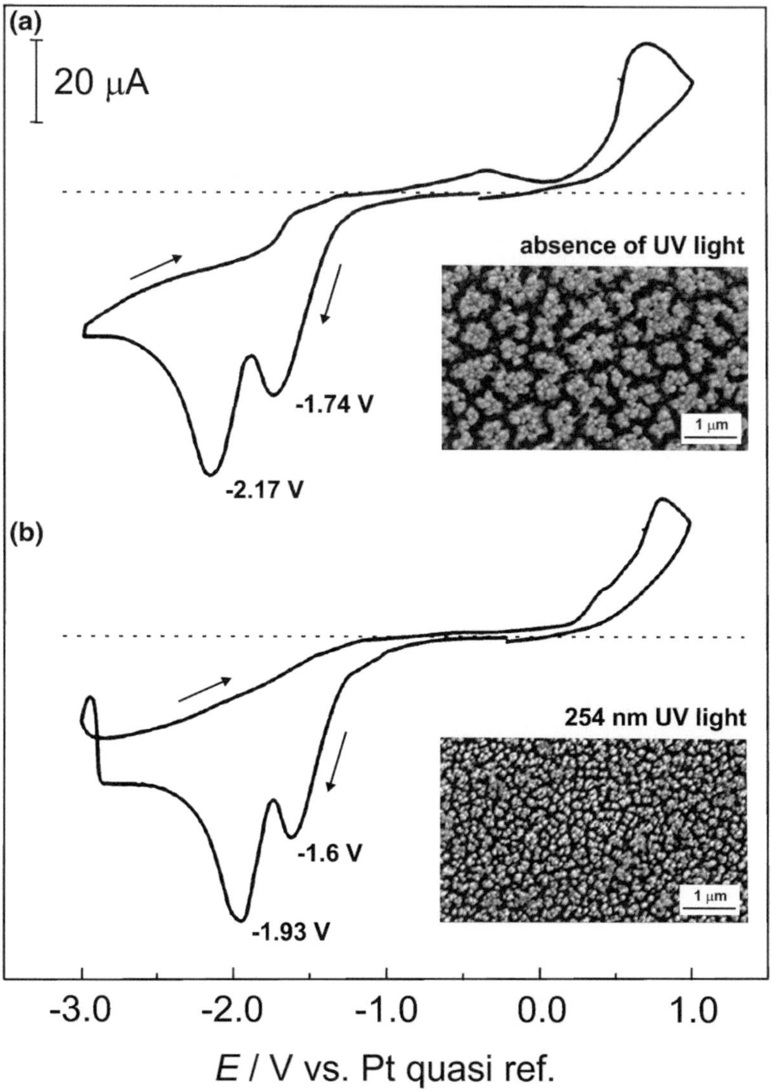

Fig. 12.3 Cyclic voltammograms of 0.1 M $GeCl_4$ in $[C_4mPyr][N(Tf)_2]$ on ITO at 25 °C: (**a**) without UV light; (**b**) in the presence of 254 nm UV light. The scan rates are 10 mV s^{-1}. *Insets* show the SEM images of Ge layers deposited potentiostatically on ITO for 30 min: (**a**) at −2.17 V, (**b**) at −1.93 V

Furthermore, it was found that micrometer thick Si_xGe_{1-x} electrodeposits can be made electrochemically at room temperature from an equimolar solution (0.1 M) of $SiCl_4$ and $GeCl_4$ in ionic liquids $[C_2mim][N(Tf)_2]$, $[C_4mPyr][N(Tf)_2]$, 1-ethyl-2,3-dimethylimidazolium bis(trifluoromethylsulfonyl)imide ($[C_2dmim][N(Tf)_2]$) and 1-butyl-1-methylpyrrolidinium tris(pentafluoroethyl)trifluorophosphate ([C_4mPyr]

[FAP]) [58]. However it was reported that the $[N(Tf)_2]$ anion is subject to an anodic decomposition at the counter electrode, and the decomposition products oxidize the Si_xGe_{1-x} deposit chemically leading to its complete dissolution after several hours. In the case of [FAP] anion the Si_xGe_{1-x} deposit remains stable even after several days. These results show again that the nature of the ionic liquid has a significant influence on the electrodeposition process. It has also been reported that the concentrations of the precursors influence the electrodeposition mechanism of Si_xGe_{1-x} from $[C_4mPyr][N(Tf)_2]$ [59]. If the Si_xGe_{1-x} is deposited from an equimolar solution of $SiCl_4$ and $GeCl_4$, a color change of the deposit from red to green is observed during the cyclic voltammetry scan. When the concentration of $GeCl_4$ is increased in the electrolyte, only a black deposit is obtained, whereas only the red colored deposit is formed by increasing of the concentration of $SiCl_4$ in the solution. In situ UV–Vis measurements revealed a layered growth mechanism for formation of Si_xGe_{1-x} [59].

12.6 Semiconductor Nanostructures

Nanostructured semiconductors are getting extensive attention due to their unique electronic and optical properties and potential applications in nanoscale sensors, catalysts, solar cells, batteries, and biomedicine.

12.6.1 Nanowires, Nanotubes, and Nanorods

In 2008, it was reported that silicon and germanium nanowires can be obtained electrochemically in $[C_4mPyr][N(Tf)_2]$ ionic liquid containing $SiCl_4$ and $GeCl_4$ using template-assisted techniques [60, 61]. Commercially available track-etched polycarbonate membranes (PC) with an average nominal pore diameter of 90 nm were used as templates. One side of such a membrane was sputtered with a 120 nm thick gold film to act as a working electrode during the electrodeposition experiment. The electrodeposition was performed at room temperature inside the pores of the membrane. After the deposition is finished, the membrane, with the nanowires embedded inside, was carefully dissolved in dichloromethane (CH_2Cl_2). The morphology of the deposited nanowires is mainly controlled by the original shape of the pores of the membrane. Si nanowires with a length of about 4–6 μm and Ge nanowires with the length of about 2 μm were made potentiostatically in $[C_4mPyr][N(Tf)_2]$ containing $SiCl_4$ or $GeCl_4$ as a source of Si or Ge, respectively [60]. Figure 12.4 shows the high-resolution SEM images of Ge and Si nanowires prepared in the commercially available M90 PC membrane. The as-deposited Si and Ge nanowires were amorphous, cylindrical, smooth, and with homogeneous diameters fixed by the pore diameter of the PC membrane. Annealing treatment enables the crystallization of Si nanowires without any modification of their shape and

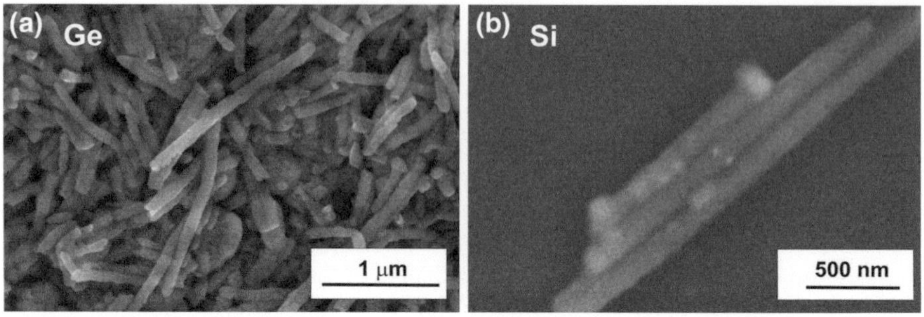

Fig. 12.4 High-resolution SEM images of (**a**) Ge and (**b**) Si nanowires after removal of the template. Reproduced from ref. [60] © 2008 RSC

composition [61]. It was demonstrated that Si_xGe_{1-x} nanowires can be obtained in $[C_4mPyr][N(Tf)_2]$ and $[C_2mim][N(Tf)_2]$ ionic liquids containing $SiCl_4$ and $GeCl_4$ by an electrochemical template synthesis [62]. Recently, Homma et al. [63] reported that amorphous Si nanopillars were electrodeposited in a UV-NIL resist template from trimethylhexylammonium bis(trifluoromethylsulfonyl)imide ($[N_{1116}][N(Tf)_2]$) ionic liquid containing $SiCl_4$.

It was reported that nanotubular structures can be grown at room temperature using template-assisted electrodeposition in ionic liquids [64, 65]. Mallet et al. showed that Si nanotubes of about 400 nm in diameter can be obtained from $SiCl_4$ in $[C_4mPyr][N(Tf)_2]$ [64]. The authors reported that depending on the deposition time, the length of the nanotubes can be adjusted from a few hundred nanometers to a few microns using the same 400 nm PC membrane. The as-deposited Si was amorphous. It was suggested that the nanotubes are formed by competition between the growth rate and the ionic diffusion inside the nanopores of the insulating template, which is quite low due to the high viscosity of $[C_4mPyr][N(Tf)_2]$ ionic liquid and the diffusion regime inside the nanopores [64]. However the growth mechanism of the semiconductor nanotubes in ionic liquids is still unclear. Free-standing open-ended Ge nanotubes can be made electrochemically in PC membranes in $[C_4mPyr][N(Tf)_2]$ containing $GeCl_4$ [65]. Figure 12.5 shows the SEM images of free-standing Ge nanotubes prepared in the PC membrane with a pore diameter of 200 nm. After the deposition of Ge on the gold sputtered membrane at −2.1 V vs. Pt-quasi ref., Au was further electrodeposited from a commercially available electrolyte to support the layer and to obtain free-standing nanotubes. Then the membrane was dissolved in CH_2Cl_2 to get the nanotubes. The diameter of the tube is about 220 nm with a wall thickness about 40 nm. The nanotubes appear to be free-standing on the Au supporting layer. It was found that increasing the time of electrodeposition leads to a decrease in the opening diameter of the nanotubes, as the deposition occurs at the inner wall of the nanotube. When Ge was deposited for 40 min, the tube wall thickness increases to about 120 nm [65]. Furthermore, it was reported that core-shell structures can be made electrochemically using this technique. In order to obtain core-shell structure, Ge nanotubes were electrodeposited followed by Cu deposition from 1-butyl-1-methylpyrrolidinium dicynamide

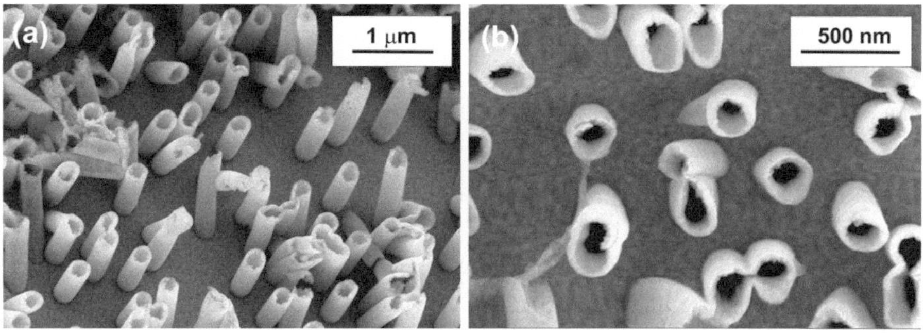

Fig. 12.5 High-resolution SEM images of free-standing Ge nanotubes after removal of the template (**a**) 20000x; (**b**) 50000x

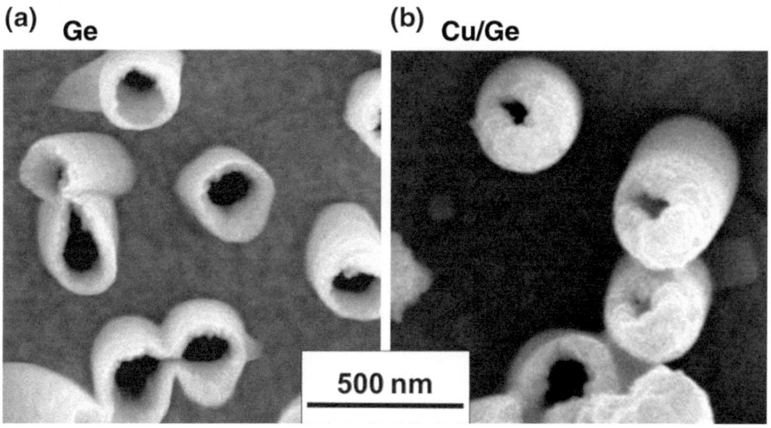

Fig. 12.6 High-resolution SEM images of (**a**) Ge and (**b**) Cu core-Ge shell nanotubes

($[C_4mPyr][N(CN)_2]$) containing CuCl [65]. The obtained free-standing micro-structure of Cu/Ge core-shell structure is presented in Fig. 12.6b. As can be seen, the nanotube hole size is reduced and the top of the nanotubes is almost completely covered compared to the open structure of Ge nanotubes shown in Fig. 12.6a. The EDX analysis revealed that the concentration of Cu is higher than the concentration of Ge. It was suggested that Cu deposition occurs both on the top and inside of the Ge nanotube leading to the formation of core-shell structures [65].

12.6.2 Macroporous Structures

Three-dimensional (3DOM) and two-dimensional (2DOM) macroporous nanostructures are materials with an ordered arrangement of pores with a high specific surface area. These materials are in particular interesting for micro- and

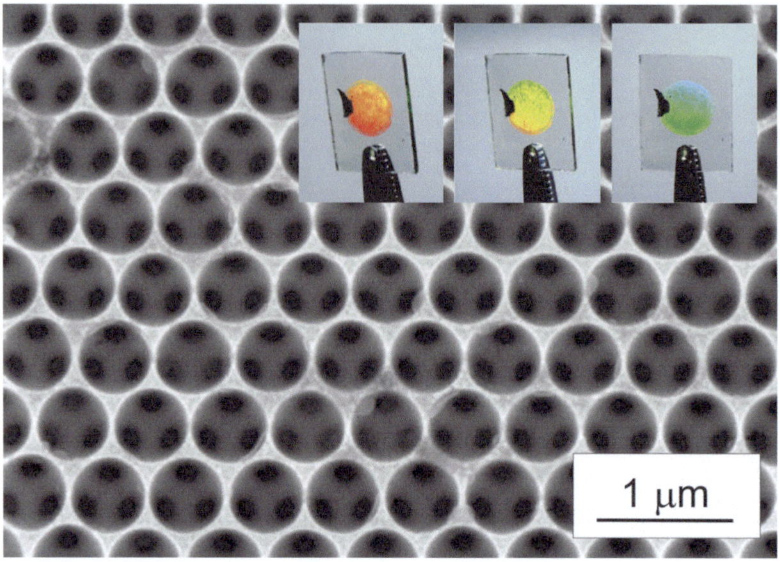

Fig. 12.7 High-resolution SEM image of a 3DOM Ge layer after removal of the PS spheres. *Inset* shows the optical photographs of the sample showing the color change by changing the angle of incident white light. Reproduced from ref. [66] © 2009 Wiley-VCH Verlag GmbH & Co. KGaA

optoelectronic technologies. In 2009, the 3DOM germanium nanostructures were synthesized electrochemically at room temperature within polystyrene colloidal crystal templates from the ionic liquids 1-hexyl-3-methylimidazolium tris (pentafluoroethyl)trifluorophosphate ([C_6mim][FAP]), and [C_2mim][N$(Tf)_2$] containing $GeCl_4$ [66, 57]. First, polystyrene colloidal crystals (PS) of about 600 nm in diameter were grown on indium tin oxide (ITO) coated glass. Then the PS template was filled with an electrolyte. The deposition occurs in the spacings between the PS spheres. After the deposition of Ge, the PS template was removed by dissolution in tetrahydrofuran (THF). As can be seen in Fig. 12.7, the deposited Ge has a well-ordered macroporous structure consisting of uniform close-packed spherical pores. Furthermore, if the angle of incident of white light is changed, the color of the 3DOM Ge deposit changes from orange to blue owing to light reflection (inset in Fig. 12.7). Reflection spectra of the 3DOM Ge electrodeposits revealed that the substrate is highly oriented, allowing ~100 % reflectance of different visible-light wavelengths [57]. Recently, the photoluminescence (PL) properties of PS, 3DOM Ge, and Ge/PS composite opal structures obtained electrochemically at room temperature in [C_6mim][FAP] containing $GeCl_4$ were reported [67]. It was found that the Ge/PS composite structure shows a strong green emission due to the disorder multiply-scattering in PS opals colloidal crystal.

The electrodeposition of silicon in PS template is considerably more difficult than that of germanium [57]. Silicon is deposited at much lower electrode potentials and has lower conductivity compared to Ge. A too negative electrode potential can

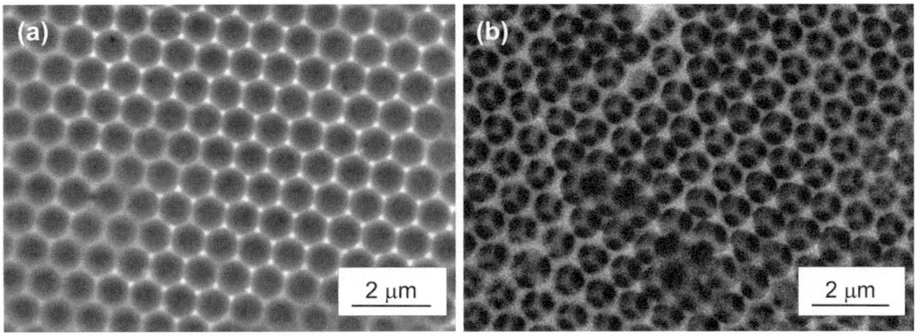

Fig. 12.8 SEM images of (**a**) 2DOM and (**b**) 3DOM Si_xGe_{1-x} after the dissolution of the PS template. Figure reprinted from ref. [57] © 2010 IUPAC

lead to disintegration of PS. It was reported that well-ordered 3DOM silicon can be electrodeposited at room temperature in $[C_4mPyr][N(Tf)_2]$ containing $SiCl_4$ using PS templates [68]. When the angle of incident of white light is changed, the 3DOM Si shows reflection colors (yellow, orange, green, and blue). The photonic band gap of the deposited 3DOM Si is found at 1,350 nm (in the near infrared regime). Furthermore, the surface of the deposit was oxidized.

The room temperature template-assisted elecrodeposition of 2DOM and 3DOM Si_xGe_{1-x} macroporous structures in an ionic liquid was reported in [57]. Figure 12.8 represents SEM images of 2DOM and 3DOM Si_xGe_{1-x} made electrochemically in $[C_2mim][N(CN)_2]$ containing $GeCl_4$ and $SiCl_4$ at -2.0 V vs. Ag for 10 and 30 min, respectively. Similar to Ge and Si opal structures, both samples show an angle-dependent reflection with changing the angle of the incident white light. Well-ordered close-packed amorphous 2DOM Si_xGe_{1-x} bowl-like and fishing-net structures were produced electrochemically from $GeCl_4$ and $SiCl_4$ in $[C_2mim][N(CN)_2]$ at room temperature and at 90 °C, respectively [69]. These results show that template-assisted electrodeposition in ionic liquids is a well-suited technique for the synthesis of 3DOM nanostructures.

12.7 Concluding Remarks

During the recent 15 years ionic liquids have attracted a rising research interest as potential electrolytes for the electrodeposition of various materials, which cannot be obtained from aqueous or organic solutions. However, despite a remarkable progress in this field, studies on the electrodeposition of semiconductors in ionic liquids are still rare. In the case of compound semiconductors to date most of the studies were performed in chlorometallate (chloroaluminates, chlorogallates, chloroindate, and chlorozincate) ionic liquids to which the required precursors were added. Chlorometallate ionic liquids are promising solvents for electrodeposition of Al-based, Ga-based, In-based, and Zn-based semiconductor materials.

They are inexpensive compared with air- and water-stable ILs. However, the main disadvantage of these salts is their extreme deliquescence. These liquids have to be prepared and stored in inert gas atmosphere. Such electrolytes require considerable safety issues as they are in part extremely aggressive. Some studies were performed in deep eutectic solvents/ILs (such as ChCl/urea). These liquids are relatively benign and inexpensive, and therefore attractive for industrial use. However, the main disadvantage of these ILs is that the electrochemical windows are significantly smaller than the ones of air- and water-stable ILs, which make them unsuitable for electrodeposition of reactive semiconductors (such as Si and Ge). In contrast to deep eutectic solvents/ILs, the electrochemical windows of ILs with $[PF_6]^-$ and $[BF_4]^-$ anions are wide enough for electrodeposition of reactive semiconductors. But, their chemical and physical properties can change upon exposure to moisture for a long time due to a slow hydrolysis of $[BF_4]^-$ and $[PF_6]^-$ liberating HF. Therefore, the use of ILs based on more hydrophobic anions such as $[N(Tf)_2]^-$ and $[FAP]^-$ is preferable.

During the last 5 years a remarkable progress has been done in electrodeposition of ZnO and CdS. However, the electrodeposition of many compound semiconductors (such as ZnS, CuS, GeSn, SiSn, $CdInSe_2$, CdHgTe, etc.) is still at its infancy. To date the electrodeposition of Si, Ge, and Si_xGe_{1-x} in ionic liquids has probably been studied in most detail. It was demonstrated that nanoscale Si, Ge, and "luminescent" Si_xGe_{1-x} can be made electrochemically at room temperature in air- and water-stable ILs containing silicon and/or germanium halides as precursors. Furthermore, ILs are well suited for the electrodeposition of free-standing semiconductor nanowires, nanotubes, core-shell structures, and macroporous structures. These materials are in particular interesting as potential anode and cathode materials for future generation Li-ion batteries. However, although the potential use of ILs for electrochemical synthesis has been demonstrated in various cases, the basic electrochemistry is yet not fully understood. Relatively little is known how the structure of the IL/electrode interface influences the deposition process and if the solutes affect the IL/electrode interfacial structure. Numerous fundamental studies are still required including the mechanism of nucleation and growth, the influence of the nature of IL on the deposition process, and the structure of the double layer to produce the semiconductor materials with required properties by electrochemical deposition in ILs. Nevertheless, ionic liquids are challenging solvents for electrochemistry!

Acknowledgments The author gratefully acknowledges Prof. Frank Endres for all his support. The author thanks Dr. Abhishek Lahiri and Dr. Giridhar Pulletikurthi for their helpful discussions.

References

1. Lincot D (2005) Electrodeposition of semiconductors. Thin Solid Films 487(1–2):40–48. doi:10.1016/j.tsf.2005.01.032
2. Pandey PK, Sahu SN, Chandra S (1996) Handbook of semiconductor electrodeposition. Marcel Dekker, New York

3. Nakamura S (2006) In: Elliot TB (ed) New research on semiconductors. Nova Science, New York, pp 159–207
4. Fulop GF, Taylor RM (1985) Electrodeposition of semiconductors. Annu Rev Mater Sci 15:197–210
5. Hodes G (2001) Electrochemistry of nanomaterials. Wiley-VCH Verlag GmbH, Weinheim
6. Schlesinger TE, Rajeshwar K, De Tacconi NR (2010) Electrodeposition of semiconductors. In: Schlesinger M, Paunovic M (eds) Modern electroplating, 5th edn. Wiley, New York, pp 383–411. doi:10.1002/9780470602638.ch14
7. Venkatasamy V, Mathe MK, Cox SM, Happek U, Stickney JL (2006) Optimization studies of HgSe thin film deposition by electrochemical atomic layer epitaxy (EC-ALE). Electrochim Acta 51(21):4347–4351. doi:10.1016/j.electacta.2005.12.012
8. Vaidyanathan R, Cox SM, Happek U, Banga D, Mathe MK, Stickney JL (2006) Preliminary studies in the electrodeposition of PbSe/PbTe superlattice thin films via electrochemical atomic layer deposition (ALD). Langmuir 22(25):10590–10595. doi:10.1021/la061625z
9. Colletti LP, Flowers BH, Stickney JL (1998) Formation of thin films of CdTe, CdSe, and CdS by electrochemical atomic layer epitaxy. J Electrochem Soc 145(5):1442–1449. doi:10.1149/1.1838502
10. Wade TL, Vaidyanathan R, Happek U, Stickney JL (2001) Electrochemical formation of a III-V compound semiconductor superlattice: InAs/InSb. J Electroanal Chem 500 (1–2):322–332. doi:10.1016/s0022-0728(00)00473-3
11. Liang X, Zhang Q, Lay MD, Stickney JL (2011) Growth of Ge nanofilms using electrochemical atomic layer deposition, with a "bait and switch" surface-limited reaction. J Am Chem Soc 133(21):8199–8204. doi:10.1021/ja109398t
12. Gaune-Escard M, Seddon KR (2010) Molten salts and ionic liquids: never the twain? Wiley, Hoboken
13. Wasserscheid P, Welton T (2007) Ionic liquids in synthesis. Wiley-VCH, Weinheim
14. Endres F, Abbott AP, MacFarlane DR (2008) Electrodeposition from ionic liquids. Wiley-VCH Verlag GmbH & Co, KGaA, Weinheim
15. Seddon KR, Plechkova N (2013) Ionic liquids uncoiled: critical expert overviews. Wiley, Hoboken
16. Borisenko N, Zein El Abedin S, Endres F (2008) Electrodeposition of semiconductors in ionic liquids. In: Electrodeposition from ionic liquids. Wiley-VCH Verlag GmbH & Co. KGaA, Weinheim, pp 147–165
17. Endres F, Zein El Abedin S (2008) Inorganic materials by electrochemical methods. In: Wasserscheid P, Welton T (eds) Ionic liquids in synthesis, vol 2. Wiley-VCH, Weinheim, pp 575–608
18. Armand M, Endres F, MacFarlane DR, Ohno H, Scrosati B (2009) Ionic-liquid materials for the electrochemical challenges of the future. Nat Mater 8(8):621–629. doi:10.1038/nmat2448
19. Borisenko N (2011) Electrodeposition of semiconductors from ionic liquids. In: Torriero AAJ, Shiddiky M (eds) Electrochemical properties and applications of ionic liquids. Nova Science, New York, pp 183–202
20. Hsiu S-I, Sun I-W (2004) Electrodeposition behaviour of cadmium telluride from 1-ethyl-3-methylimidazolium chloride tetrafluoroborate ionic liquid. J Appl Electrochem 34 (10):1057–1063. doi:10.1023/B:JACH.0000042670.84645.c5
21. Chauhan KR, Burgess IJ, Chang GS, Mukhopadhyay I (2014) Preparation of CdTe thin film by electrodeposition in butyl methyl imidazolium bath at 80 °C. J Electroanal Chem 713:70–76
22. Golgovici F, Visan T (2012) Electrodeposition behaviour of cadmium telluride from choline chloride-urea ionic liquids. Chalcogenide Lett 9(4):165–174
23. Lin M-C, Chen P-Y, Sun I-W (2001) Electrodeposition of zinc telluride from a zinc chloride-1-ethyl-3-methylimidazolium chloride molten salt. J Electrochem Soc 148(10):C653–C658. doi:10.1149/1.1396337

24. Dale PJ, Samantilleke AP, Shivagan DD, Laurence MP (2007) Synthesis of cadmium and zinc semiconductor compounds from an ionic liquid containing choline chloride and urea. Thin Solid Films 515(15):5751–5754. doi:10.1016/j.tsf.2006.12.072
25. Izgorodin A, Winther-Jensen O, Winther-Jensen B, MacFarlane DR (2009) CdS thin-film electrodeposition from a phosphonium ionic liquid. Phys Chem Chem Phys 11 (38):8532–8537. doi:10.1039/b906995j
26. Azaceta E, Tena-Zaera R, Marcilla R, Fantini S, Echeberria J, Pomposo JA, Grande H, Mecerreyes D (2009) Electrochemical deposition of ZnO in a room temperature ionic liquid: 1-butyl-1-methylpyrrolidinium bis(trifluoromethane sulfonyl)imide. Electrochem Commun 11 (11):2184–2186. doi:10.1016/j.elecom.2009.09.026
27. Azaceta E, Marcilla R, Mecerreyes D, Ungureanu M, Dev A, Voss T, Fantini S, Grande H-J, Cabañero G, Tena-Zaera R (2011) Electrochemical reduction of O_2 in 1-butyl-1-methylpyrrolidinium bis(trifluoromethanesulfonyl)imide ionic liquid containing Zn^{2+} cations: deposition of non-polar oriented ZnO nanocrystalline films. Phys Chem Chem Phys 13(29):13433–13440. doi:10.1039/c1cp20718k
28. Harati M, Love D, Lau WM, Ding Z (2012) Preparation of crystalline zinc oxide films by one-step electrodeposition in reline. Mater Lett 89:339–342. doi:10.1016/j.matlet.2012.08.136
29. Tulodziecki M, Tarascon J-M, Taberna PL, Guéry C (2012) Electrodeposition growth of oriented ZnO deposits in ionic liquid media. J Electrochem Soc 159(12):D691–D698. doi:10.1149/2.027212jes
30. Wicelinski SP, Gale RJ (1986) In: Saboungi M-L, Johnson K, Newman DS, Inman D (eds) Proceedings of the fifth international symposium on molten salts, vol 86-1. Electrochemical Society, Inc., Pennington, NJ, p 144
31. Carpenter MK, Verbrugge MW (1990) Electrochemical codeposition of gallium and arsenic from a room temperature chlorogallate melt. J Electrochem Soc 137(1):123–129. doi:10.1149/1.2086346
32. Carpenter MK, Verbrugge MW (1994) Electrochemical codeposition of indium and antimony from a chloroindate molten salt. J Mater Res 9(10):2584–2591
33. Yang M-H, Yang M-C, Sun I-W (2003) Electrodeposition of indium antimonide from the water-stable 1-ethyl-3-methylimidazolium chloride/tetrafluoroborate ionic liquid. J Electrochem Soc 150(8):C544–C548. doi:10.1149/1.1590328
34. Mann O, Aravinda CL, Freyland W (2006) Microscopic and electronic structure of semimetallic Sb and semiconducting AlSb fabricated by nanoscale electrodeposition: an in situ scanning probe investigation. J Phys Chem B 110(43):21521–21527. doi:10.1021/jp064384c
35. Aravinda CL, Freyland W (2006) Nanoscale electrocrystallisation of Sb and the compound semiconductor AlSb from an ionic liquid. Chem Commun 16:1703–1705. doi:10.1039/b517243h
36. Gandhi T, Raja KS, Misra M (2008) Room temperature electrodeposition of aluminum antimonide compound semiconductor. Electrochim Acta 53(24):7331–7337. doi:10.1016/j.electacta.2008.04.014
37. Tsuda T, Hussey CL (2008) Electrodeposition of photocatalytic AlInSb semiconductor alloys in the lewis acidic aluminum chloride-1-ethyl-3-methylimidazolium chloride room-temperature ionic liquid. Thin Solid Films 516(18):6220–6225. doi:10.1016/j.tsf.2007.11.114
38. Steichen M, Djemour R, Gütay L, Guillot J, Siebentritt S, Dale PJ (2012) Direct synthesis of single-phase p-type SnS by electrodeposition from a dicyanamide ionic liquid at high temperature for thin film solar cells. J Phys Chem C 117(9):4383–4393. doi:10.1021/jp311552g
39. Mann O, Freyland W (2007) Mechanism of formation and electronic structure of semiconducting ZnSb nanoclusters electrodeposited from an ionic liquid. Electrochim Acta 53(2):518–524. doi:10.1016/j.electacta.2007.07.034
40. Katayama Y, Yokomizo M, Mura T, Kishi T (2001) Preparation of a novel fluorosilicate salt for electrodeposition of silicon at low temperature. Electrochemistry 69(11):834–837

41. Zein El Abedin S, Borissenko N, Endres F (2004) Electrodeposition of nanoscale silicon in a room temperature ionic liquid. Electrochem Commun 6(5):510–514. doi:10.1016/j.elecom.2004.03.013
42. Borisenko N, Zein El Abedin S, Endres F (2006) In situ STM investigation of gold reconstruction and of silicon electrodeposition on Au(111) in the room temperature ionic liquid 1-butyl-1-methylpyrrolidinium bis(trifluoromethylsulfonyl)imide. J Phys Chem B 110 (12):6250–6256
43. Al-Salman R, Zein El Abedin S, Endres F (2008) Electrodeposition of Ge, Si and $Si_{(x)}Ge_{(1-x)}$ from an air- and water-stable ionic liquid. Phys Chem Chem Phys 10(31):4650–4657. doi:10.1039/b806996b
44. Bebensee F, Borissenko N, Frerichs M, Höfft O, Maus-Friedrichs W, Zein El Abedin S, Endres F (2008) Surface analysis of nanoscale aluminium and silicon films made by electrodeposition in ionic liquids. Z Phys Chem 222:671–686. doi:10.1524/zpch.2008.5315
45. Komadina J, Akiyoshi T, Ishibashi Y, Fukunaka Y, Homma T (2013) Electrochemical quartz crystal microbalance study of Si electrodeposition in ionic liquid. Electrochim Acta 100:236–241. doi:10.1016/j.electacta.2012.07.043
46. Pulletikurthi G, Lahiri A, Carstens T, Borisenko N, Zein El Abedin S, Endres F (2013) Electrodeposition of silicon from three different ionic liquids: possible influence of the anion on the deposition process. J Solid State Electrochem 17(11):2823–2832. doi:10.1007/s10008-013-2185-1
47. Scherrer P (1918) Bestimmung der Größe und der inneren Struktur von Kolloidteilchen mittels Röntgenstrahlen. Göttinger Nachrichten Math Phys 2:98–100
48. Lahiri A, Olschewski M, Höfft O, Zein El Abedin S, Endres F (2013) In situ spectroelectrochemical investigation of Ge, Si, and Si_xGe_{1-x} electrodeposition from an ionic liquid. J Phys Chem C 117(4):1722–1727. doi:10.1021/jp309646r
49. Endres F, Zein El Abedin S (2002) Nanoscale electrodeposition of germanium on Au(111) from an ionic liquid: an in situ STM study of phase formation—part I. Ge from $GeBr_4$. Phys Chem Chem Phys 4(9):1640–1648. doi:10.1039/b110268k
50. Endres F, Zein El Abedin S (2002) Nanoscale electrodeposition of germanium on Au(111) from an ionic liquid: an in situ STM study of phase formation—part II. Ge from $GeCl_4$. Phys Chem Chem Phys 4(9):1649–1657. doi:10.1039/b110560b
51. Endres F, Schrodt C (2000) In situ STM studies on germanium tetraiodide electroreduction on Au(111) in the room temperature molten salt 1-butyl-3-methylimidazolium hexafluorophosphate. Phys Chem Chem Phys 2(24):5517–5520. doi:10.1039/b007897m
52. Endres F (2001) Electrodeposition of a thin germanium film on gold from a room temperature ionic liquid. Phys Chem Chem Phys 3(15):3165–3174. doi:10.1039/b102232f
53. Endres F (2002) Electrodeposition of nanosized germanium from $GeBr_4$ and $GeCl_4$ in an ionic liquid. Electrochem Solid State Lett 5(3):C38–C40. doi:10.1149/1.1448185
54. Murugesan S, Kearns P, Stevenson KJ (2012) Electrochemical deposition of germanium sulfide from room-temperature ionic liquids and subsequent Ag doping in an aqueous solution. Langmuir 28(13):5513–5517. doi:10.1021/la300551z
55. Wu MX, Brooks NR, Schaltin S, Binnemans K, Fransaer J (2013) Electrodeposition of germanium from the ionic liquid 1-butyl-1-methylpyrrolidinium dicyanamide. Phys Chem Chem Phys 15(14):4955–4964. doi:10.1039/c3cp44554b
56. Lahiri A, Zein El Abedin S, Endres F (2012) UV-assisted electrodeposition of germanium from an air- and water-stable ionic liquid. J Phys Chem C 116(33):17739–17745. doi:10.1021/jp3062543
57. Al-Salman R, Meng XD, Zhao JP, Li Y, Kynast U, Lezhnina MM, Endres F (2010) Semiconductor nanostructures via electrodeposition from ionic liquids. Pure Appl Chem 82 (8):1673–1689. doi:10.1351/pac-con-09-09-25
58. Al-Salman R, Al Zoubi M, Endres F (2011) Evidence for an influence of anodic decomposition products of ionic liquids on the electrodeposition of Si_xGe_{1-x} semiconductor. J Mol Liq 160 (2):114–118. doi:10.1016/j.molliq.2011.03.003

59. Lahiri A, Olschewski M, Höfft O, Zein El Abedin S, Endres F (2013) Insight into the electrodeposition of Si_xGe_{1-x} thin films with variable compositions from a room temperature ionic liquid. J Phys Chem C 117(49):26070–26076. doi:10.1021/jp408042e
60. Al-Salman R, Mallet J, Molinari M, Fricoteaux P, Martineau F, Troyon M, Zein El Abedin S, Endres F (2008) Template assisted electrodeposition of germanium and silicon nanowires in an ionic liquid. Phys Chem Chem Phys 10(41):6233–6237. doi:10.1039/b809075k
61. Mallet J, Molinari M, Martineau F, Delavoie F, Fricoteaux P, Troyon M (2008) Growth of silicon nanowires of controlled diameters by electrodeposition in ionic liquid at room temperature. Nano Lett 8(10):3468–3474. doi:10.1021/nl802352e
62. Al-Salman R, Endres F (2009) Template-assisted electrodeposition of Si_xGe_{1-x} nanowires with varying length and composition from two different ionic liquids. J Mater Chem 19 (39):7228–7231. doi:10.1039/b909265j
63. Homma T, Komadina J, Nakano Y, Ouchi T, Akiyoshi T, Iichibashi Y, Nishimura Y, Nishida T, Fukunaka Y (2012) Templated electrodeposition of silicon nanowires from ionic liquid. ECS Trans 41(46):9–15. doi:10.1149/1.4729177
64. Mallet J, Martineau F, Namur K, Molinari M (2013) Electrodeposition of silicon nanotubes at room temperature using ionic liquid. Phys Chem Chem Phys 15(39):16446–16449. doi:10.1039/c3cp51522b
65. Lahiri A, Willert A, Zein El Abedin S, Endres F (2014) A simple and fast technique to grow free-standing germanium nanotubes and core-shell structures from room temperature ionic liquids. Electrochim Acta 121:154–158. doi:10.1016/j.electacta.2013.12.084
66. Meng XD, Al-Salman R, Zhao JP, Borissenko N, Li Y, Endres F (2009) Electrodeposition of 3D ordered macroporous germanium from ionic liquids: a feasible method to make photonic crystals with a high dielectric constant. Angew Chem Int Ed 48(15):2703–2707. doi:10.1002/anie.200805252
67. Meng XD, Zhao JP, Li HB, Endres F, Li Y (2012) Enhanced photoluminescence of ordered macroporous germanium electrochemically prepared from ionic liquids. Opt Express 20 (9):9421–9430
68. Liu X, Zhang Y, Ge DT, Zhao JP, Li Y, Endres F (2012) Three-dimensionally ordered macroporous silicon films made by electrodeposition from an ionic liquid. Phys Chem Chem Phys 14(15):5100–5105. doi:10.1039/c2cp23236g
69. Xin W, Zhao J, Ge D, Ding Y, Li Y, Endres F (2013) Two-dimensional Si_xGe_{1-x} films with variable composition made via multilayer colloidal template-guided ionic liquid electrodeposition. Phys Chem Chem Phys 15(7):2421–2426. doi:10.1039/c2cp43983b

Chapter 13
A Disproportionation Reaction-Driven Electroless Deposition of Metals in RTILs

Da Zhang and Takeo Ohsaka

13.1 Introduction

Room-temperature ionic liquids (RTILs) are molten salts composed entirely of cations and anions, which have attracted intensive interest as new possible media for electrodeposition of metals, because RTILs have many advantages such as a wide electrochemical potential window, high ionic conductivity, and good thermal stability [1–4]. Also, there are a rising number of publications [5–15] demonstrating that anions and cations of RTILs can influence chemical and physicochemical reactions. RTILs are not only liquids with interesting physical properties, but also they can totally alter reaction pathways and products [5, 6]. Previous attempts on RTILs have revealed that, as a new kind of solvents, they are favorable templates for the preparation of predictable chemical nanostructures [7–10]. RTILs stabilize metal nanoparticles due to their high ionic and dielectric properties, high polarity, and the ability to form supramolecular networks. A protective shell can be fabricated because anions will coordinate with the prepared metal nanoparticles and cations form hydrogen bridges with anions [7–10]. The controlled and reproducible chemical nanostructures can be prepared and kept stable for several months without any extra stabilizing molecules. Many metal nanoparticles with unusual catalytic properties such as Co [16], Al [17], Mg [18], Ta [19], Ag [20], Au [21], Pt [22], Ir [23], Rh [24], Te [25], and Ru [26] have been prepared in RTILs and applied in extensive chemical and electrochemical reactions, e.g., in

D. Zhang
School of Environmental and Materials Engineering, Shanghai Second Polytechnic University, Jinhai Road 2360, Pudong New Area, Shanghai 201209, China

T. Ohsaka (✉)
Department of Electronic Chemistry, Interdisciplinary Graduate School of Science and Engineering, Tokyo Institute of Technology, 4259 Nagatsuta Midori-ku, Mail Box G1-5, Yokohama 226-8502, Japan
e-mail: ohsaka@echem.titech.ac.jp

A.A.J. Torriero (ed.), *Electrochemistry in Ionic Liquids*,
DOI 10.1007/978-3-319-15132-8_13

the production of hydrogen from methane [27], oxygen reduction [28], and formic acid oxidation [29].

Several methods have been employed for the fabrication of metal nanoparticles, such as vapor deposition and sputtering [30, 31]. However, these physical methods are not so satisfactory to control the morphology and crystallographic orientations. Compared with them, electrodeposition is of great use because of the facial control of the characteristics of metal nanoparticles (e.g., size, shape, and thickness), which may result in various functions of metal nanoparticles-electrodeposited electrodes by adjusting the electrolysis conditions (current density, electrolysis potential, electrolysis time, etc.), bath chemistry (solvent, pH, additives), electrode substrate, and temperature. The electrodeposition of Pt and Pt-Au alloy from RTILs has been reported [32–35]. He et al. have found that the catalytic performance and utilization efficiency for methanol oxidation of the Pt film prepared in RTILs would be much higher than that prepared in aqueous solution because the Pt nanoclusters formed in RTILs would be smaller [32]. Yu et al. have reported a potential-controllable synthesis and deposition of Pt particles in RTIL [15]. On the other hand, electroless deposition is technically convenient to operate and occurs through chemically promoted reduction of metal ions without an externally applied potential. It is effective for arbitrarily controlling the morphology and size of the nanoparticles by suitably choosing the preparation conditions such as the reducing agents, stabilizers, and temperature [21, 36], and has been extensively used in fields such as metal film coating, corrosion, and preparation of bimetallic catalyst particles [37–39]. Of them, disproportionation reactions of different complexes in RTILs have received great attention over the past few years. Disproportionations of non-metal are attracting an increasing interest in diverse fields of chemistry because they are likely to have a significant impact upon the mechanism research in RTILs. Carter et al. have first reported the irreversible oxygen reduction reaction (ORR) in imidazolium chloride-aluminum chloride molten salts and they have considered that the electrogenerated superoxide ($O_2^{\bullet -}$) undergoes a disproportionation reaction with the residual impurities in the used molten salts (e.g., HCl) to form hydrogen peroxide [40], which is same as the suggestion by Islam et al. that electrogenerated $O_2^{\bullet -}$ undergoes disproportionation in the presence of a proton donor in 1-ethyl-3-methylimidazolium tetrafluoroborate ([C_2mim][BF_4]) [41]. Furthermore, Pozo-Gonzalo et al. discovered the generation of $O_2^{\bullet -}$ in the presence of large quantities of water in trihexyl(tetradecyl)phosphonium chloride ([$P_{6\,6\,6\,14}$][Cl]) [42], which is most likely due to a strong ion-pairing phenomenon between the superoxide anion and the phosphonium cation. As a result, reversibility of this redox reaction is observed for the first time in the presence of water. The width of electrochemical window, vital for electrochemical measurement, is decided by the electrochemical stability of RTILs. So the disproportionation of two 1-butyl-3-methylimidazolium radicals, formed in the electrochemical breakdown of 1-butyl-3-methylimidazolium tetrafluoroborate ([C_4mim][BF_4]), was studied and found to react with each other in a radical–radical coupling reaction by Kroon et al. [43]. Lagrost et al. also found that the electron-transfer disproportionation between a hydroxyl adduct $(Ant\text{-}H_2O)^{\bullet +}$, produced from the electrochemical oxidation of anthracene (Ant) and reaction with H_2O, and $Ant^{\bullet +}$ occurs to give Ant and a hydroxy dication adduct $(Ant\text{-}H_2O)^{2+}$ in RTILs [44]. Besides, Ninie et al. reported the disproportionation of polyanions $S_4{}^{2-}$ and $S_8{}^{2-}$ in 1-butyl-3-methyl-imidazolium dicyanamide ($[C_4mim][N(CN)_2]$) [45]. About metal nanoparticles, there are many investigations concerning disproportionation of gold in RTILs, including the fabrication of Au nanoparticles via the disproportionation of two-electron reduction product ($[Au^ICl_2]^-$) of $[Au^{III}Cl_4]^-$ [5, 46] and dendrite-like hierarchical flat Au-Ag nanostructures via the galvanic replacement reactions [47]. Recently, electroless deposition of Au was carried out by Aldous et al. by dissolving Au complex ($H[AuCl_4]$) into 1-butyl-3-methylimidazolium bis(trifluoromethyl)sulfonylimide ($[C_4mim][N(Tf)_2]$), demonstrating that it is a more straightforward and effective method for preparing Au nanoparticles [48]. Zhang et al. found the disproportionation reaction of $[Pt^{II}Cl_4]^{2-}$ will largely influence the size, morphology, and crystalline facets of Pt nanoparticles prepared via electro- and electroless deposition and discussed the feasibility of the mechanism for disproportionation of $[Pt^{II}Cl_4]^{2-}$ in RTILs [6, 49]. In addition, the slow disproportionation of Cu(I) was found to occur in protic ionic liquids and strongly affected by water content [50]. Mukhopadhyay et al. observed the violet color during the electrochemical measurement, which is possibly due to the conversion of Ti(IV) to Ti(III) via a comproportionation reaction, i.e., $Ti^{2+} + Ti^{4+} \rightarrow 2Ti^{3+}$ [51]. Based on previous studies, it is expected that the preparation of metal nanoparticles of interest via disproportionation reaction in RTILs is novel and promising.

The disproportionation reactions of metal nanoparticles could be observed in not only RTILs but also other electrolytes such as organic and aqueous solutions, so it is a significant issue to study the mechanisms in different media and obtain the possible interpretations. This chapter focuses upon understanding the effect and probable mechanism of disproportionation reactions in the formation of metal nanoparticles, especially Au, Pt, and Cu in RTILs.

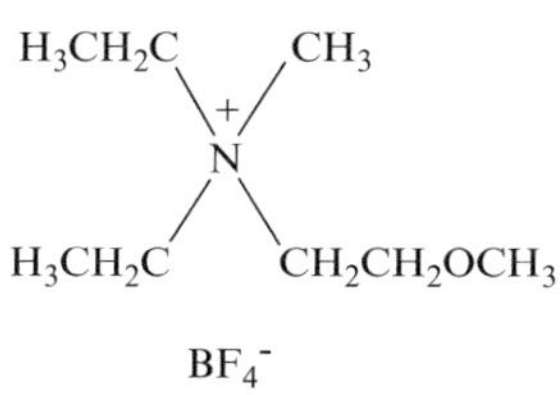

[DEME][BF4]

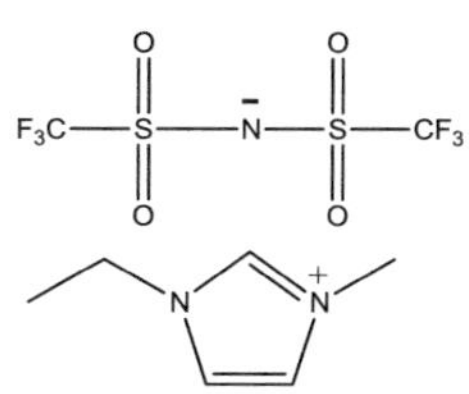

$[C_2mim][N(Tf)_2]$

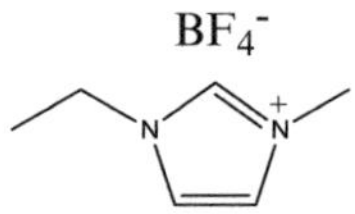

$[C_2mim][BF_4]$

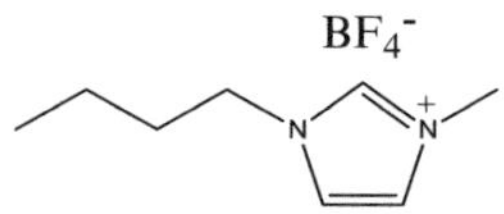

$[C_2mim][BF_4]$

13.2 Disproportionation Reaction

A disproportionation reaction can take place with an element in one oxidation state but produce products in which that element appears in two different oxidation states as shown typically by Eq. 13.1, when the standard potential ($E^0{}_{A'/A}$) of the A′/A redox couple is more positive than that ($E^0{}_{A/A''}$) of the A/A″ redox couple and thus the standard Gibbs free energy (ΔG^0) < 0 for the reaction of Eq. 13.1.

$$2A \rightarrow A' + A'' \tag{13.1}$$

where A, A′, and A″ are all different chemical species with different oxidation states and the oxidation states of A, A′, and A″ are in the order $A'' > A > A'$.

For example, the standard potentials of the complexes of Au [52], Pt [53], and Cu [54] in aqueous solution have been reported (Eqs. 13.2–13.7), confirming that their disproportionation reactions (Eqs. 13.8–13.10) are thermodynamically possible.

$$\left[Au^{III}Cl_4\right]^- + 2e^- \rightleftharpoons \left[Au^{I}Cl_2\right]^- + 2Cl^- \quad E^0 = 0.926\,\mathrm{V\ vs.\ SHE} \tag{13.2}$$

$$\left[Au^{I}Cl_2\right]^- + e^- \rightleftharpoons Au^0 + 2Cl^- \quad E^0 = 1.154\,\mathrm{V\ vs.\ SHE} \tag{13.3}$$

$$\left[Pt^{IV}Cl_6\right]^{2-} + 2e^- \rightleftharpoons \left[Pt^{II}Cl_4\right]^{2-} + 2Cl^- \quad E^0 = 0.68\,\mathrm{V\ vs.\ SHE} \tag{13.4}$$

$$\left[Pt^{II}Cl_6\right]^{2-} + 2e^- \rightleftharpoons Pt^0 + 2Cl^- \quad E^0 = 0.755\,\mathrm{V\ vs.\ SHE} \tag{13.5}$$

$$Cu^{2+} + e^- \rightleftharpoons Cu^+ \quad E^0 = 0.159\,\mathrm{V\ vs.\ SHE} \tag{13.6}$$

$$Cu^+ + e^- \rightleftharpoons Cu^0 \quad E^0 = 0.52\,\mathrm{V\ vs.\ SHE} \tag{13.7}$$

$$3\left[Au^{I}Cl_2\right]^- \rightarrow 2Au^0 + \left[Au^{III}Cl_4\right]^- + 2Cl^- \tag{13.8}$$

$$2\left[Pt^{II}Cl_4\right]^{2-} \rightarrow Pt^0 + \left[Pt^{IV}Cl_6\right]^{2-} + 2Cl^- \tag{13.9}$$

$$2Cu^+ \rightarrow Cu^0 + Cu^{2+} \tag{13.10}$$

13.2.1 Disproportionation Reaction of Gold in RTILs

Gold nanoparticles have very often been synthesized via disproportionation and characterized, but the details of the reduction process in many cases are not elucidated and only a few molecular mechanisms have been proposed. For instance, Eustis et al. suggested that the disproportionation reaction of $[Au^{I}X_2]^-$ (X: Cl or Br) in ethylene glycol involves electron transfer between two gold ions while halide ions are attached to each gold ion [55]. This bimolecular complex contains a gold–gold bond as the following equation.

$$2\left[Au^{I}X_2\right]^- \rightarrow \begin{matrix}\left[{}^{-}X - Au^{I} - X^{-}\right] \\ | \\ \left[{}^{-}X - Au^{I} - X^{-}\right]\end{matrix} \rightarrow X^- + Au^0 + \left[Au^{II}X_3\right]^- \tag{13.11}$$

They found that no gold nanoparticle formation is observed when tetrabromoauric acid is used as the starting material, while the rate of disproportionation and gold nanoparticles formation will increase if replacing Br^- by Cl^-. There will be a large energy barrier to the formation of the gold–gold dimer with Br^- due to the larger size of Br^- ion than that of Cl^- ion [56]. Furthermore, it will be relatively unstable due to the Br^- Br^- repulsion even if the dimer is formed. These two factors decrease the concentration of the dimer complex and in turn decrease the rate of formation of Au^0 and gold nanoparticles. This mechanism is also useful in explaining the effect of the silver ion. The silver ions can react with the halide ions, produced in the disproportionation reaction, to form insoluble AgCl(s) or AgBr(s) (Eq. 13.12). The addition of Ag+ ions to the solution removes the halide ions from the solution, thus driving the equilibrium to the right, which accelerates the formation rate of gold nanoparticles in Eq. 13.11 and results in an increase in the amount of Au^0 produced. Figure 13.1 shows the changes in the absorbance of $HAuCl_4$ in $[C_4mim][BF_4]$ as a function of the irradiation time with and without the addition of silver ions [47].

$$Ag^+ + X^- \rightarrow AgX \downarrow \tag{13.12}$$

Much attention has been devoted to the formation of Au particles in RTILs via a disproportionation of the 2*e*-reduction product of $[Au^{III}Cl_4]^-$ (i.e., $[Au^{I}Cl_2]^-$) to $[Au^{III}Cl_4]^-$ and Au. Xu et al. observed the disproportionation reaction of $[Au^{I}Cl_2]^-$ in aluminum chloride-1-methyl-3-ethylimidazolium, which could be reasonably understood based on the obtained formal potentials ($E^{0\prime}$) of the Au(I)/Au^0 and Au(I)/Au(III) redox couples, i.e., 0.374 ± 0.002 and 0.310 ± 0.002 V, respectively [57]. Oyama et al. [5] found that two reduction waves were obtained at ca. 0.2 and -0.5 V vs. Ag wire in the potential scan from 1.0 to -1.0 V in $[C_2mim][BF_4]$, and two oxidation waves appeared at ca. 0.5 and 1.2 V in the reverse potential scan (Fig. 13.2). The first and second cathodic waves are considered to correspond to the reduction of

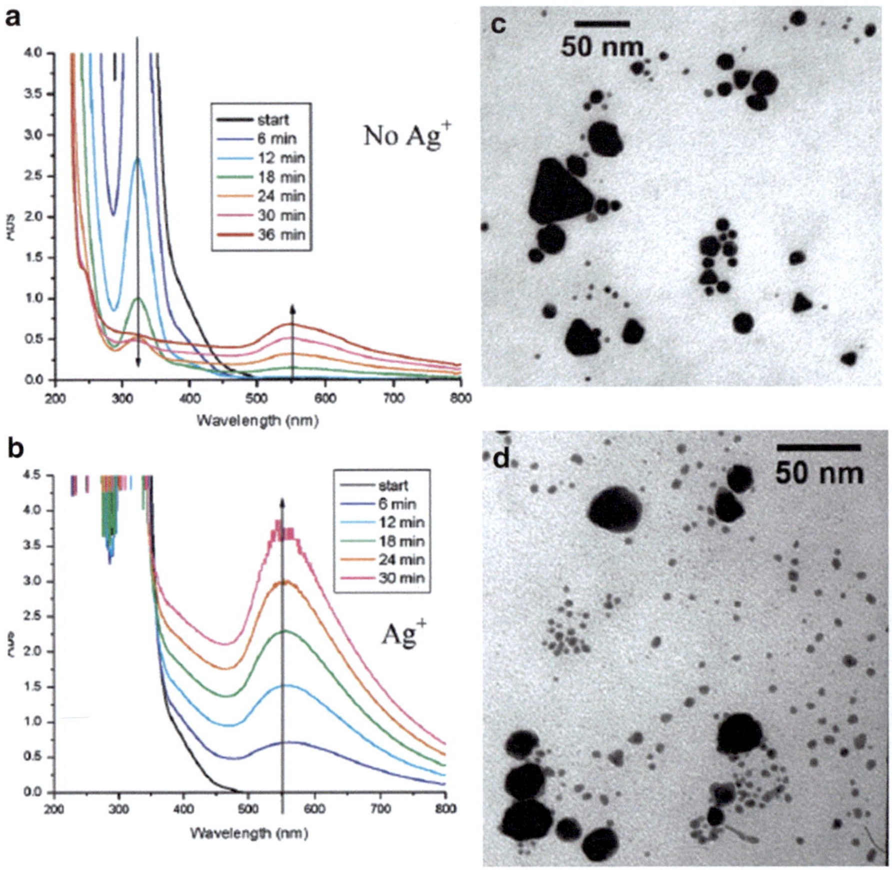

Fig. 13.1 Effect of adding Ag^+ ions on the rate of gold nanoparticle formation in $[C_4mim][BF_4]$: time dependence of the absorption spectra of $HAuCl_4$ solutions during irradiation (**a**) without silver and (**b**) with silver nitrate (3.5×10^{-4} M). TEM images of solutions of $HAuCl_4$ after 36 min of irradiation (**c**) without silver and (**d**) with silver nitrate (3.5×10^{-4} M). The addition of Ag^+ to the solution accelerates the rate of formation of the gold nanoparticles. Reprinted from Ref. [47], Copyright (2010) with permission from the Royal Society of Chemistry

$[Au^{III}Cl_4]^-$ to $[Au^{I}Cl_2]^-$ and further reduction of $[Au^{I}Cl_2]^-$ to Au metal, respectively and the anodic waves at ca. 0.5 and 1.2 V are ascribable to the oxidations of the deposited Au metal to $[Au^{I}Cl_2]^-$ and $[Au^{I}Cl_2]^-$ to $[Au^{III}Cl_4]^-$, respectively. The anodic peak current at 0.5 V was found to increase, resulting from the disproportionation reaction of $[Au^{I}Cl_2]^-$ to $[Au^{III}Cl_4]^-$ and Au, when the electrode potential was held at 0 V for 200 s and then was scanned in the positive direction of potential [5]. They also used an in situ electrochemical quartz crystal microbalance (EQCM) technique based on a Pt film-coated quartz crystal electrode. A series of two-electron (2*e*) and one-electron (1*e*) reductions of the $[Au^{III}Cl_4]^-$ to $[Au^{I}Cl_2]^-$ and $[Au^{I}Cl_2]^-$ to Au metal were recognized at the Pt surface [46]. Besides, the disproportionation

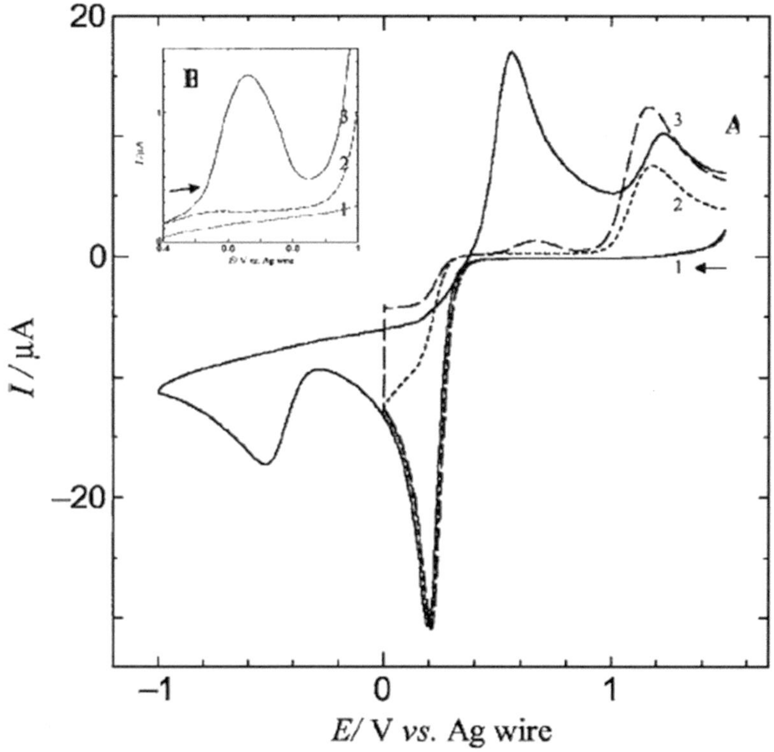

Fig. 13.2 (*A*) CVs obtained at GC electrode ($\Phi = 3$ mm) in N_2-saturated [C_2mim][BF_4] containing 10 mM $[Au^{III}Cl_4]^-$. Potential scan range: (1, *solid line*) 1.5 to −1.0 V vs. Ag wire; (2, *dashed line*) 1.5–0 V. Potential scan rate 10 mV s^{-1}. (*B*) Linear sweep voltammograms obtained in [C_2mim][BF_4] at 10 mV s^{-1} in the (*1*) absence and (*2,3*) the presence of 10 mM $[Au^{III}Cl_4]^-$. Voltammogram 2 is the same as that (*enlarged one*) shown by the *dashed line* in Fig. 13.1a. Voltammogram 3 was measured in the anodic scan after the electrode potential was held at 0 V for 200 s. Reprinted from Ref. [5], Copyright (2007) with permission from the Electrochemical Society

reaction of $[Au^{I}Cl_2]^-$ (i.e., the 2*e*-reduction product of $[Au^{III}Cl_4]^-$) to $[Au^{III}Cl_4]^-$ and Au metal was also observed and the results obtained from the EQCM measurements strongly support the data obtained by the CV measurements [5].

The replacement mechanism is complicated due to simultaneous co-occurrence of multiple reactions in a complex equilibrium environment, as proposed in Scheme 13.1 by Pearson et al. [47]. During the course of Ag oxidation by $[Au^{III}Br_4]^-$ in [C_4mim][BF_4], $[Au^{I}Br_2]^-$, Ag^+ and Br^- ions are initially generated (step 1). The Ag^+ and Br^- ions can undergo a series of reactions to form a soluble $[C_4mim]^+[AgBr_2]^-$ complex [58]. The possible formation of AgBr onto Ag nanoparticles surface (step 2) and its further dissolution by forming a soluble complex (step 3) might be partly responsible for the Ag template size reduction as observed by TEM. Additionally, considering the known slow nature of the $[Au^{I}Br_2]^-$ disproportionation reaction [56] and its strong dependence on the relative overpotential available for the nucleation of resulting Au^0 required to drive the

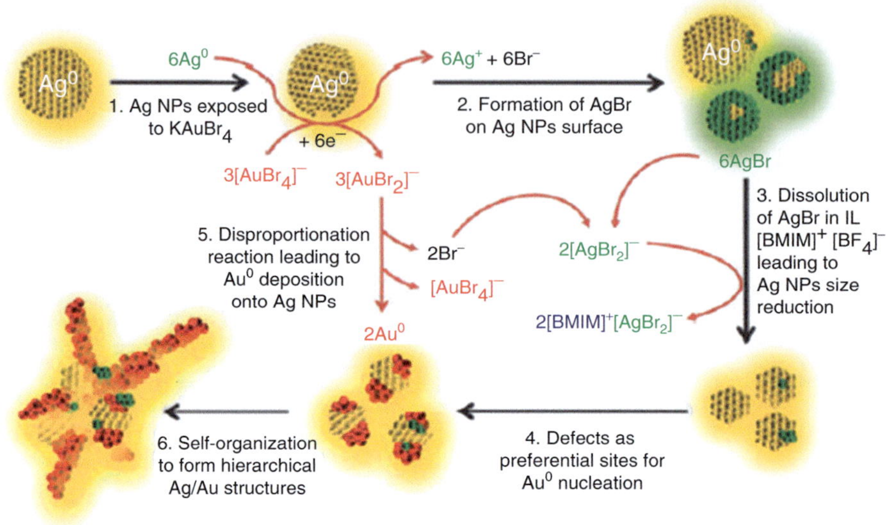

Scheme 13.1 Reaction scheme explaining the proposed mechanism for synthesis of hierarchical dendritic structures in the IL $[C_4mim][BF_4]$. Reprinted from Ref. [47], Copyright (2010) with permission from the Royal Society of Chemistry

forward reaction [59], it is most likely that the defect sites generated on Ag nanoparticles after AgBr dissolution can provide a facile surface for the nucleation of Au^0 generated via disproportionation (step 4). Moreover, new $[Au^{III}Br_4]^-$ ions will be continuously regenerated through this disproportionation reaction (step 5), that will again undergo a series of the aforementioned reaction steps.

Aldous et al. observed electroless deposition of gold on GC when it was immersed in $[C_4mim][N(Tf)_2]$ containing $H[AuCl_4]$ [48]. They suggested that the deposition of gold is not due to the electrochemical instability of $[N(Tf)_2]^-$ but the active sites on the carbon surface and the disproportionation of $[Au^ICl_2]^-$ in the presence of proton. It is probable that protons can activate the surface defects or impurities on the GC electrode and thus facilitate the reduction process of the adsorbed $[Au^{III}Cl_4]^-$ to $[Au^ICl_2]^-$. The presence of H^+ then encourages the deposition of gold via the disproportionation of $[Au^ICl_2]^-$.

13.2.2 Disproportionation Reaction of Platinum in RTILs

13.2.2.1 Redox Behavior of $[Pt^{IV}Cl_6]^{2-}$ and $[Pt^{II}Cl_4]^{2-}$ in $[DEME][BF_4]$ [6]

Typical CVs at GC electrode in $[DEME][BF_4]$ are shown as insets in Fig. 13.3a, b. $[DEME]$ $[BF_4]$ will be reduced when the potential is more negative than −3.0 V and

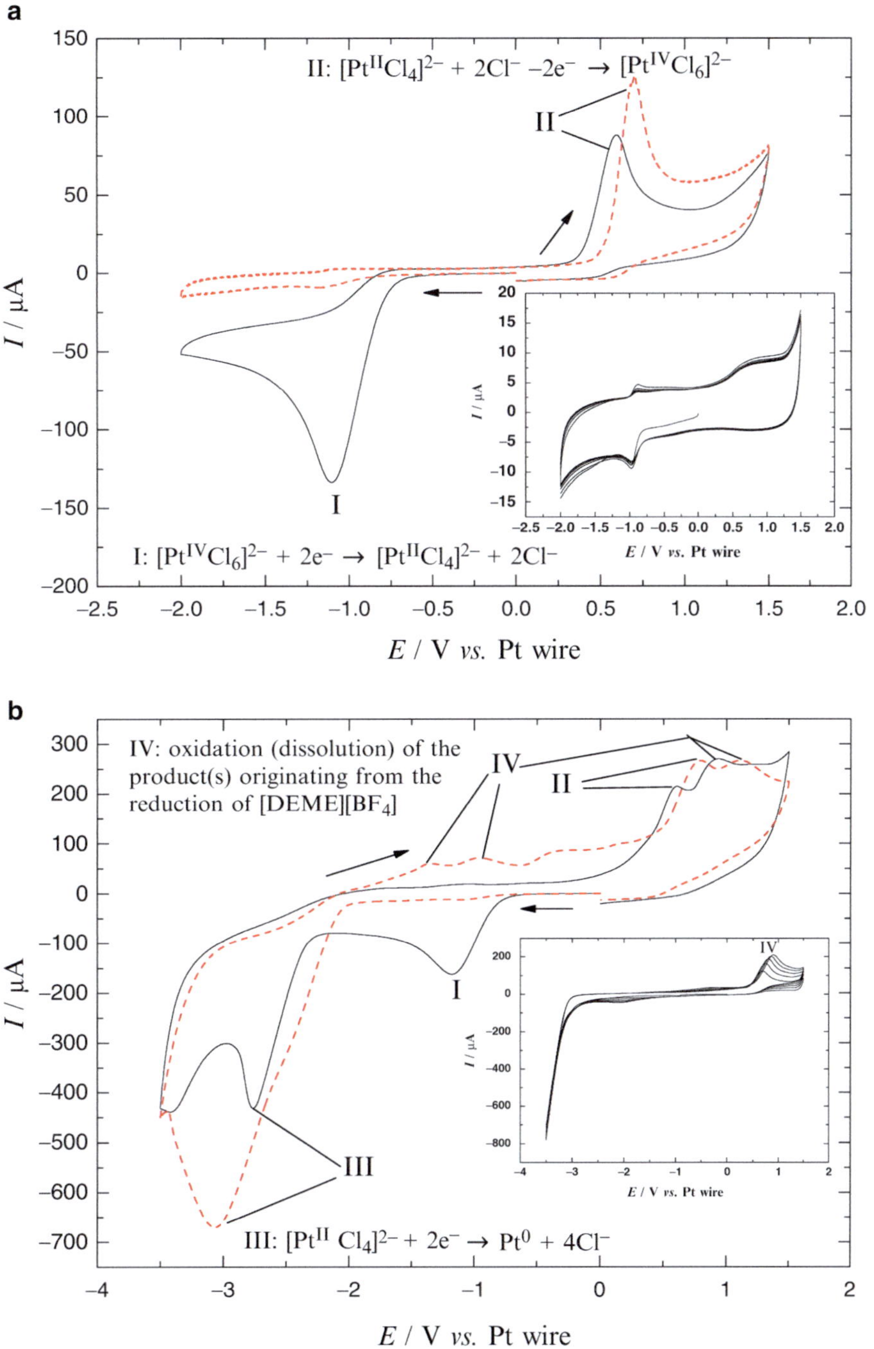

Fig. 13.3 CVs obtained at GC electrode in [DEME][BF_4] containing 70 mM $K_2[PtCl_6]$ (*solid line*) and $K_2[PtCl_4]$ (*dashed line*). Potential scan range: (**a**) −2.0 to 1.5 V and (**b**) −3.5 to 1.5 V vs. Pt wire. *Insets* in (**a**) and (**b**) are CVs obtained at GC electrode in [DEME][BF_4]. Potential scan rate: 500 mV s^{-1}. Reprinted from Ref. [6], Copyright (2011) with permission from the American Chemistry Society

the oxidation peak, IV, is observed at ca. 0.7 V which may correspond to the oxidation of the reduction product of $[DEME]^+$.

Parts a and b of Fig. 13.3 depict typical CVs obtained at GC electrode in N_2-saturated $[DEME][BF_4]$ containing 70 mM $[Pt^{IV}Cl_6]^{2-}$ or $[Pt^{II}Cl_4]^{2-}$. From the comparison of these CVs, the cathodic and anodic peaks observed at ca. −1.1 and 0.6–0.7 V, i.e., I and II, are considered to correspond to the reduction of $[Pt^{IV}Cl_6]^{2-}$ to $[Pt^{II}Cl_4]^{2-}$ and the reverse reaction, respectively (i.e., Eqs. 13.13 and 13.14).

$$\left[Pt^{IV}Cl_6\right]^{2-} + 2e^- \rightarrow \left[Pt^{II}Cl_4\right]^{2-} + 2Cl^- \tag{13.13}$$

$$\left[Pt^{II}Cl_4\right]^{2-} + 2Cl^- - 2e^- \rightarrow \left[Pt^{IV}Cl_6\right]^{2-} \tag{13.14}$$

When the potential was scanned to −3.5 V in the $[Pt^{IV}Cl_6]^{2-}$ containing solution, the second cathodic peak, III, was observed at ca. −2.8 V in agreement with the first cathodic peak observed in the $[Pt^{II}Cl_4]^{2-}$ solution (Fig. 13.3b), indicating that this peak corresponds to the reduction of $[Pt^{II}Cl_4]^{2-}$ to Pt^0 (Eq. 13.15).

$$\left[Pt^{II}Cl_4\right]^{2-} + 2e^- \rightarrow Pt^0 + 4Cl^- \tag{13.15}$$

13.2.2.2 Disproportionation Reaction of $[Pt^{II}Cl_4]^{2-}$ in $[DEME][BF_4]$

As mentioned above, Zhang et al. have found the 2-step reduction of $[Pt^{IV}Cl_6]^{2-}$ to Pt in $[DEME][BF_4]$ [6], i.e., the reduction of $[Pt^{IV}Cl_6]^{2-}$ to $[Pt^{II}Cl_4]^{2-}$ and further reduction of $[Pt^{II}Cl_4]^{2-}$ to Pt^0. The $E^{0\prime}$ of the former is more negative than that of the latter in aqueous solution [54] and thus, in general, the formation of Pt particles via a disproportionation of $[Pt^{II}Cl_4]^{2-}$ to $[Pt^{IV}Cl_6]^{2-}$ and Pt in $[DEME][BF_4]$ can be also expected to take place as Eq. 13.9. Recently, Yu et al. [15] have examined the formation of Pt particles at indium tin oxide (ITO) electrode at a high and low overpotential (i.e., −2.8 and −1.8 V vs. Ag/AgCl(KCl-sat.)) in $[C_4mim][PF_6]$ containing 20 mM H_2PtCl_6 and have demonstrated the following remarkable points: When a high overpotential was applied, the nanoparticles prefer to deposit onto the ITO electrode surface since the formation rate of so-called Pt adatoms $[Pt^0]_{ad}$ is fast and, as a result, the stabilization and thus solubilization of $[Pt^0]_{ad}$ by RTIL molecules through the steric and/or electrostatic interaction(s) are suppressed. Conversely, when a low overpotential was applied, the $[Pt^0]_{ad}$ tend to form clusters that are efficiently stabilized by RTIL molecules through the steric and/or electrostatic interaction(s) between both components, and the nanoparticles prefer to solubilize into RTIL solvent to form a homogeneous dispersion. In this case, by taking the potentials used for the deposition of Pt particles into account, as mentioned above, it is thought that the overall 4-electron reduction of $[Pt^{IV}Cl_6]^{2-}$ to Pt metal occurs at a high overpotential of −2.8 V, while at a low overpotential of −1.8 the disproportionation of $[Pt^{II}Cl_4]^{2-}$ takes place to form Pt metal and $[Pt^{IV}Cl_6]^{2-}$.

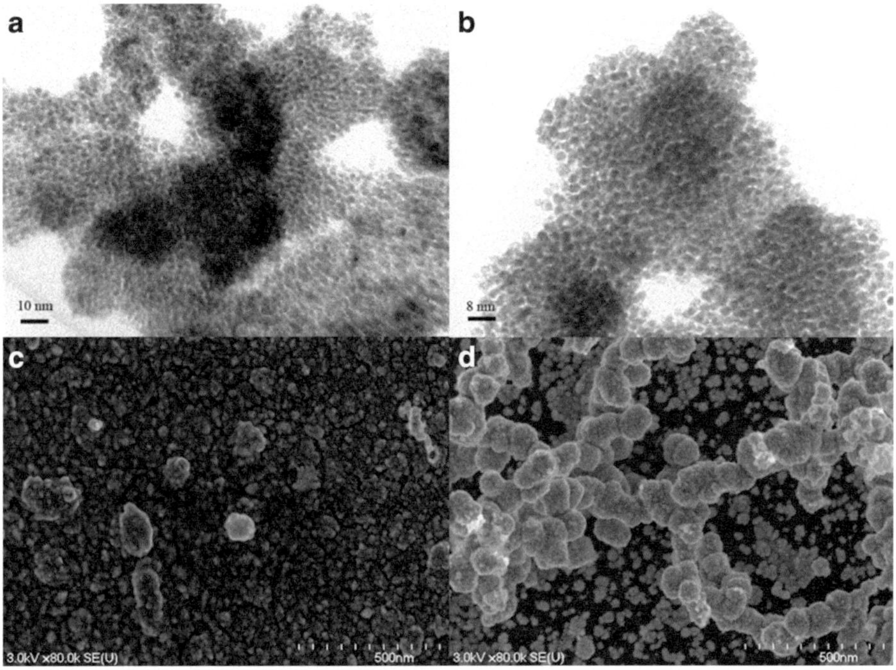

Fig. 13.4 TEM (**a**, **b**) and SEM (**c**, **d**) images of the Pt particles electrodeposited on GC electrodes. Pt particles were electrodeposited in N_2-saturated [DEME][BF_4] containing 70 mM $[PtCl_6]^{2-}$ by holding the potential at (**a**, **c**) −2.0 and (**b**, **d**) −3.5 V for 480 s. Reprinted from Ref. [6], Copyright (2011) with permission from the American Chemistry Society

Typical TEM and SEM images of Pt particles electrodeposited by holding the potential at (a) −2.0 and (b) −3.5 V are shown in Fig. 13.4 [6]. From the comparison of the TEM images (Fig. 13.4a, b), we can see that the sizes of the Pt nanoparticles prepared by holding the electrode potential at −2.0 and −3.5 V, at which the Pt particles are formed via the disproportionation of $[Pt^{II}Cl_4]^{2-}$ to $[Pt^{IV}Cl_6]^{2-}$ and Pt and the 4*e*-reduction of $[Pt^{IV}Cl_6]^{2-}$ to Pt, respectively, are almost the same in the range of ca. 1–2 nm. These small nanoparticles are “grown” to form bigger particles with different morphologies as seen from the SEM images (Fig. 13.4c, d): In the case of the deposition at −2.0 V the GC electrode surface is totally, relatively compactly covered with Pt particles of relatively uniform size of ca. 10–50 nm. On the other hand, in the case of the electrodeposition at −3.5 V, small particles of ca. 50–100 nm and the grown-up particles of ca. 100–200 nm cover the GC surface irregularly and coarsely. These different deposition morphologies may originate from the different deposition mechanisms. In the former case, after Pt nuclei are formed on the GC surface, the subsequent deposition of Pt may occur on the bare GC surface rather than on the previously formed Pt nuclei based on the equilibrium of the disproportionation

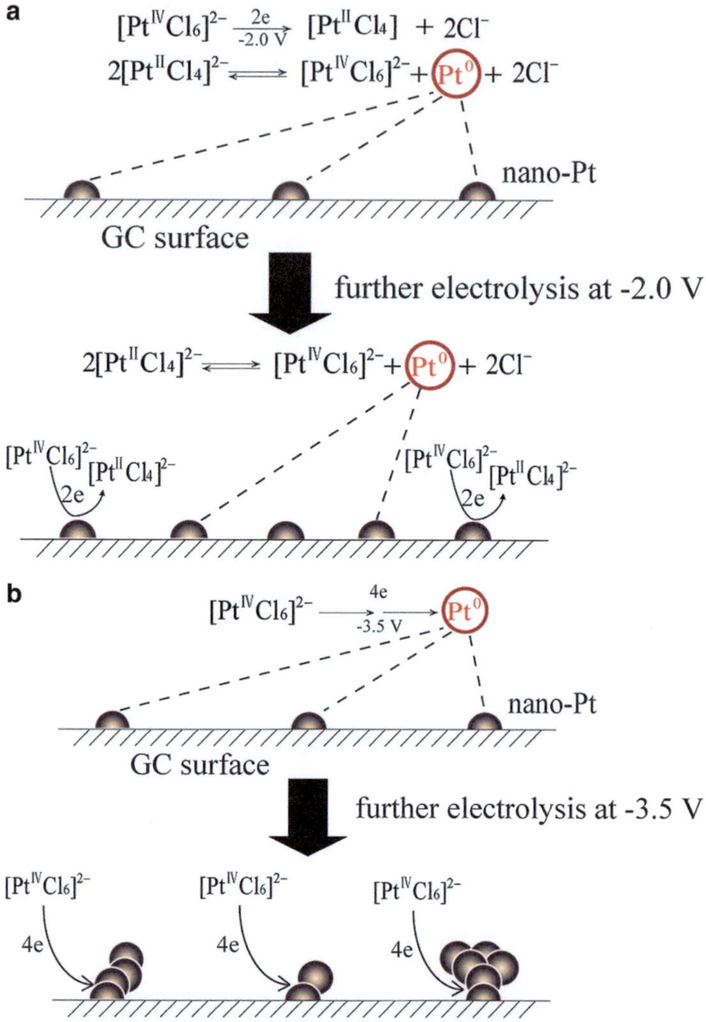

Scheme 13.2 Schematic illustration of Pt deposition via (**a**) the disproportionation reaction of $[Pt^{II}Cl_4]^{2-}$ to $[Pt^{IV}Cl_6]^{2-}$ and Pt and (**b**) the 4*e*-electroreduction of $[Pt^{IV}Cl_6]^{2-}$ to Pt. Reprinted from Ref. [6], Copyright (2011) with permission from the American Chemistry

reaction (Eq. 13.9). Thus, Pt particles of relatively uniform sizes relatively uniformly and compactly cover the whole of the GC surface (Scheme 13.2a, see Fig. 13.4c). On the other hand, in the latter case, i.e., in the case of the 4*e*-electrodeposition of $[Pt^{IV}Cl_6]^{2-}$ to Pt, after Pt nuclei are formed on the GC surface, the subsequent 4*e*-reduction of $[Pt^{IV}Cl_6]^{2-}$ may favorably take place on the previously formed Pt nuclei, because the overpotential for the 4*e*-reduction of $[Pt^{IV}Cl_6]^{2-}$ to Pt is less on the Pt particles than on the GC surface. Thus, the favorable deposition growth of Pt continuously occurs on Pt particles, resulting in the morphologically irregularly grown-up Pt particles (Scheme 13.2b, see Fig. 13.4d). Interestingly, the Pt nanoparticles prepared by holding the potential at −2.0 and −3.5 V are found to be relatively enriched in Pt(100) and Pt(110) facets, respectively.

13.2.2.3 Effects of Proton, Anion, and Temperature on the Disproportionation Reaction of $[Pt^{II}Cl_4]^{2-}$ [49]

No Pt deposition was observed when GC electrodes were immersed in $[C_2mim]$ $[N(Tf)_2]$ containing $K_2[PtCl_4]$ for 1 day, but the Pt particles were prepared by adding organic acids ($H[N(Tf)_2]$ or CH_3SO_3H) into $[C_2mim][N(Tf)_2]$ containing $K_2[PtCl_4]$ under N_2 gas atmosphere. The deposited amount of Pt nanoparticles onto GC electrodes gradually increased with increasing temperature when the GC electrodes were immersed in $[C_2mim][N(Tf)_2]$ containing $K_2[PtCl_4]$ and $H[N(Tf)_2]$. When $K_2[PtCl_6]$ was used instead of $K_2[PtCl_4]$, the deposition of Pt was observed actually when the GC electrode was immersed in $[C_2mim][N(Tf)_2]$ containing $K_2[PtCl_6]$ and $H[N(Tf)_2]$ above 80 °C and the amount increased gradually with increasing temperature. When $[C_2mim][N(Tf)_2]$ was replaced by $[C_2mim][BF_4]$, no deposition of Pt particles could be observed in the absence of CH_3SO_3H and a little bit of the deposition was observed in its presence. The amount of Pt deposition increased only slightly with increasing temperature and no Pt deposition could be observed when $K_2[PtCl_6]$ was used instead of $K_2[PtCl_4]$. Thus, it may be concluded that the formation of a large quantity of Pt particles took place when $[C_2mim][N(Tf)_2]$ contained $K_2[PtCl_4]$ and protons, while a very small quantity of Pt particles was formed when $[C_2mim][BF_4]$ was used instead of $[C_2mim][N(Tf)_2]$ keeping the other conditions constant. When $K_2[PtCl_6]$ was used instead of $K_2[PtCl_4]$, a large amount of Pt particles could be formed only in $[C_2mim][N(Tf)_2]$ containing $H[N(Tf)_2]$ at high temperature ($\geq$80 °C).

13.2.2.4 Mechanism of Electroless Deposition of Pt

The formation of Au metal via the disproportionation of $[Au^{I}Cl_2]^-$ to $[Au^{III}Cl_4]^-$ and Au in $[C_4mim][N(Tf)_2]$ containing protons was proposed by Aldous et al. [48], who suggested the "proton-induced" disproportionation. The fact that the electroless deposition of Pt metal could be observed at 25 °C in $[C_2mim][N(Tf)_2]$ and $[C_2mim][BF_4]$ containing $[Pt^{II}Cl_4]^{2-}$ and protons, but not in the absence of protons may suggest that the proton-induced disproportionation reaction of $[Pt^{II}Cl_4]^{2-}$ might have occurred [49]. The Pt formation was also found to take place in the presence of Li^+ and K^+, although their effect on the Pt formation is significantly small compared with that of H^+ [49].

The disproportionation reaction of $[Pt^{II}Cl_4]^{2-}$ is a divalent anion-divalent anion reaction and thus this bimolecular reaction could be largely effected electrostatically by cations which may effectively approach the "reaction domain" of the disproportionation of $[Pt^{II}Cl_4]^{2-}$ through the so-called ionic liquid organized network [7–10]. As mentioned above, this disproportionation reaction does not occur actually in the $[C_2mim][N(Tf)_2]$ containing $K_2[PtCl_4]$ in the absence of H^+, suggesting that $[Pt^{II}Cl_4]^{2-}$ anions cannot be brought so close together that their disproportionation reaction does not take place, probably because of their electrostatic repulsion. Therefore, the addition of small, freely mobile cations in the $[C_2mim][N(Tf)_2]$ containing $K_2[PtCl_4]$ can be expected to enable this disproportionation, because they can effectively approach the "reaction domain" of the disproportionation of $[Pt^{II}Cl_4]^{2-}$ and consequently reduce the electrostatic repulsion between $[Pt^{II}Cl_4]^{2-}$ anions. The diffusion coefficients of cations can be considered as a reasonable parameter reflecting their mobile degree. The diffusion coefficient of H^+ is larger than that of Li^+ [49, 60]. The diffusion coefficient of K^+ is larger than that of Li^+ in *N*-butyl-*N*-methylpyrrolidinium bis(trifluoromethylsulfonyl)imide ($[C_4mPyr][N(Tf)_2]$) as reported by Wibowo et al. [61]. Thus, it can be thought that the diffusion coefficients of H^+, K^+, and Li^+ in $[C_2mim][N(Tf)_2]$ increase in the order of $H^+ > K^+ > Li^+$. The fact that H^+ is much more effective than K^+ and Li^+ on the disproportionation of $[Pt^{II}Cl_4]^{2-}$ ions and the effectiveness is in the order of $H^+ > K^+ > Li^+$ in agreement with that of their diffusion coefficients. Here it should be noted that the disproportionation of $[Pt^{II}Cl_4]^{2-}$ ions was observed in $[DEME][BF_4]$ without adding these small cations [6], probably due to the weak hydrogen-bond strength between $[DEME]^+$, containing only aliphatic C-H groups, and BF_4^-, namely, the ionic liquid network structure in $[DEME][BF_4]$ is not so rigid compared with that in $[C_2mim][N(Tf)_2]$ and $[C_2mim][BF_4]$, where the imidazolium ring protons (e.g., aromatic C-H groups) in $[C_2mim]^+$ possess a preferential affinity as hydrogen-bond donors to F atoms of $[N(Tf)_2]^-$ or $[BF_4]^-$ [62–64], and thus $[DEME]^+$ ions may easily approach the "reaction domain" of the disproportionation of $[Pt^{II}Cl_4]^{2-}$. In addition, the fact that, in the presence of proton, Pt-NPs formation in $[C_2mim][BF_4]$ is much less than in $[C_2mim][N(Tf)_2]$ may be understood based on the much smaller diffusion coefficient of proton in RTIL containing $[BF_4]^-$ anion than that in RTIL containing $[N(Tf)_2]^-$ anion [65]. The Pt deposition from $[Pt^{II}Cl_4]^{2-}$ and $[Pt^{IV}Cl_6]^{2-}$ was also found to occur via their reduction by carbon monoxide (CO) gas generated from the decomposition of $H[N(Tf)_2]$ at high temperatures (typically $\geq$80 °C) [49].

In summary, the electroless Pt deposition from $[Pt^{II}Cl_4]^{2-}$ in $[C_2mim][N(Tf)_2]$ containing H^+, K^+, or Li^+ takes place via a small cation-assisted disproportionation of $[Pt^{II}Cl_4]^{2-}$ to $[Pt^{IV}Cl_6]^{2-}$ and Pt, that is, this disproportionation reaction is enhanced in the order of $H^+ > K^+ > Li^+$ as well as via the reduction of $[Pt^{II}Cl_4]^{2-}$ by CO gas generated by the decomposition of $H[N(Tf)_2]$ at high temperature ($\geq$80 °C), while the Pt deposition from $[Pt^{IV}Cl_6]^{2-}$ occurs via its reduction by the CO gas (Scheme 13.3).

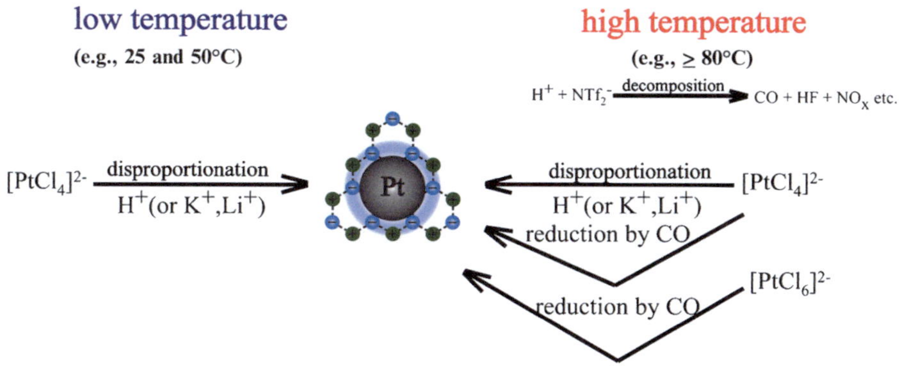

Scheme 13.3 Electroless deposition of Pt-NPs from $[Pt^{II}Cl_4]^{2-}$ and $[Pt^{IV}Cl_6]^{2-}$ in $[C_2mim][N(Tf)_2]$ at low and high temperatures. Reprinted from Ref. [49], Copyright (2013) with permission from the American Chemistry Society

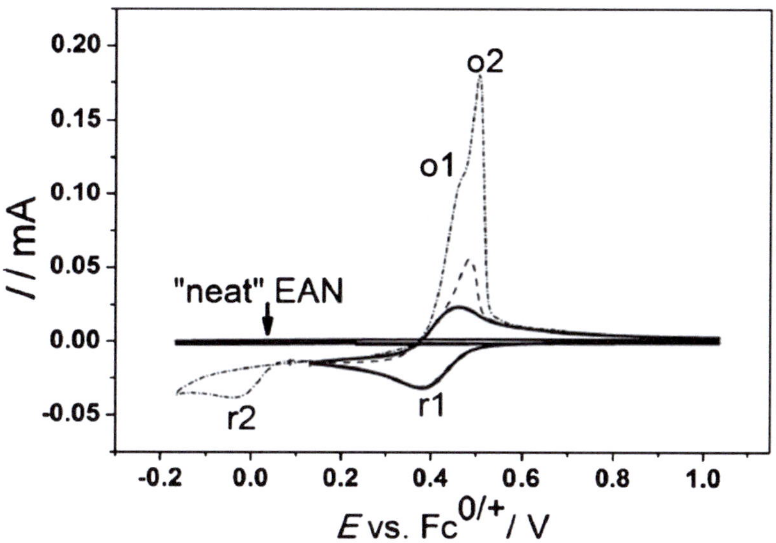

Fig. 13.5 CVs of 10 mM Cu^{2+} in EAN obtained at a scan rate of 0.1 V s^{-1} with a GC electrode. Reprinted from Ref. [50], Copyright (2012) with permission from the Electrochemical Society

13.3 Disproportionation Reaction of Copper in RTILs

The slow disproportionation of Cu(I), strongly affected by water content, was found to occur in protic ionic liquids by Gunawan et al. [50]. Figure 13.5 shows the representative cyclic voltammograms obtained at a GC electrode at a scan rate of 0.1 V s^{-1} for reduction of Cu(II) in protic ionic liquid (PIL), ethylammonium nitrate (EAN). The reduction of Cu(II) in "dried" EAN (<100 ppm H_2O) proceeds with two well-defined mass transport-limited processes ($r1$, $r2$) with the peak potentials at $E_{p,\ r1} = 0.38$ V and $E_{p,\ r2} = -0.017$ V, respectively. These processes are attributed to the two-step, one-electron reduction mechanism (Eqs. 13.6 and 13.7).

On reversing the scan direction at 0.13 V, the reduction of Cu(II) to Cu(I) is close to an ideal one-electron reversible process. Interestingly, scanning the potential to more negative values than the $r1$ process, but prior to the onset of the $r2$ process, an oxidation stripping peak is detected at 0.50 V on the anodic scan. This voltammetric behavior is distinctly different from that in aprotic ionic liquids (AILs), organic solvents, or aqueous media [66–70], and is attributed to the disproportionation of Cu(I) in EAN via Eq. 13.10.

On reversing the scan direction at −0.17 V, a current loop is detected at 0.034 and 0.13 V, indicative of the nucleation and growth mechanism of Cu deposition via Eq. 13.7. When electrolysis is carried out at $E_{dep} = 0.33$ V where Cu is deposited via the disproportionation reaction (Eqs. 13.6 and 13.10), smooth and shiny thin films are generated which contain nanocrystals having diameters 50–100 nm, with some microcrystals sparsely present. Gunawan et al. [50] also suggested that different types of the PILs clearly play a role upon deposition process, presumably owing to different properties such as surface tensions, electrical conductivity, viscosity, diffusion constant, and adsorption.

The disproportionation of Cu(I) to Cu and Cu(II), undergoing rapid reaction, in aqueous media is due to the much higher solvation enthalpy of Cu(II) in comparison to Cu(I) and the aquacomplexed Cu(I) species, e.g., $[Cu(H_2O_2)]^+$ are very unstable [68–72]. As a result, electrodeposition of copper from aqueous electrolytes, carried out with Cu(II) salts, usually follows a one-step reduction mechanism of Cu(II) to Cu^0 [68, 69]. However, studies in ILs such as 1-ethyl-3-methylimidazolium dicyanamide ($[C_2mim][N(CN)_2]$) and $[C_2mim][BF_4]$ show that Cu(I) species are stable [70, 71] and undergo one-electron reduction to Cu metal [72]. The fact that Cu (I) disproportionates at a low rate in the RTILs suggests that the formation of Cu^0 may take place via a slow cation-assisted disproportionation of Cu(I) to Cu(II) and Cu^0.

13.4 Conclusion

A disproportionation reaction of $[Au^IBr_2]^-$, $[Pt^{II}Cl_4]^{2-}$, and Cu^+ as well as a disproportionation reaction-driven electroless deposition of Au, Pt, and Cu nanoparticles in RTILs have been introduced. The voltammograms corresponding to the reductions of $[Au^IBr_2]^-$ to Au^0, $[Pt^{II}Cl_4]^{2-}$ to Pt^0, and Cu^+ to Cu^0 together with those corresponding to the reductions of $[Au^{III}Br_4]^-$ to $[Au^IBr_2]^-$, $[Pt^{IV}Cl_6]^{2-}$ to $[Pt^{II}Cl_4]^{2-}$ and Cu^{2+} to Cu^+ are observed separately and clearly, being different from the electrode reactions of $[Au^{III}Br_4]^-$, $[Pt^{IV}Cl_6]^{2-}$, and Cu^{2+} in aqueous solutions and common organic solvents. Thus, $[Au^IBr_2]^-$, $[Pt^{II}Cl_4]^{2-}$, and Cu^+ can be electrogenerated in situ via the $2e^-$, $2e^-$, and e^- reductions of $[Au^{III}Br_4]^-$,

$[Pt^{IV}Cl_6]^{2-}$, and Cu^{2+}, respectively, resulting in the individual metal nanoparticles, the size and morphology, and the relative ratio of the crystalline facets of which are unique and different from those of the corresponding nanoparticles electrodeposited conventionally. Some factors controlling the disproportionations and their mechanisms are also discussed. A disproportionation reaction-driven electroless deposition in RTILs is expected as a promising procedure to develop novel nanostructures such as electrocatalysts.

Acknowledgments The present work was supported by Grant-in-Aid for Scientific Research (A) (No. 19206079) to T.O. from the Ministry of Education, Culture, Sports, Science and Technology (MEXT), Japan and to D. Z. from Tokyo Institute of Technology Global COE Program for Energy Science.

References

1. Sheldon R (2001) Catalytic reactions in ionic liquids. Chem Commun (Camb) (23):2399–2407
2. Antonietti M, Kuang D, Smarsly B, Zhou Y (2004) Ionic liquids for the convenient synthesis of functional nanoparticles and other inorganic nanostructures. Angew Chem Int Ed 43:4988–4992
3. Wasserscheid P, Welton T (2002) Ionic liquids in synthesis. Wiley-VCH, Weinheim
4. Rogors RD, Seddon KR (2002) Ionic liquids industrial applications to green chemistry. American Chemical Society, Washington, DC
5. Oyama T, Okajima T, Ohsaka T (2007) Electrodeposition of gold at glassy carbon electrodes in room-temperature ionic liquids. J Electrochem Soc 154:D322–D327
6. Zhang D, Chang W, Okajima T, Ohsaka T (2011) Electrodeposition of platinum nanoparticles in a room-temperature ionic liquid. Langmuir 27:14662–14668
7. Redel E, Kramer J, Thomann R, Janiak C (2009) Synthesis of Co, Rh and Ir nanoparticles from metal carbonyls in ionic liquids and their use as biphasic liquid-liquid hydrogenation nanocatalysts for cyclohexene. J Organomet Chem 694:1069–1075
8. Redel E, Thomann R, Janiak C (2008) Use of ionic liquids (ILs) for the IL-anion size-dependent formation of Cr, Mo and W nanoparticles from metal carbonyl $M(CO)_6$ precursors. Chem Commun 15:1789–1791
9. Kramer J, Redel E, Thomann R, Janiak C (2008) Use of ionic liquids for the synthesis of iron, ruthenium, and osmium nanoparticles from their metal carbonyl precursors. Organometallics 27:1976–1978
10. Redel E, Thomann R, Janiak C (2008) First correlation of nanoparticle size-dependent formation with the ionic liquid anion molecular volume. Inorg Chem 47:14–16
11. Park JW, Yamauchi K, Takashima E, Tachikawa N, Ueno K, Dokko K, Watanabe M (2013) Solvent effect of room temperature ionic liquids on electrochemical reactions in lithium-sulfur batteries. J Phys Chem C 117:4431–4440
12. Ueno K, Inaizumi S, Hata K, Watanabe M (2009) Colloidal interaction in ionic liquids: effects of ionic structures and surface chemistry on rheology of silica colloidal dispersions. Langmuir 25:825–831
13. Ueno K, Inaba A, Kondoh M, Watanabe M (2008) Colloidal stability of bare and polymer-grafted silica nanoparticles in ionic liquids. Langmuir 24:5253–5259
14. Yu P, Yan J, Zhang J, Mao L (2007) Cost-effective electrodeposition of platinum nanoparticles with ionic liquid droplet confined onto electrode surface as micro-media. Electrochem Commun 9:1139–1144

15. Yu P, Qian Q, Wang X, Cheng H, Ohsaka T, Mao L (2010) Potential-controllable green synthesis and deposition of metal nanoparticles with electrochemical method. J Mater Chem 20:5820–5822
16. Fukui R, Katayama Y, Miura T (2005) Electrodeposition of cobalt from a hydrophobic room-temperature molten salt system. Electrochemistry 73:567–569
17. Stafford GR, Kongstein OE, Haarberg GM (2006) In situ stress measurements during aluminum deposition from $AlCl_3$-EtMeImCl ionic liquid. J Electrochem Soc 153:C207–C212
18. NuLi Y, Wang J, Wang J, Xu J, Wang P (2005) Electrochemical magnesium deposition and dissolution with high efficiency in ionic liquid. Electrochem Solid State Lett 8:C166–C169
19. Fl Abedin SZ, Farag HK, Moustafa EM, Welz-Biermann U, Endres F (2005) Electroreduction of tantalum fluoride in a room temperature ionic liquid at variable temperatures. Phys Chem Chem Phys 7:2333–2341
20. Borissov D, Aravinda CL, Freyland W (2005) Comparative investigation of underpotential deposition of Ag from aqueous and ionic electrolytes: an electrochemical and in situ STM study. J Phys Chem B 109:11606–11615
21. Zhang H, Cui H (2009) Synthesis and characterization of functionalized ionic liquid-stabilized metal (gold and platinum) nanoparticles and metal nanoparticle/carbon nanotube hybrids. Langmuir 25:2604–2612
22. Scheeren CW, Machado G, Dupont J, Fichtner PFP, Texeira SB (2003) Nanoscale Pt(0) particles prepared in imidazolium room temperature ionic liquids: synthesis from an organometallic precursor, characterization, and catalytic properties in hydrogenation reactions. Inorg Chem 42:4738–4742
23. Dupont J, Fonseca GS, Umpierra AP, Fichtner PFP, Texeira SB (2002) Transition-metal nanoparticles in imidazolium ionic liquids: recyclable catalysts for biphasic hydrogenation reactions. J Am Chem Soc 124:4228–4229
24. Fonseca GS, Umpierre AP, Fichtner PF (2003) The use of imidazolium ionic liquids for the formation and stabilization of Ir^0 and Rh^0 nanoparticles: efficient catalysts for the hydrogenation of arenes. Chemistry 9:3263–3269
25. Zhu YJ, Wang WW, Qi RJ, Hu XL (2004) Microwave-assisted synthesis of single-crystalline tellurium nanorods and nanowires in ionic liquids. Angew Chem Int Ed 43:1410–1414
26. Silveira ET, Umpierre AP, Rossi LM, Machado G, Morais J, Soares GV, Baumvol IJR, Texeira SR, Finchtner PFP, Dupont J (2004) The partial hydrogenation of benzene to cyclohexene by nanoscale ruthenium catalysts in imidazolium ionic liquids. Chemistry 10:3734–3740
27. Qu Y, Sutherland AM, Lien J, Suarez GD, Guo T (2010) Probing site activity of monodisperse Pt nanoparticle catalysts using steam reforming of methane. J Phys Chem Lett 1(1):254–259
28. Tritsaris GA, Greeley J, Rossmeisl J, Norskov JK (2011) Atomic-scale modeling of particle size effects for the oxygen reduction reaction on Pt. Catal Lett 141:909–913
29. Chan KY, Ren JW, Cheng SA, Tsang KY (2004) Supported mixed metal nanoparticles as electrocatalysts in low temperature fuel cells. J Mater Chem 14:505–516
30. Fiordaliso EM, Dahl S, Chorkendorff I (2012) Strong metal support interaction of Pt and Ru nanoparticles deposited on HOPG probed by the H-D exchange reaction. J Phys Chem C 116:5773–5780
31. Yoo SJ, Cho Y-H, Park H-S, Lee JK, Sung Y-E (2008) High utilization of Pt nanocatalysts fabricated using a high-pressure sputtering technique. J Power Sources 178:547–553
32. He P, Liu H, Li Z, Li J (2005) Electrodeposition of platinum in room-temperature ionic liquids and electrocatalytic effect on electro-oxidation of methanol. J Electrochem Soc 152(4):E146–E153
33. Chen H, Dong S (2007) Self-assembly of ionic liquids-stabilized Pt nanoparticles into two-dimensional patterned nanostructures at the air-water interface. Langmuir 23:12503–12507

34. Xiao F, Mo Z, Zhao F, Zeng B (2008) Ultrasonic-electrodeposition of gold-platinum alloy nanoparticles on multi-walled carbon nanotubes-ionic liquid composite film and their electrocatalysis towards the oxidation of nitrite. Electrochem Commun 10:1740–1743
35. Xiao F, Zhao F, Zhang Y, Guo G, Zeng B (2009) Ultrasonic electrodeposition of gold-platinum alloy nanoparticles on ionic liquid-chitosan composite film and their application in fabricating nonenzyme hydrogen peroxide sensors. J Phys Chem C 113:849–855
36. Redel E, Walter M, Thomann R, Vollmer C, Hussein L, Scherer H, Kruger M, Janiak C (2009) Synthesis, stabilization, functionalization and, DFT calculations of gold nanoparticles in fluorous phases (PTFE and ionic liquids). Chemistry 15:10047–10059
37. Lu P, Shi Z, Walker AV (2011) Selective electroless deposition of copper on organic thin films with improved morphology. Langmuir 27:13022–13028
38. Vitry V, Sens AAF, Kanta AF, Delaunois F (2012) Wear and corrosion resistance of heat treated and as-plated duplex NiP/NiB coatings on 2024 aluminum alloys. Surf Coat Technol 206:3421–3427
39. Chio I, Ahn SH, Kim JJ, Kwon OJ (2011) Preparation of Pt_{shell}-Pd_{core} nanoparticle with electroless deposition of copper for polymer electrolyte membrane fuel cell. Appl Catal Environ 102:608–613
40. Carter MT, Hussey CL, Strubinger SKD, Osteryoung RA (1991) Electrochemical reduction of dioxygen in room-temperature imidazolium chloride-aluminum chloride molten salts. Inorg Chem 30:1149–1151
41. Islam MM, Ferdousi BN, Okajima T, Ohsaka T (2005) A catalytic activity of a mercury electrode towards dioxygen reduction in room-temperature ionic liquids. Electrochem Commun 7:789–795
42. Pozo-Gonzalo C, Torriero AAJ, Forsyth M, Howlett PC (2013) Redox chemistry of the superoxide ion in a phosphonium-based ionic liquid in the presence of water. J Phys Chem Lett 4:1834–1837
43. Kroon MC, Buijs W, Peters CJ, Witkamp GJ (2005) Decomposition of ionic liquids in electrochemical processing. Green Chem 8:241–245
44. Lagrost C, Carrie D, Vaultier M, Hapiot P (2003) Reactivities of some electrogenerated organic cation radicals in room-temperature ionic liquids: toward an alternative to volatile organic solvents? J Phys Chem A 107:745–752
45. Manan NSA, Aldous L, Alias Y, Murray P, Yellowlees LJ, Lagunas MC, Hardacre C (2011) Electrochemistry of sulfur and polysulfides in ionic liquids. J Phys Chem B 115:13873–13879
46. Oyama T, Yamaguchi S, Rahman MR, Okajima T, Ohsaka T, Oyama N (2010) EQCM study of the $[Au^{III}Cl_4]^-$-$[Au^{I}Cl_2]^-$-Au(0) redox system in 1-ethyl-3-methylimidazolium tetrafluoroborate room-temperature ionic liquid. Langmuir 26(11):9069–9075
47. Pearson A, O'Mullane AP, Bansal V, Bhargava SK (2010) Galvanic replacement mediated transformation of Ag nanospheres into dendritic Au-Ag nanostructures in the ionic liquid $[BMIM][BF_4]$. Chem Commun 46:731–733
48. Aldous L, Silvester DS, Pitner WR, Compton RG, Lagunas MC, Hardacre C (2007) Voltammetric studies of gold, protons, and $[HCl_2]^-$ in ionic liquids. J Phys Chem C 111:8496–8503
49. Zhang D, Okajima T, Lu D, Ohsaka T (2013) Electroless deposition of platinum nanoparticles in room-temperature ionic liquids. Langmuir 29:11931–11940
50. Gunawan CA, Suryanto BHR, Zhao C (2012) Electrochemical study of copper in room temperature protic ionic liquids ethylammonium nitrate and triethyammonium methylsulfonate. J Electrochem Soc 159(10):D611–D615
51. Mukhopadhyay I, Aravinda CL, Borissov D, Freyland W (2005) Electrodeposition of Ti from $TiCl_4$ in the ionic liquid 1-methyl-3butyl-imidazolium bis(trifluoromethylsulfone) imide at room temperature: study on phase formation by in situ electrochemical scanning tunneling microscopy. Electrochim Acta 50:1275–1281

52. Lingane JJ (1962) Standard potentials of half-reactions involving +1 and +3 gold in chloride medium. Equilibrium constant of the reaction $AuCl_4^- + 2Au + 2Cl^- = 3AuCl_2^-$. J Electroanal Chem 4:332–342
53. El Sawy EN, Birss VI (2013) A comparative study of the electrodeposition of nanoporous Ir and Pt thin films. J Electrochem Soc 160(9):D386–D393
54. Bertocci U, Wagman DD (1985) Standard potential of copper. In: Bard AJ, Parsons R, Jordan J (eds) Standard potentials in aqueous solution. Marcel Dekker, New York, pp 287–293
55. Eustis S, El-Sayed MA (2006) Molecular mechanism of the photochemical generation of gold nanoparticles in ethylene glycol: support for the disproportionation mechanism. J Phys Chem B 110:14014–14019
56. Cotton FA, Wilkinson G (1988) Advanced inorganic chemistry. Wiley, New York
57. Xu XH, Hussey CL (1992) The electrochemistry of gold at glassy carbon in the basic aluminum chloride-1-methyl-3-ethylimidazolium chloride molten salt. J Electrochem Soc 139(11):3103–3108
58. Grishina EP, Vladimirova TV, Ramenskaya LM, Shilovskii KS (2007) Anodic oxidation of silver in 1-butyl-3-methylimidazolium bromide ionic liquid. Russ J Electrochem 43:234–237
59. Aldous L, Silvester DS, Villagran C, Pitner WR, Compton RG, Lagunas MC, Hardacre C (2006) Electrochemical studies of gold and chloride in ionic liquids. New J Chem 30:1576–1583
60. Meng Y, Aldous L, Belding SR, Compton RG (2012) The hydrogen evolution reaction in a room temperature ionic liquid: mechanism and electrocatalyst trends. Phys Chem Chem Phys 14:5222–5228
61. Wibowo R, Aldous L, Jones SEW, Compton RG (2010) The group I alkali metals in ionic liquids: electrodeposition and determination of their kinetic and thermodynamic properties. ECS Trans 33(7):523–535
62. Mele A, Tran CD, Lacerda SHD (2003) The structure of a room-temperature ionic liquid with and without trace amounts of water: the role of C–H···O and C–H···F interactions in 1-*n*-butyl-3-methylimidazolium tetrafluoroborate. Angew Chem Int Ed 42:4364–4366
63. Wu B, Liu Y, Zhang YM, Wang H (2009) Probing intermolecular interactions in ionic liquid-water mixtures by near-infrared spectroscopy. Chemistry 15:6889–6893
64. Noda A, Hayamizu K, Watanabe M (2001) Pulsed-gradient spin-echo ^1H and ^{19}F NMR ionic diffusion coefficient, viscosity, and ionic conductivity of non-chloroaluminate room-temperature ionic liquids. J Phys Chem B 105:4603–4610
65. Meng Y, Aldous L, Belding SR, Compton RG (2012) The formal potentials and electrode kinetics of the proton/hydrogen couple in various room temperature ionic liquids. Chem Commun 48:5572–5574
66. Naskar JP, Chowdhury S, Drew MGB, Datta D (2002) Chemistry of the copper(I)-water bond. Some new observations. New J Chem 26:170–175
67. Moen A, Nicholson DG (1995) Reduction of copper(II) with subsequent disproportionation of copper(I) during the hydrothermal syntheses of microporous silicoaluminum phosphates SAPO-5 and -11. J Chem Soc Faraday Trans 91:3529–3535
68. Hu CC, Wu CM (2003) Effects of deposition modes on the microstructure of copper deposits from an acidic sulfate bath. Surf Coat Technol 176:75–83
69. Grujicic D, Pesic B (2002) Electrodeposition of copper: the nucleation mechanisms. Electrochim Acta 47:2901–2912
70. Chen PY, Sun IW (1999) Electrochemical study of copper in a basic 1-ethyl-3-methylimidazolium tetrafluoroborate room temperature molten salt. Electrochim Acta 45:441–450
71. Leong TI, Sun IW, Deng Mj WCM, Chen PY (2008) Electrochemical study of copper in the 1-ethyl-3-methylimidazolium dicyanamide room temperature ionic liquid. J Electrochem Soc 155:F55–F60
72. Chen PY, Deng MJ, Zhuang DX (2009) Electrochemical codeposition of copper and manganese from room-temperature N-butyl-N-methylpyrrolidinium bis(trifluoromethyl-sulfonyl) imide ionic liquid. Electrochim Acta 54:6935–6940

Part IV
Reactions in Ionic Liquids

Chapter 14
Voltammetry of Adhered Microparticles in Contact with Ionic Liquids: Principles and Applications

Shu-Feng Zhao, Cameron L. Bentley, Michael D. Horne, Alan M. Bond, and Jie Zhang

14.1 Introduction

Historically, electrochemists have focused primarily on studying the electrochemical properties of conductive solid materials (e.g., battery electrodes, metallic deposition and dissolution), and compounds dissolved in molecular solvents such as acetonitrile (CH_3CN), dimethylformamide, and water. In order to study the electrochemistry of insulating or poorly conductive solid phases by voltammetry, elaborate chemically modified electrode configurations were introduced, such as composite electrodes, carbon paste electrodes with an organic binder or an electrolyte solution serving as a binder, suspensions of solid particles, solid particles immobilized on an electrode in a polymer film, or solid compounds sandwiched between solid electrodes. These methods are often limited by the high level of undesirable ohmic (IR_u) distortion and poor resolution resulting from the low electrical conductivity of the solids.

S.-F. Zhao
School of Chemistry, Monash University, Clayton, VIC 3800, Australia

Mineral Resources Flagship, CSIRO, Box 312, Clayton South, VIC 3169, Australia

C.L. Bentley
School of Chemistry, Monash University, Clayton, VIC 3800, Australia

Energy Flagship, CSIRO, Box 312, Clayton South, VIC 3169, Australia

M.D. Horne
Mineral Resources Flagship, CSIRO, Box 312, Clayton South, VIC 3169, Australia

A.M. Bond • J. Zhang (✉)
School of Chemistry, Monash University, Clayton, VIC 3800, Australia
e-mail: jie.zhang@monash.edu

A.A.J. Torriero (ed.), *Electrochemistry in Ionic Liquids*,
DOI 10.1007/978-3-319-15132-8_14

An alternative method for studying the voltammetry of non- or poorly conducting solid phases involves adhering microparticles of the compound of interest to the electrode surface [1]. This method, known as the "voltammetry of adhered microparticles", has been utilized and developed extensively in recent years [2–5]. With this method, useful information can be obtained from the electrochemical response generated at three-phase (electrode/electrolyte/microparticle interface) junctions [2, 4, 6]. For example, it is possible to determine the composition of metal alloys and assess their electrochemical corrosion stability, with notable examples being lead-antimony [1], tin-bismuth [7], mercury-tin, and tin-silver alloys [8]. This method has also proven to be very helpful in understanding the redox chemistry of solids such as Prussian blue [9], Dawson molybdate anion salts [10], and microcrystals of $[Bu_4N][Cr(CO)_5I]$ [11]. In other examples using this method [12], thermodynamic data have been obtained from the measurement of voltammetric peak potentials under diffusion-controlled conditions following dissolution of oxidized or reduced solid in aqueous solutions and ionic liquid media. The use of minimal sample amounts, especially when using microelectrodes, also makes it possible to explore the use of relatively high scan rates because the small current magnitude reduces the severity of the ohmic (IR_u) drop [13]. On the basis of these studies, it may be concluded this technique provides a very worthy addition to established voltammetric methods, as it allows electrochemical studies to be performed with extremely small masses (e.g., single crystals or arrays of microparticles), regardless of electronic conductivity of the solid phase.

In recent years, ionic liquids have received substantial attention in sustainable chemistry applications [14–18] because they typically possess desirable properties, such as negligible volatility, high thermal and chemical stability, low toxicity, nonflammability, and the ability to dissolve a wide range of organic and inorganic compounds. For electrochemical applications, ionic liquids have the additional advantages of intrinsically high conductivity and electrochemical stability, as detailed in recent review articles [5, 19]. Furthermore, electrochemical kinetics, mass transfer and mechanisms associated with charged species are often unique in ionic liquids, in part due to the high ionic strength of this type of media compared to conventional molecular solvent (electrolyte) media. Nevertheless, despite having a range of attractive properties, ionic liquids also have some significant drawbacks when compared to conventional molecular solvents [20], such as high viscosity, high production costs, and difficulties in purification. In addition, trace impurities present in ionic liquids such as halides, dissolved gases, and water, can have a pronounced effect on the physical properties (e.g., viscosity, density, conductivity, solubility, electrochemical window, and electroactive species stability), which may be problematic in some applications. Furthermore, although many compounds possess high thermodynamic solubility in ionic liquids, the rate of their dissolution in such media is frequently extremely slow. As a consequence, it can take several hours to prepare the solutions needed for electrochemical studies, even when the dissolution process is assisted by sonication or heating. A potential complication with slow dissolution kinetics is that the compound may decompose or react with the ionic liquid over these lengthy time scales. Fortunately, the adhered microparticle method often can avoid the need for solute dissolution and therefore provides a powerful method of giving access to electrochemical data in ionic liquid media. Advances in using this method for electrochemical measurements in ionic liquids are reviewed herein.

14.2 Electrode Preparation and Experimental Setup

The simple procedure of mechanical attachment of microcrystalline particles onto electrode surfaces has been described in detail by Scholz and Meyer [1, 2]. This method of immobilizing solid particles on the surface of an electrode has the great advantage that no chemical bonding or modification of the solid particle or the electrode surface is required. In brief, a chemically modified electrode based on adhered microparticles can be prepared by placing a few milligrams of microparticles on high purity weighing or filter paper. Subsequently, a polished working electrode is pressed onto the paper substrate and then rubbed over the solid to ensure that some microcrystalline solid adheres to the electrode surface. Typically, when a 1-mm-diameter electrode is used, microgram-to-nanogram quantities of sample become adhered to the electrode. Finally, after attachment of the solid compound, the working electrode is immersed in the required media and voltammograms are recorded.

Schematic representations of chemically modified working electrodes of the kind described above in contact with either bulk ionic liquid or an ionic liquid layer are provided in Fig. 14.1. In one experiment described below, adhered ferrocene (Fc) is oxidized to Fc^+ (see Sect. 14.3.1), using the experimental arrangement given in Scheme 1 of Fig. 14.1, while the microchemical type of experiment with the experimental setup outlined in Scheme 2 of Fig. 14.1 was used to investigate the oxidation of adhered *trans*-[Mn-(CN)$(CO)_2${P$(OPh)_3$}$(Ph_2PCH_2PPh_2)$] (*trans*-Mn) to *trans*-Mn^+ (see Sect. 14.3.2). The arrangement in Scheme 2 is achieved by initially coating the microparticle-modified electrode (prepared as described above) with a thin layer of ionic liquid using a micro syringe, before the modified electrode is placed in contact with an aqueous electrolyte solution. This microchemical approach produces an electrode configuration that is analogous to ones used for the investigation of redox reactions that occur across organic electrolyte/aqueous electrolyte interfaces [21]. In Scheme 1, the reference and counter electrodes are in direct contact with bulk ionic liquid, whilst in Scheme 2 these electrodes are in contact with the aqueous electrolyte phase. The fact that both the reference and counter electrodes are located in the aqueous phase (in Scheme 2) requires that transport of ions must occur across the ionic liquid-aqueous electrolyte interface. This configuration also introduces an additional junction potential term, which is described by the Nernst-Donnan [22] equation under conditions where a common ion exists between the ionic liquid thin layer and aqueous electrolyte phases. For example, when 1-butyl-3-methylimidazolium hexafluorophosphate ([C_4mim][PF_6]) is the ionic liquid and

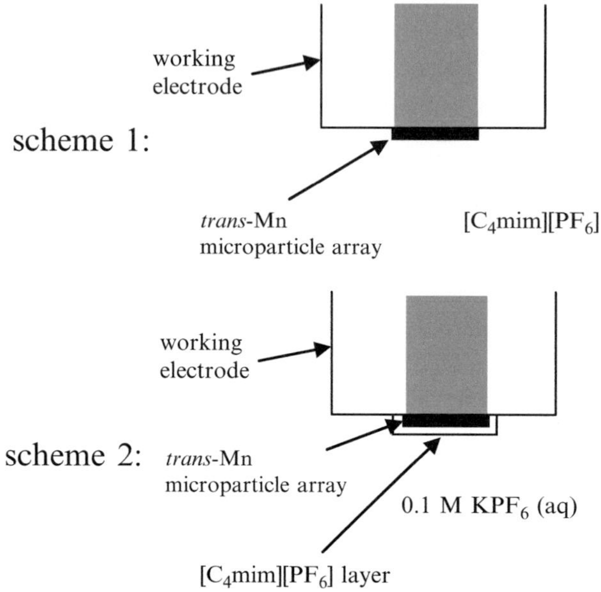

Fig. 14.1 Schematic illustration of two chemically modified working electrode-solution interface configurations used in ionic liquids studies. *Scheme 1*: An array of microparticles adhered to an electrode in contact with bulk ionic liquid. *Scheme 2*: An electrode-microparticle array-ionic liquid layer-aqueous electrolyte microchemical configuration. Adapted with permission from Zhang et al., *Anal. Chem.* 2003, *75*, 6938–6948 [23]. Copyright 2013, American Chemical Society

KPF_6 is used as aqueous electrolyte or when presaturation of the aqueous phase with the ionic liquid is achieved (e.g., $[C_4mim][PF_6]$, although regarded as a hydrophobic ionic liquid, is slightly soluble in water).

14.3 Basic Principles

The basic principles of the voltammetry of adhered microparticles in contact with ionic liquids in both bulk and thin-layer ionic liquid configurations are demonstrated in this section for the oxidation of Fc and *trans*-Mn (structure shown in Fig. 14.2).

14.3.1 Voltammetry of Adhered Microparticles in Contact with Bulk Ionic Liquids

In studies [24] on the Fc oxidation process in $[C_4mim][PF_6]$, it was found that despite the high thermodynamic solubility of Fc (~10 mM), the rate of dissolution

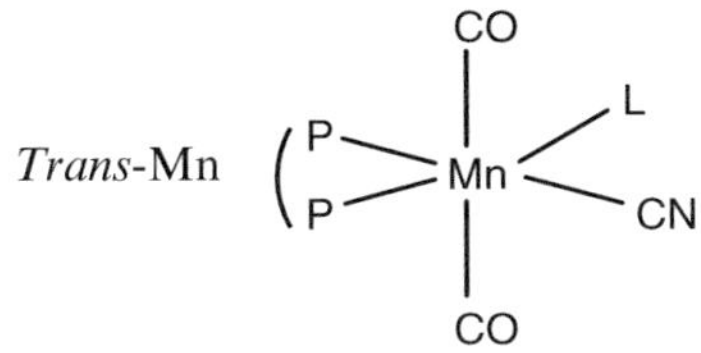

Fig. 14.2 Structure of *trans*-Mn: [L = $P(OPh)_3$ and P-P = $Ph_2PCH_2PPh_2$]

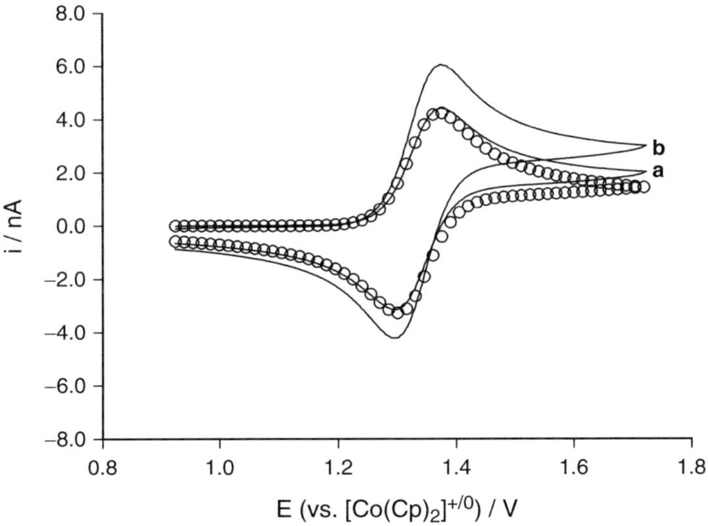

Fig. 14.3 Experimental (*solid lines*) and simulated (*open circles*) voltammograms obtained when solid Fc particles are adhered to a 25 μm diameter Pt microdisk electrode in contact with $[C_4mim]$ $[PF_6]$ in the absence (*a*) and presence (*b*) of 10 mM Fc. The simulated voltammogram (*open circles*) was calculated using parameter values of $k^0 = 0.0025$ cm s^{-1}, $D = 3.0 \times 10^{-8}$ cm^2 s^{-1} for both Fc and Fc^+, $n = 1$, $\alpha = 0.5$, and an electrode area of 4.91×10^{-6} cm^2 and normalized to the peak current in (*b*) in order to demonstrate the shape and peak potential equivalence of voltammograms obtained from dissolved and adhered material. (k^0, D, n, and α are standard heterogeneous electron-transfer rate constants for the electrode reactions, diffusion coefficient, the number of electrons, and electron-transfer rate constant). Adapted with permission from Zhang et al., *Anal. Chem.* 2003, *75*, 2694–2702 [24]. Copyright 2013, American Chemical Society

of Fc was very slow. In order to obtain a 10 mM solution, solid Fc particles had to be left in contact with $[C_4mim][PF_6]$ in an ultrasonic bath for ~3 h. However, it is just the slow kinetics of dissolution that facilitates voltammetric studies of adhered Fc microparticles [24, 25].

Cyclic voltammograms for oxidation of solid Fc mechanically adhered to a 25 μm diameter Pt microdisk electrode in $[C_4mim][PF_6]$ are shown in Fig. 14.3a. The potential of the $Fc^{0/+}$ process was converted to the cobaltocenium $[Co(Cp)_2]^{+/0}$ reference scale using dissolved $[Co(Cp)_2]^+$ as an internal reference standard. The voltammogram obtained under these conditions exhibits the classical behavior associated with a diffusion-controlled solution-based process, provided the mass of the adhered solid is carefully controlled. Importantly, under adhered solid-state conditions, although the current magnitude is enhanced, the oxidation/reduction peak potential and scan rate dependence of the peak current are indistinguishable from voltammetric measurements made with dissolved Fc. Also shown in Fig. 14.3 is a cyclic voltammogram obtained from the oxidation of adhered Fc microparticles in the presence of 10 mM dissolved Fc (see Fig. 14.3b). The voltammetric characteristics (i.e., shapes and positions) are essentially identical to those obtained from dissolved Fc or adhered Fc microparticles, with the current magnitudes of mixtures being additive. Finally, the simulated voltammogram shown in Fig. 14.3 (normalized with respect to the peak current) illustrates the equivalence of the characteristics (i.e., shapes and peak potentials) of voltammograms obtained from dissolved Fc or adhered Fc microparticles.

In order to probe the nature of charge neutralization and dissolution processes that accompany the voltammetry of adhered solid Fc in $[C_4mim][PF_6]$, electrochemical quartz crystal microbalance (EQCM) measurements were undertaken (Fig. 14.4). There is no weight loss prior to the onset of the oxidation process, as evident from the lack of change in mass-E data in Fig. 14.4, consistent with the extremely slow dissolution kinetics of Fc. Upon sweeping the potential, in the positive potential direction, an increase in mass is observed as soon as Fc starts to be oxidized to Fc^+, a process which is assumed to be accompanied by insertion of $[PF_6]^-$ into the Fc solid, as required to maintain charge balance. At more positive potentials, this initial mass increase region is followed by a much larger (and ongoing) loss of mass. Finally, it can be noted that by the time the potential has returned to its initial value, a significant net loss of mass is evident. These EQCM observations indicate that following oxidation of Fc, a significant level of dissolution of the product, $[Fc][PF_6]$ (solid), occurs on the voltammetric time scale. Accordingly, the oxidation of Fc at the particle/ionic liquid interface can be represented by the reaction scheme shown below:

$$\begin{array}{ccc} \text{Fc(solid)} + \text{PF}_6^- \text{(ionic liquid)} & \rightleftharpoons & [\text{Fc}^+][\text{PF}_6^-]\text{(solid)} + \text{e}^- \\ & & \updownarrow \\ & & \text{Fc}^+\text{(ionic liquid)} + \text{PF}_6^- \text{(ionic liquid)} \end{array} \tag{14.1}$$

$$\text{Fc(ionic liquid)} \rightleftharpoons \text{Fc}^+\text{(ionic liquid)} + \text{e}^- \tag{14.2}$$

For convenience, this reaction can be written in terms of two different one electron transfer pathways:

$$\text{Fc(solid)} \overset{E^{0'}_{\text{solid}}}{\rightleftharpoons} \text{Fc}^+\text{(ionic liquid)} + \text{e}^- \tag{14.3}$$

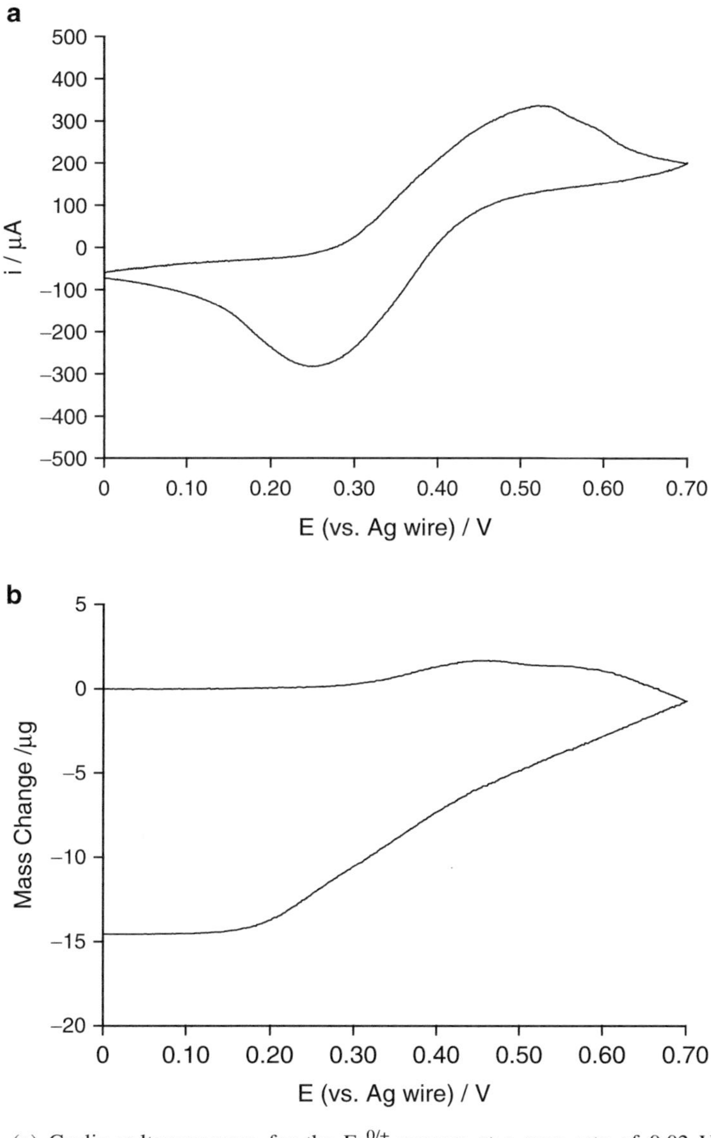

Fig. 14.4 (**a**) Cyclic voltammogram for the $Fc^{0/+}$ process at a scan rate of 0.02 V s^{-1} and (**b**) EQCM data obtained when Fc particles are adhered to an Au electrode surface in contact with $[C_4mim][PF_6]$. Adapted with permission from Zhang et al., *Anal. Chem.* 2003, *75*, 2694–2702 [24]. Copyright 2013, American Chemical Society

$$\text{Fc(ionic liquid)} \overset{E_{\text{il}}^{0'}}{\rightleftharpoons} \text{Fc}^{+}\text{(ionic liquid)} + \text{e}^{-} \tag{14.4}$$

where $E_{\text{solid}}^{0'}$ and $E_{\text{il}}^{0'}$ are the formal potentials of processes involving solid- and solution-phase species (Eq. 14.3) and solution-phase species only (Eq. 14.4), respectively.

With the aim of establishing the relationship between the formal potentials for the parallel electron-transfer pathways given by Eqs. 14.3 and 14.4, a thermodynamic analysis of the voltammetry of Red(solid) adhered to an electrode surface can be undertaken:

$$\begin{array}{ccc} \text{Red(solid)} + \text{PF}_6^- \text{(ionic liquid)} & \xrightleftharpoons{E^{0'}_{\text{solid1}}} & [\text{Ox}][\text{PF}_6]\text{(solid)} + \text{e}^- \\ \Big\updownarrow K_2 & & \Big\updownarrow K_1 \\ \text{Red(ionic liquid)} & & \text{Ox}^+ \text{(ionic liquid)} + \text{PF}_6^- \text{(ionic liquid)} \end{array} \quad (14.5)$$

$$\text{Red(ionic liquid)} \xrightleftharpoons{E^{0'}_{\text{il}}} \text{Ox}^+\text{(ionic liquid)} + \text{e}^- \quad (14.6)$$

where the general terms Red and Ox are now used instead of Fc and Fc^+ respectively. In this scheme, $K_1 = [\text{Ox}^+\text{(ionic liquid)}][\text{PF}_6^-\text{(ionic liquid)}]/[[\text{Ox}][\text{PF}_6]\text{(solid)}]$ is the equilibrium constant for the process involving the dissolved species (Ox^+(ionic liquid) and PF_6^-(ionic liquid)) and [Ox][PF_6](solid) at the particle/ionic liquid interface, $K_2 = [\text{Red(ionic liquid)}]/[\text{Red(solid)}]$ is the equilibrium constant for the dissolution processes involving Red(solid) and Red(ionic liquid), and $E^{0'}_{\text{solid1}}$ is the formal potential of the process involving a solid redox transformation. It follows that:

$$E^{0'}_{\text{solid}} = E^{0'}_{\text{solid1}} - (RT/F)\ln K_1 \quad (14.7)$$

$$E^{0'}_{\text{il}} = E^{0'}_{\text{solid1}} + (RT/F)\ln K_2/K_1 \quad (14.8)$$

where $E^{0'}_{\text{solid}}$, $E^{0'}_{\text{solid1}}$ and $E^{0'}_{\text{il}}$ are as defined above. Therefore:

$$E^{0'}_{\text{il}} - E^{0'}_{\text{solid}} = (RT/F)\ln K_2 \quad (14.9)$$

From the EQCM data (see Fig. 14.4), it may be concluded that the thermodynamic analysis in Eqs. 14.7–14.9 is likely to be applicable on the voltammetric time scale due to the slow rate of the dissolution/reprecipitation process involving Red (solid) and Red(ionic liquid). As per Eqs. 14.5 and 14.6, the processes that contribute to the voltammetry are assumed to be: (1) oxidation of Red(solid) to Ox^+(solid), which is accompanied by charge neutralization involving the insertion of $[PF_6]^-$; (2) dissolution of [Ox][PF_6](solid) with the equilibrium relationship between the dissolved species (Ox^+(ionic liquid)) and $[PF_6]^-$(ionic liquid), and [Ox][PF_6](solid) at the particle/ionic liquid interface being governed by the equilibrium constant K_1; and (3) reduction of solution-phase Ox^+ to solution-phase Red. Thus, overall the processes to be modelled are as follows:

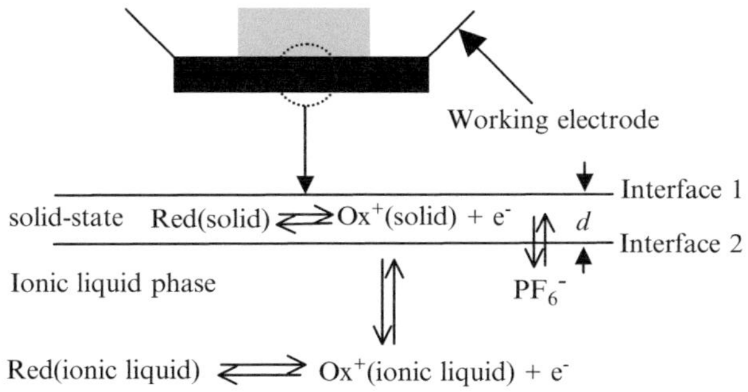

Fig. 14.5 Schematic diagram of the voltammetry of Red(solid) adhered to an electrode surface. *d* is the thickness of the adhered solid. Adapted with permission from Zhang et al., *Anal. Chem.* 2003, *75*, 2694–2702 [24]. Copyright 2013, American Chemical Society

$$\text{Red(solid)} \rightleftharpoons \text{Ox}^{+}\text{(ionic liquid)} + \text{e}^{-} \tag{14.10}$$

$$\text{Red(ionic liquid)} \rightleftharpoons \text{Ox}^{+}\text{(ionic liquid)} + \text{e}^{-} \tag{14.11}$$

which give rise to the following thermodynamic relationships:

$$E_{\text{solid},1/2} = E^{0'}_{\text{solid}} - (RT/F)\ln(D_{\text{il,Ox}}/D_{\text{solid,Red}})^{1/2} \tag{14.12}$$

$$E_{\text{il},1/2} = E^{0'}_{\text{il}} - (RT/F)\ln(D_{\text{il,Ox}}/D_{\text{il,Red}})^{1/2} \tag{14.13}$$

where $E_{\text{solid},1/2}$ and $E_{\text{il},1/2}$ are the half-wave potentials of the processes given in Eqs. 14.10 and 14.11 respectively, ($E_{1/2} = \left(E_{\text{p}}^{\text{ox}} + E_{\text{p}}^{\text{red}}\right)/2$ with E_{p}^{ox} and $E_{\text{p}}^{\text{red}}$ being peak potentials for the oxidation and reduction component of the experiment respectively), and $D_{\text{il,Ox}}$, $D_{\text{il,Red}}$, and $D_{\text{solid,Red}}$ are the diffusion coefficients of the oxidized or reduced forms in the ionic liquid and solid phase respectively. In modelling the processes described in Eqs. 14.10 and 14.11, it was assumed that both the solid- and solution-phase Red/Ox^{+} processes occur at Interface 2 shown in Fig. 14.5 and exhibit diffusion-controlled, reversible, one-electron reactions indicating that all relevant species at the particle/ionic liquid interface are governed by the Nernst equation [26]. Furthermore, the model assumes the absence of reactions between any oxidized and reduced species, and the theoretical analysis does not take into account any change of film thickness arising from dissolution. Studies on other systems have suggested that a hopping-type electron-transfer process in the solid may be treated as an overall diffusion problem having an apparent diffusion coefficient [27, 28]. That is, the electrode reaction and mass transport mechanisms of a dissolved species in a thick-layer format are adopted here for the surface-confined species.

To characterize the conditions under which the shape and position of voltammograms of dissolved species and adhered microparticles are identical, as

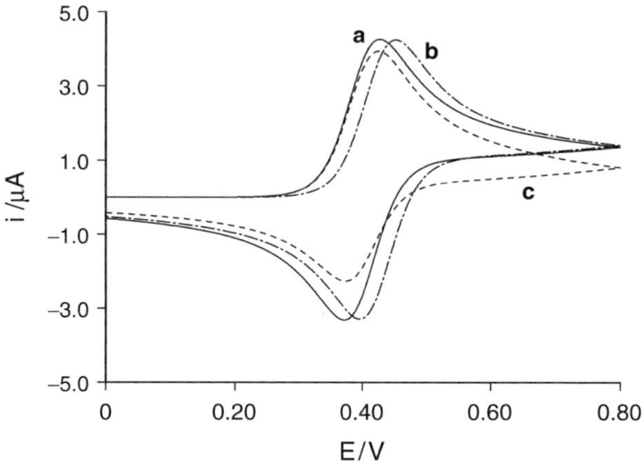

Fig. 14.6 Influence of $E^{0'}_{\text{solid}}$ and d (defined in Fig. 14.5) on voltammograms simulated according to the mechanism postulated in Fig. 14.5. Voltammogram a: $E^{0'}_{\text{solid}} = 0.3$ V, $d = 1.0$ μm (effectively undistinguishable from purely solution-phase voltammogram with $E_{1/2} = E^{0'}_{\text{il}} = 0.4$ V). Voltammogram b: $E^{0'}_{\text{solid}} = 0.5$ V and $d = 1.0$ μm. Voltammogram c: $E^{0'}_{\text{solid}} = 0.3$ V and $d = 0.1$ μm. Adapted with permission from Zhang et al., *Anal. Chem.* 2003, *75*, 2694–2702 [24]. Copyright 2013, American Chemical Society

found experimentally, simulations based on the model described in Eqs. 14.10 and 14.11 [24] have been undertaken to give the simulated cyclic voltammograms shown in Fig. 14.6. These simulations are in agreement with experimental observations, and demonstrate that a voltammogram obtained from adhered particles can be indistinguishable from that of the dissolved species if the solid layer is sufficiently thick, the product dissolution process is fast enough, and $E_{\text{solid1,1/2}}$ is not too much more positive than $E_{\text{il,1/2}}$. The simulations also predict that a nonsymmetrical-shaped voltammogram will result if the mass of the solid particles on the electrode surface is not sufficient to provide a continuous supply of dissolved (product) material. This was observed experimentally for the oxidation of Fc and the reduction of Fc^+ [24]. The model also predicts that if the mass of Fc or $[Fc][PF_6]$ on the electrode surface is very large, the solution forms of either Fc or Fc^+ will be reprecipitated on the electrode surface under conditions where their solubility is exceeded during the course of the reduction or oxidation process respectively. Under these high-mass conditions, voltammograms with surface-confined characteristics are expected. Therefore, in practice, it is important to control the loading of electroactive species, in order to obtain well-defined diffusion-controlled voltammograms.

14.3.2 Voltammetry of Adhered Microparticles in Contact with a Thin Layer of Ionic Liquid

The experimental arrangement outlined in Scheme 2 of Fig. 14.1 is relevant when a microparticle-modified electrode is coated with a thin layer of ionic liquid before being placed in contact with aqueous solution. The results of modelling of the [*trans*-Mn]$^{0/+}$ process in both the ionic liquid dissolved and adhered states have been described in detail by Zhang and Bond [23] and compared to results obtained by microchemical methods at an electrode/ionic liquid/aqueous electrolyte interface under the conditions of outlined in Scheme 2.

A cyclic voltammogram derived from the oxidation of solid *trans*-Mn microparticles adhered to a glassy carbon (GC) electrode in [C_4mim][PF_6] is shown in Fig. 14.7. A well-defined, reversible, one-electron process is observed under these conditions and aside from the magnitude of the current, the voltammogram is essentially indistinguishable from that derived from dissolved *trans*-Mn in this media. By analogy with on the $Fc^{0/+}$ process (Eqs. 14.5 and 14.6), the reactions occurring at the electrode/microparticle/ionic liquid interface under these conditions can be summarized as follows:

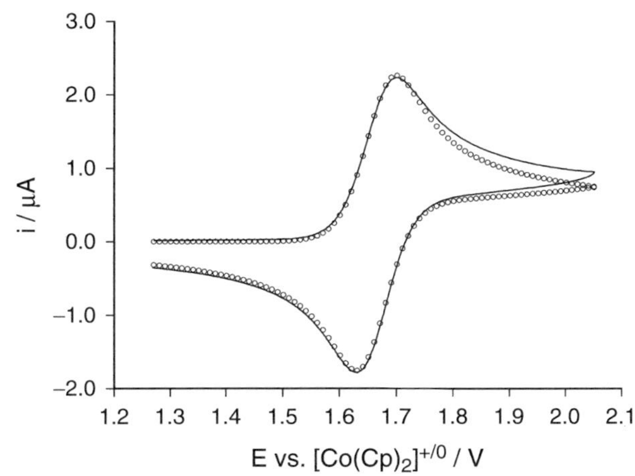

Fig. 14.7 Comparison of an experimental (*solid lines*) (scan rate 0.1 V s^{-1}) cyclic voltammogram obtained when an array of solid *trans*-Mn microparticles is adhered to a 1 mm diameter GC disk electrode (high-mass case) that is placed in contact with [C_4mim][PF_6] and simulated data (*open circle*) that represents a cyclic voltammogram for the case where *trans*-Mn is dissolved in bulk ionic liquid. The simulated voltammogram was calculated for a reversible one-electron-transfer process using $D = 9.1 \times 10^{-9}$ cm^2 s^{-1} for both *trans*-Mn and [*trans*-Mn]$^+$, electrode area 0.00857 cm^2, uncompensated resistance (R_u) = 4,500 Ω, and temperature (T) = 293 K and has been normalized to the peak current of the experimentally obtained voltammogram in order to demonstrate the shape and peak potential equivalence of voltammograms obtained from adhered and dissolved material. Adapted with permission from Zhang et al., *Anal. Chem.* 2003, *75*, 6938–6948 [23]. Copyright 2013, American Chemical Society

$$trans\text{-}Mn(solid) + PF_6^-(ionic\ liquid) \rightleftharpoons [trans\text{-}Mn][PF_6](solid) + e^-$$
$$\downarrow$$
$$[trans\text{-}Mn]^+(ionic\ liquid) + PF_6^-(ionic\ liquid) \quad (14.14)$$

$$trans\text{-}Mn(ionic\ liquid) \rightleftharpoons [trans\text{-}Mn]^+(ionic\ liquid) + e^- \quad (14.15)$$

Cyclic voltammetric data obtained (low- and high-mass cases) using the experimental setup outlined in Scheme 2 of Fig. 14.1 and with microparticles of *trans*-Mn and coated with a layer of ionic liquid (thick 1 μL, or thin $\leq$0.1 μL) in an aqueous (0.1 M KPF_6) electrolyte are presented in Fig. 14.8. Examination of

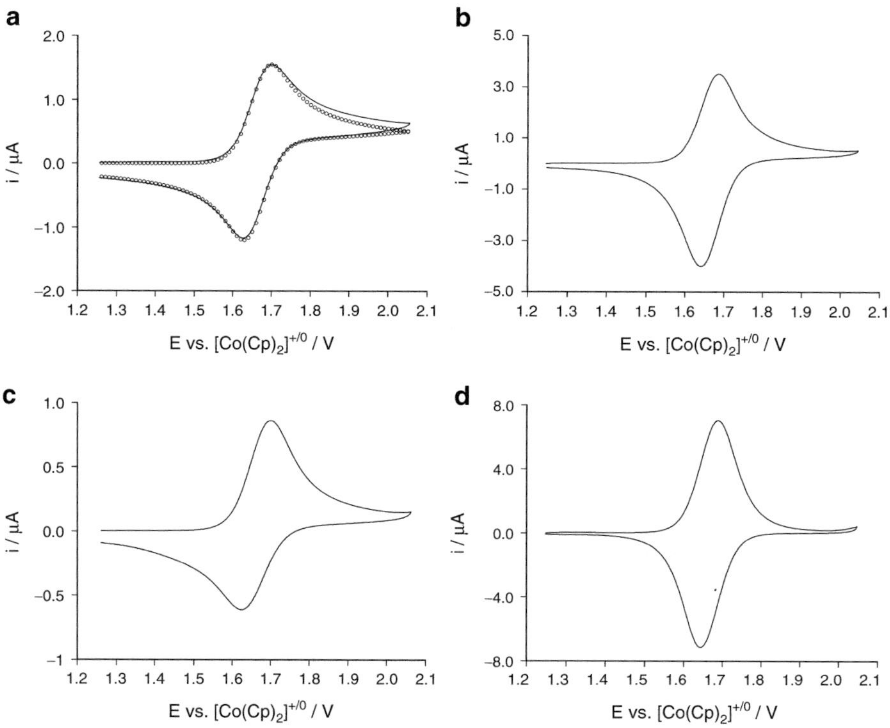

Fig. 14.8 Cyclic voltammograms obtained when an array of solid *trans*-Mn microparticles is adhered to a 1 mm diameter GC electrode surface that is placed in contact with a layer of $[C_4mim][PF_6]$ (Scheme 2, Fig. 14.1; scan rate 0.1 V s^{-1}). (**a**) Comparison of data obtained with a high mass of *trans*-Mn and thick layer of $[C_4mim][PF_6]$ (*solid lines*) and a simulated cyclic voltammogram (*open circle*) representing oxidation of *trans*-Mn dissolved in the ionic liquid. The simulated used the parameters given in the caption to Fig. 14.7 and was normalized to the experimental peak current. (**b**) High mass of *trans*-Mn and thin layer of $[C_4mim][PF_6]$. (**c**) Low mass of *trans*-Mn and thick layer of $[C_4mim][PF_6]$. (**d**) Low mass of *trans*-Mn (Sub-μg) and thin layer of $[C_4mim][PF_6]$. Adapted with permission from Zhang et al., *Anal. Chem.* 2003, *75*, 6938–6948 [23]. Copyright 2013, American Chemical Society

Fig. 14.8a reveals that when the electrode is modified with a high mass of *trans*-Mn microparticles and a thick layer of [C_4mim][PF_6], a well-defined, diffusion-controlled, one-electron reversible oxidation process is observed. Once again, aside from the current magnitude, the voltammogram obtained under microchemical conditions (Fig. 14.8a) is indistinguishable from that obtained from the adhered *trans*-Mn microparticles in contact with bulk [C_4mim][PF_6] (Fig. 14.7).

Keeping the quantity of *trans*-Mn constant, while reducing the volume of the [C_4mim][PF_6] layer to submicroliter level changes the predominant mode of mass transport from semi-infinite diffusion to finite diffusion, as applies in a thin-layer cell, to give the voltammograms shown in Fig. 14.8b. Under these conditions, the peak height of the reduction component is larger than the oxidation one and the overall current magnitude is enhanced as a result of the more efficient electrolysis than achieved in a thin-layer ionic liquid configuration. Conversely, keeping the volume of the [C_4mim][PF_6] layer constant, while reducing the quantity of *trans*-Mn (e.g., low mass/thick-layer conditions, see Fig. 14.8c), results in a decrease of current (as expected). Furthermore, the peak current for oxidation is now significantly larger than that for the reduction component. A combination of a low mass of *trans*-Mn and a thin layer of [C_4mim][PF_6] results in extensive electrolysis. Under these conditions, mass transport is governed almost solely by finite diffusion, evidenced by the almost symmetrical shape of the voltammogram shown in Fig. 14.8d. Clearly the microchemical method is beneficial in the context of "green chemistry", resulting in faster electrolysis times with minimal quantities of solvents and chemicals. This experimental setup has the additional advantage in that it allows a stable aqueous reference electrode such as silver/silver chloride (Ag/AgCl) to be used, although calibration against an internal standard is still necessary as discussed below.

On the basis of results reviewed above, the principal processes involved in the oxidation of *trans*-Mn microparticles under the microchemical conditions outlined in Scheme 2 of Fig. 14.1 may be represented schematically as in Fig. 14.9, again using symbols Red (reduced form) and Ox^+ (oxidized form). In the model employed, the rates of both dissolution and reprecipitation of Red were assumed to be negligibly slow on the voltammetric time scale. The important initial processes in the voltammetry are the oxidation of Red (solid) to Ox^+ (solid), accompanied by charge neutralization (insertion of $[PF_6]^-$ into the solid) and the dissolution of [Ox][PF_6] (solid) at the particle/ionic liquid interface. These processes are assumed to be established rapidly on the voltammetric time scale. It was also assumed that the reduction of Ox^+ (ionic liquid) to solution-phase Red is confined to the ionic liquid layer, and therefore a charge neutralization process involving the transfer of $[PF_6]^-$ across the [C_4mim][PF_6]/0.1 M KPF_6 aqueous electrolyte interface must occur for the duration of the voltammetric measurement. Additionally, the transfer of ions across the interface is required for charge neutralization because the working and counter electrodes are situated in different phases. Ion transfer across two immiscible liquid/liquid interfaces is usually a rapid process [29] and is therefore unlikely to be rate-limiting. This assumption is consistent with the experimental data shown in Fig. 14.8a, since a reversible $[trans\text{-}Mn]^{0/+}$ response would not occur if the $[PF_6]^-$ anion transfer across the interface was slow.

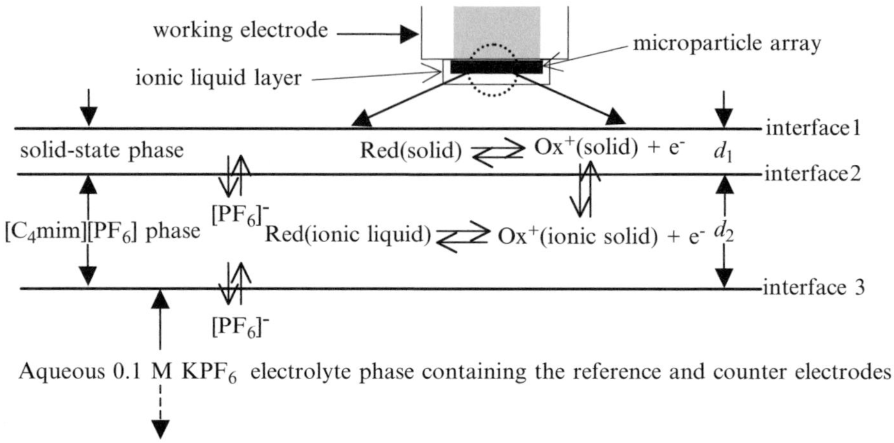

Fig. 14.9 Schematic representation of the dominant processes assumed to influence the voltammetric response when a chemically modified electrode consisting of a redox-active microparticle adhered to an electrode and coated with a layer of ionic liquid is placed in contact with aqueous electrolyte. d_1 and d_2 are the thicknesses of solid-state phase and ionic liquid phase respectively. Adapted with permission from Zhang et al., *Anal. Chem.* 2003, *75*, 6938–6948 [23]. Copyright 2013, American Chemical Society

On the basis of the above findings, the aqueous electrolyte phase can be assumed to simply provide a phase boundary whose junction potential contribution is removed by reporting potentials against the reversible couple $[Co(Cp)_2]^{+/0}$ dissolved in the ionic liquid phase used as an internal reference standard. From examination of the experimental voltammograms shown in Fig. 14.8, it is evident that the relative thickness of the solid (i.e., quantity or mass) and ionic liquid phases influence whether mass transport to the electrode surface is governed by semi-infinite or finite diffusion, with the data reported under conditions of Fig. 14.7 (semi-infinite) and Fig. 14.8d (predominantly finite) representing the extremes. Further experimental results and theoretical analysis (simulations) [23] have confirmed that in order to obtain a voltammogram which is indistinguishable from that in the dissolved state using the experimental setup outlined in Scheme 2 of Fig. 14.1, the ionic liquid layer must be thick enough to facilitate mass transport by semi-infinite diffusion and the solid layer must be "thick enough" (i.e., contain enough mass) to provide a continuous supply of dissolved (product) material on the voltammetric timescale.

14.4 Applications

14.4.1 Measurement of Reversible Potentials for Compounds with Slow Dissolution Kinetics

Unlike in aqueous media, where there are well established, stable reference electrodes such as the Ag/AgCl and saturated Calomel electrode (SCE), it is common in organic solvents to calibrate the potential vs. the reversible potential of an internal standard. Typically, the reference scale is derived by the analysis of voltammograms obtained from the oxidation of Fc or reduction of the cobaltocenium cation (Cc^+). In practice, this involves converting the reference potential scale vs. Ag/AgCl or a quasi-reference (e.g., Ag or Pt wire) potential to the $Fc^{0/+}$ or $[Co(Cp)_2]^{+/0}$ scale. The $Fc^{0/+}$ and $[Co(Cp)_2]^{+/0}$ processes are used for reference electrode calibration because they provide almost ideal electrochemically and chemically reversible voltammograms, with the assumption being made that their reversible potentials are independent of the medium [25, 30, 31].

As applies in organic solvents, voltammetric studies in ionic liquids are often carried out using quasi-reference electrodes, in particular Pt wire [32–35] or Ag wire [35–37]. Unfortunately, the use of quasi-reference electrodes makes it very difficult to compare voltammetric data from laboratory to laboratory, and furthermore the potentials observed cannot be directly related to those obtained in conventional molecular solvents. The I^-/I_3^- [38] and Li^+/Li couple [39] have been used as reference electrodes in ionic liquids, but not so commonly in conventional solvents, so again comparisons of potential data in the different media are difficult. Bond and co-workers [25] have presented detailed studies at Au, GC, and Pt macrodisk electrodes to assess the feasibility of using the voltammetry of dissolved cobaltocenium hexafluorophosphate ($[Co(Cp)_2][PF_6]$) or Fc to provide a reliable reference scale in $[C_4mim][PF_6]$. Unfortunately, as highlighted above (see Sect. 14.3.1), although the thermodynamic solubility of Fc and its derivatives in ionic liquids is more than adequate for this purpose, the rate of dissolution is very slow. Therefore, it is not very convenient to use dissolved Fc for potential scale calibration. However, the slow kinetics of dissolution of Fc in $[C_4mim][PF_6]$ allowed assessment of its voltammetry in the form of adhered solid (see above) under conditions where the shape of the voltammograms is identical to that found for dissolved Fc. This equivalence is very useful, as it allows the IUPAC recommended $Fc^{0/+}$ process to be used as a reference potential scale in ionic liquids [25] via the use of the simple adhered solid-state method.

Many other compounds also have similar solubility properties to Fc and exhibit a reaction scheme analogous to the $Fc^{0/+}$ case in ionic liquids (see schematic diagram in Fig. 14.5), which allows voltammograms from adhered microparticles to be essentially indistinguishable from that obtained with the dissolved material. Notable examples include Fc derivatives [12, 24, 25, 40], 1,3,5-tris(3-((ferrocenylmethyl) amino)pyridiniumyl)-2,4,6-triethylbenzene hexafluorophosphate ($[PD][PF_6]_3$) [12, 41] and *trans*-Mn [23]. Thus, voltammetric studies using arrays of adhered solid microparticles are frequently able to provide an efficient method for obtaining the formal potentials of redox processes in ionic liquids.

In the case of Fc derivatives 1,1′-diacetylferrocene $[Fe(CpCOCH_3)_2]$, 1,1′-dimethylferrocene $[Fe(CpCH_3)_2]$, ferrocenecarboxaldehyde [CpFeCpCOH], and decamethylferrocene $[Fe(Cp(CH_3)_5)_2]$ [25], cyclic voltammograms obtained at a gold electrode chemically modified with adhered microparticles in contact with $[C_4mim][PF_6]$ produced the cyclic voltammograms shown in Fig. 14.10. Parameters obtained from these Fc derivatives from the cyclic voltammograms are summarized in Table 14.1. In each case, a chemically reversible, one-electron oxidation

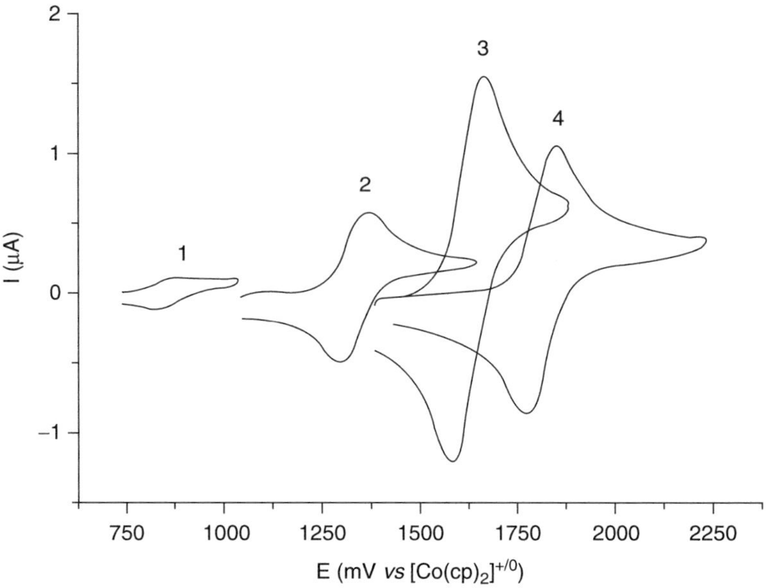

Fig. 14.10 Cyclic voltammograms of (*1*) $[Fe(Cp(CH_3)_5)_2]$, (*2*) Fc, (*3*) [CpFeCpCOH], and (*4*) $[Fe(CpCOCH_3)_2]$ solid adhered to a Au electrode in $[C_4mim][PF_6]$ ($\nu = 0.1$ V s^{-1}). Adapted with permission from Hultgren et al., *Anal. Chem.* 2002, *74*, 3151–3156 [25]. Copyright 2013, American Chemical Society

Table 14.1 Cyclic voltammetric data for Fc and Fc derivatives adhered to a Au electrode ($\nu = 100$ mV s^{-1}) in $[C_4mim][PF_6]$

Compounds	$E_f^{0'}$ *(mV)*[a]	ΔE_p (mV)	$\lvert I_p^{ox}/I_p^{red}\rvert$	$E_f^{0'}$ *(mV)*[a] CH_3CN[b]
Fc	+1,332	70	1.04	+1,335
$[Fe(CpCH_3)_2]$	+1,233	134	0.99	+1,219
$[Fe(CpCOCH_3)_2]$	+1,813	77	1.19	+1,800
[CpFeCpCOH]	+1,627	83	1.06	+1,618
$[Fe(Cp(CH_3)_5)_2]$	+856	60	0.88	+825

Adapted with permission from Hultgren et al., *Anal. Chem.* 2002, *74*, 3151–3156 [25]. Copyright 2013, American Chemical Society
[a]mV vs. $[Co(Cp)_2]^{+/0}$ reference potential scale
[b]Solutions contain 0.1 M $[Bu_4N][PF_6]$ as the supporting electrolyte

process was observed, with the reversible potential being dependent on the substituent on the cyclopentadienyl ring. $E_f^{0'}$ is the formal potential of the reversible processes involving Fc and Fc derivatives. In most cases, the $E_f^{0'}$ values vs. $[Co(Cp)_2]^{+/0}$ obtained in $[C_4mim][PF_6]$ were similar to those found when the molecule was dissolved in CH_3CN. $[Fe(Cp(CH_3)_5)_2]$ provided the largest deviation from equivalence, with an $E_f^{0'}$ value 31 mV more positive than that found in CH_3CN. The peak-to-peak potential (ΔE_p) values for the solid-state processes varied over the range of 60 mV for $[Fe(Cp(CH_3)_5)_2]$ to 134 mV for $[Fe(CpCH_3)_2]$ with a scan rate of 100 mV s^{-1} while the ratio $|I_p^{red}/I_p^{ox}|$ (where I_p^{red} and I_p^{ox} are the reductive and oxidative peak currents, respectively) was close to unity in all instances. The formal potentials reported in Table 14.1 are expected to be very similar to those found if studies had been undertaken with samples dissolved in [C_4mim][PF_6].

14.4.2 Measurements of the Kinetics of First-Order Homogeneous Reactions Coupled to Heterogeneous Electron Transfer

Voltammetry with adhered microparticles can also be usefully applied when a heterogeneous electron transfer process is followed by a homogeneous chemical reaction [13]. Straightforward quantitative evaluation is restricted to first-order-coupled chemical reactions, where data analysis is independent of the unknown concentration of dissolved material derived from the initially adhered solid. Two processes that have been well studied in organic solvents (electrolyte) have also been considered in ionic liquid medium: the one-electron oxidation of *cis*-[Mn(CN)$(CO)_2$[P$(OPh)_3$] (DPM)] (DPM = $Ph_2PCH_2PPh_2$) (*cis*-Mn) and *cis*-[W$(CO)_2$ $(DPE)_2$] (DPE = $Ph_2PCH_2CH_2PPh_2$) (*cis*-W) (see structures in Fig. 14.11) to their *cis*$^+$ isomeric forms followed by first-order isomerization to their *trans*$^+$ configuration. In [C_4mim] [PF_6], 1-ethyl-3-methylimidazolium bis(trifluoromethylsulfonyl)imide ([C_2mim] [N$(Tf)_2$]) or 1-butyl-1-methylpyrrolidinium bis(trifluoromethylsulfonyl)imide ([C_4mpyr][N$(Tf)_2$]), these processes are known to follow an *ECE* mechanism [42–44]:

$$\textit{cis}\text{-Mn (solid)} \rightleftharpoons \textit{cis}\text{-Mn}^+\text{(ionic liquid)} + e^- \qquad (14.16)$$

$$\begin{array}{c} \textit{cis}\text{-Mn (ionic liquid)} \overset{E^{0'}_{il,1}}{\rightleftharpoons} \textit{cis}\text{-Mn}^+\text{(ionic liquid)} + e^- \\ \downarrow k_f \\ \textit{trans}\text{-Mn (ionic liquid)} \overset{E^{0'}_{il,2}}{\rightleftharpoons} \textit{trans}\text{-Mn}^+\text{(ionic liquid)} + e^- \end{array} \qquad (14.17)$$

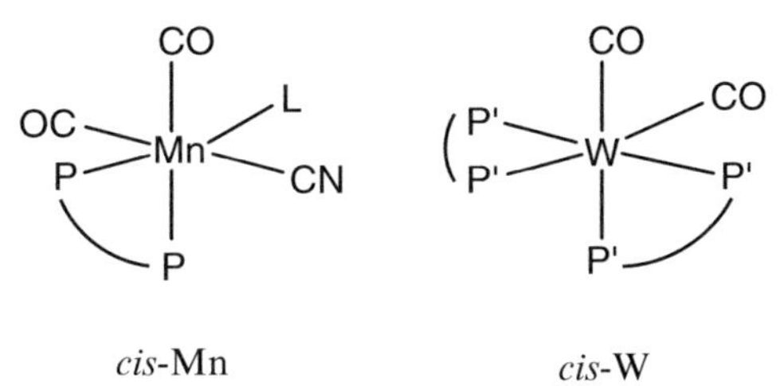

Fig. 14.11 Structure of *cis*-Mn and *cis*-W. [L = P$(OPh)_3$, P-P = $Ph_2PCH_2PPh_2$, and P′-P′ = $Ph_2PCH_2CH_2PPh_2$]

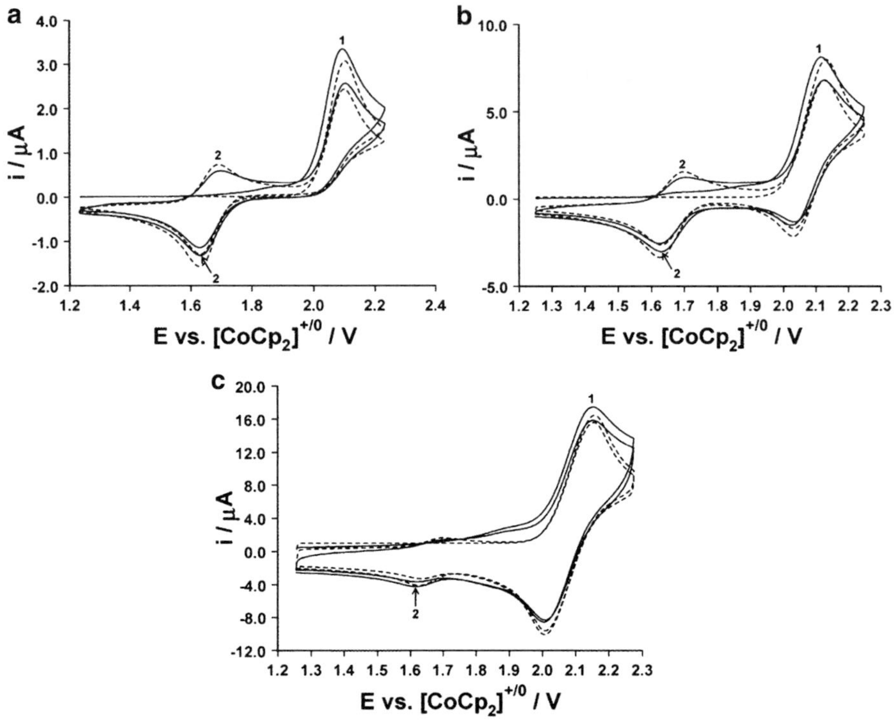

Fig. 14.12 A comparison of experimental (*solid lines*) cyclic voltammograms obtained for the oxidation of *cis*-Mn microparticles mechanically adhered to a 1 mm diameter GC electrode surface prior to being placed in contact with $[C_2mim][N(Tf)_2]$ and normalized simulated (*dashed lines*) data obtained as a range of scan rates according to the mechanism described by Eq. 14.17 with $k_f = 4\ s^{-1}, E^{0'}_{il,1} = 2.075\ V$ and $E^{0'}_{il,2} = 1.665\ V$ (vs. $[Co(Cp)_2]^{+/0}$). Scan rates are 0.1 (**a**), 1.0 (**b**), and 10.0 (**c**) V s^{-1}. Numbers 1 and 2 in the figure designate the first and second cycle of potential, respectively. Adapted with permission from Zhang et al., *J Phys Chem B*. 2004, *108*, 7363–7372 [13]. Copyright 2013, American Chemical Society

where $E^{0'}_{il,1}$, $E^{0'}_{il,2}$ and k_f are the formal potentials of the process involving solely solution-phase species and the rate constant of the forward reaction (Eq. 14.17) [13], respectively.

A comparison of experimental and simulated cyclic voltammograms obtained from oxidation of *cis*-Mn microparticles mechanically adhered to a GC electrode in $[C_2mim][N(Tf)_2]$ and simulated data is shown in Fig. 14.12. Once again, aside from the larger magnitudes of the measured currents in the ionic liquid medium, the voltammetric characteristics obtained under dissolved or solid-state conditions are almost identical. The formal potentials ($E^0_{il,1}$ and $E^0_{il,2}$), k_f, and the ratio of equilibrium constants (K_1/K_2) ($K_1 = k_f/k_b$, $K_2 = k_{f1}/k_{b1}$, k_f and k_{f1} are the rate constants of the forward reactions described in Fig. 14.13. k_b and k_{b1} are the rate constants of the backward reactions) derived from the voltammetric data were identical under dissolved and adhered solid-state conditions, as shown in Table 14.2.

Table 14.2 Summary of the thermodynamic, kinetic, and other parameters derived from the voltammetric oxidation of *cis*-Mn in $[C_4mim][PF_6]$, $[C_2mim][N(Tf)_2]$, and $[C_4mPry][N(Tf)_2]$

	Dissolved state					Adhered solid state			
Solvent	$E^{0'}_{il,1}/V^a$	$E^{0'}_{il,2}/V^a$	k_f/s^{-1}	10^{-7} $K_1/K_2{}^b$	10^8 $D/$ $cm^2\ s^{-1}$	$E^{0'}_{il,1}/V^a$	$E^{0'}_{il,2}/V^a$	k_f/s^{-1}	10^{-7} $K_1/K_2{}^b$
$[C_4mim]$ $[PF_6]$	2.085	1.675	4	1.13	1.0	2.085	1.675	4	1.13
$[C_2mim]$ $[N(Tf)_2]$	2.075	1.665	4	1.13	10.0	2.075	1.665	4	1.13
$[C_4mPry]$ $[N(Tf)_2]$	2.050	1.640	4	1.13	3.9	2.050	1.640	4	1.13

Adapted with permission from Zhang et al., *J Phys Chem B*. 2004, *108*, 7363–7372 [13]. Copyright 2013, American Chemical Society

[a]vs. $[Co(Cp)_2]^{+/0}$

[b]Calculated from the relationship $K_1/K_2 = \exp\left[(F/RT)\ \left(E^{0'}_{il,1} - E^{0'}_{il,2}\right)\right]$ that can be derived from the reaction scheme shown in Fig. 14.13

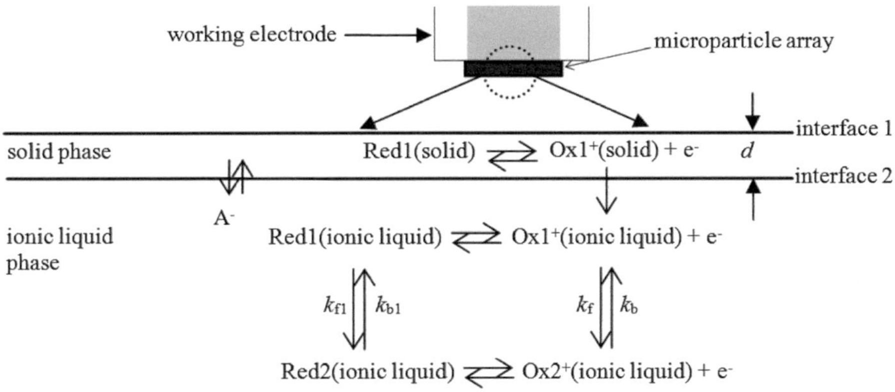

Fig. 14.13 Schematic representation of the principal processes assumed to influence the voltammetric response when a microparticle-modified electrode is placed in contact with an ionic liquid and when dissolved electrogenerated species Ox^+(ionic liquid) undergoes a square reaction scheme. Adapted with permission from Zhang et al., *J Phys Chem B*. 2004, *108*, 7363–7372 [13]. Copyright 2013, American Chemical Society

Analogous results were obtained with *cis*-W. A schematic diagram that illustrates the principal processes believed to be involved in the oxidation of microparticles of *cis*-Mn or *cis*-W adhered to an electrode surface in contact with an ionic liquid is provided in Fig. 14.13. The mechanism is now considerably more complicated than for the previously described systems, as dissolved electrogenerated species $Ox1^+$(ionic liquid), now undergoes a square reaction scheme. Nevertheless, despite the greater complexity in the *ECE* reaction mechanism, the voltammetry of adhered microparticles method can be used to determine the relevant thermodynamic and kinetic parameters when step *C* is a first-order homogeneous reaction, as applies in the case of an isomerization reaction.

14.4.3 Voltammetric Measurements on Reactive Species

14.4.3.1 Studies with Compounds that React with the Ionic Liquid or an Impurity such as Water Present in the Ionic Liquid

Polyoxometalate anions have been studied for many years due to their wide range of interesting structural, redox, and catalytic properties [45]. The voltammetry of adhered polyoxometalate microparticles may provide an efficient method for obtaining their reversible potentials, even when product of reduction reacts with the ionic liquid itself or adventitious water [46–51].

A detailed study on the reduction of $[SiW_{12}O_{40}]^{4-}$ (α, β, γ* isomers), $[S_2W_{18}O_{62}]^{4-}$ (α, γ* isomers), and $[S_2Mo_{18}O_{62}]^{4-}$ (α isomer) microparticles (as Bu_4N^+ salts) adhered to an electrode surface in contact with the "distillable ionic liquid" dimethylammonium dimethylcarbamate (DIMCARB, an equilibrium mixture largely composed of *N,N*-dimethylammonium, *N′,N′*-dimethylcarbamate, and the corresponding *N′,N′*-dimethylcarbamic acid as shown in Fig. 14.14) has been carried out [47]. Reduction by four or more electron equivalents leads to rapid dissolution of the adhered solid and the subsequent observation of a series of well-defined, solution-phase processes under conditions of cyclic voltammetry. Examples of well-defined cyclic voltammograms obtained from microparticles of $[Bu_4N]_4[\alpha\text{-}SiW_{12}O_{40}]_{(s)}$ or $[Bu_4N]_4[\beta\text{-}SiW_{12}O_{40}]_{(s)}$ adhered to a GC electrode in contact with DIMCARB are shown in Fig. 14.15a, b. Essentially the same characteristics are observed with solutions of these salts in CH_3CN. Interestingly, reduction of $[\gamma\text{-}SiW_{12}O_{40}]^{4-}$ dissolved in DIMCARB could not be detected due to reaction with this medium, while the voltammetry from adhered $[Bu_4N]_4[\gamma\text{-}SiW_{12}O_{40}]_{(s)}$ is well-defined (Fig. 14.15c). Similar observations have been made with $[Bu_4N]_4[S_2W_{18}O_{62}]$ (α, γ* isomers) and $[Bu_4N]_4[S_2Mo_{18}O_{62}]$ (α isomer), and again have been attributed to a reaction between the polyoxometalate anion and

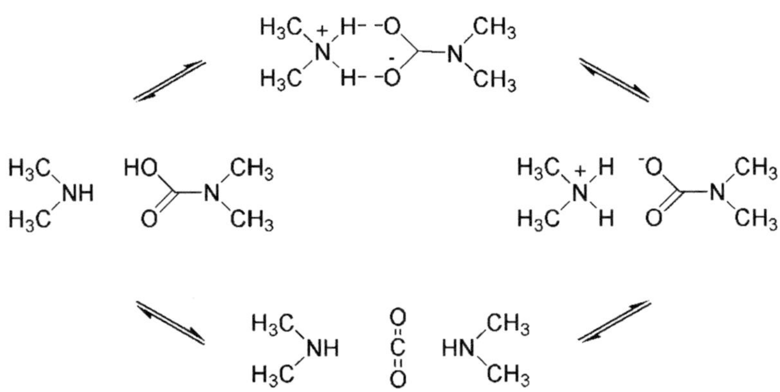

Fig. 14.14 Dynamic equilibrium of DIMCARB proposed on the basis of NMR measurement. Adapted with permission from Zhang et al., *Analyst*. 2005, *130*, 1132–1147 [5]. Copyright 2013, The Rotary Society of Chemistry

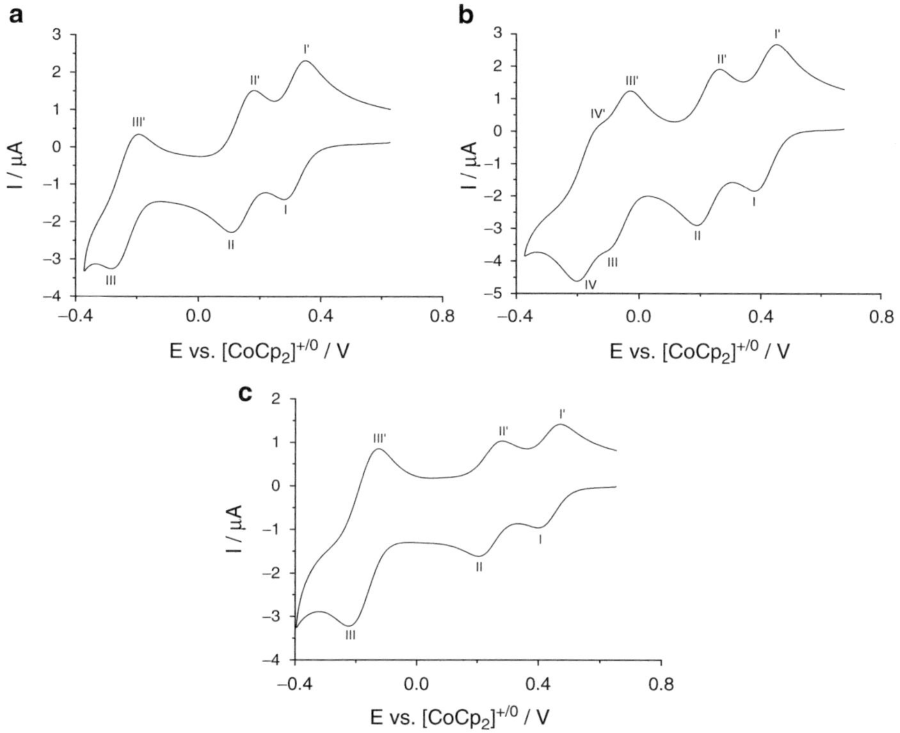

Fig. 14.15 Cyclic voltammograms for the reduction of: (**a**) $[Bu_4N]_4[\alpha\text{-}SiW_{12}O_{40}]$; (**b**) $[Bu_4N]_4[\beta\text{-}SiW_{12}O_{40}]$, and (**c**) $[Bu_4N]_4[\gamma\text{-}SiW_{12}O_{40}]$ microparticles adhered mechanically to a 1.5 mm diameter GC electrode in contact with DIMCARB, $v = 0.1$ V s^{-1}. Adapted with permission from Zhang et al., *Electrochem. Commun.* 2005, *7*, 1283–1290 [47], Copyright 2013, Elsevier

DIMCARB upon preparation of the polyoxometalate solutions to form an electroinactive species. Fortunately, the adhered microparticles technique avoids prolonged contact with DIMCARB, thereby allowing well-defined voltammetry to be observed. The initial process involved in the reduction of adhered microparticles of $[Bu_4N]_4[SiW_{12}O_{40}]_{(s)}$ (α, β, and γ* isomers) in DIMCARB is summarized as follows for the first reduction process:

$$[Bu_4N]_4[SiW_{12}O_{40}]_{(s)} + [Me_2NH_2]^+_{(DIMCARB)} + e^- \rightleftharpoons [Me_2NH_2][Bu_4N]_4[SiW_{12}O_{40}]_{(s)}$$

$$\downarrow \text{dissolution}$$

$$[Me_2NH_2]^+_{(DIMCARB)} + 4[Bu_4N]^+_{(DIMCARB)} + [Siw_{12}O_{40}]^{5-}_{(DIMCARB)} \qquad (14.18)$$

A comparison of the formal potentials, $E^{0'}$ values found for the $[SiW_{12}O_{40}]^{4-}$ isomers in DIMCARB, $[C_4mim][PF_6]$ and CH_3CN (0.1 M $[Bu_4N][ClO_4]$) is

Table 14.3 Reversible potential data for the reduction of $[SiW_{12}O_{40}]^{4-}$ isomers

Isomeric form	State	Solvent/ electrolyte	$E^{0'}$/V (vs. $[Co(Cp)_2]^{+/0}$) 4−/5−	5−/6−	6−/7−	7−/8−	Reference
α	Dissolved	DIMCARB	0.310	0.147	−0.239	−0.433	[47]
	Solid	DIMCARB	0.313	0.145	−0.238	−0.431	
β	Dissolved	DIMCARB	0.412	0.224	−0.069		
	Solid	DIMCARB	0.415	0.225	−0.068		
γ*	Solid	DIMCARB	0.435	0.244			
α	Dissolved	$[C_4mim][PF_6]$	0.614	0.263			[51]
	Solid	$[C_4mim][PF_6]$	0.621	0.272	−0.244	−0.605	
α	Dissolved	CH_3CN/0.1 M $[Bu_4N][ClO_4]$	0.240	−0.265	−0.930		[46]
β	Dissolved	CH_3CN/0.1 M $[Bu_4N][ClO_4]$	0.315	−0.215	−0.860		
γ*	Dissolved	CH_3CN/0.1 M $[Bu_4N][ClO_4]$	0.435	−0.040	−0.710		

provided in Table 14.3 [47]. The values for α and β isomers obtained from dissolved samples or adhered solid salts are essentially identical. The same order of difficulty of reduction of the isomers, α > β > γ*, applies in all media. Interestingly, the difference in $E^{0'}$ values between the 4−/5− and 5−/6− one-electron reduction processes is significantly less in DIMCARB than in $[C_4mim][PF_6]$. The reversible potentials of polyoxometalate and other highly charged compounds in organic solvents correlate with the polarity [48–50]. Data reported in a review [51] on aprotic ionic liquids suggested that $E^{0'}$ values can provide information on the polarity of the ionic liquids. The $E^{0'}$ order for the 4−/5− process of 0.240, 0.310, and 0.614 V vs. $[Co(Cp)_2]^{+/0}$ in CH_3CN, DIMCARB, and $[C_4mim][PF_6]$, respectively, suggests that the polarity of DIMCARB may be closer to CH_3CN than to $[C_4mim][PF_6]$. Evidently, the adhered microparticles method can be employed to establish the voltammetric characteristics of species that react slowly in solution.

14.4.3.2 Studies with Compounds with Products from Reduction or Oxidation That React with the Ionic Liquid or an Impurity Such as Water Present in the Ionic Liquid

The voltammetric reduction of dissolved and adhered solid forms of benzophenone and 1,4-benzoquinone has been investigated in the ionic liquid $[C_4mpyr][N(Tf)_2]$ under “wet” and “dry” conditions [52] and found to exhibit a high sensitivity to proton availability [53, 54] from water. These compounds are significant in green chemistry and biochemical contexts [55]. Cyclic voltammograms obtained from the reduction of dissolved benzophenone and 1,4-benzoquinone in extensively dried

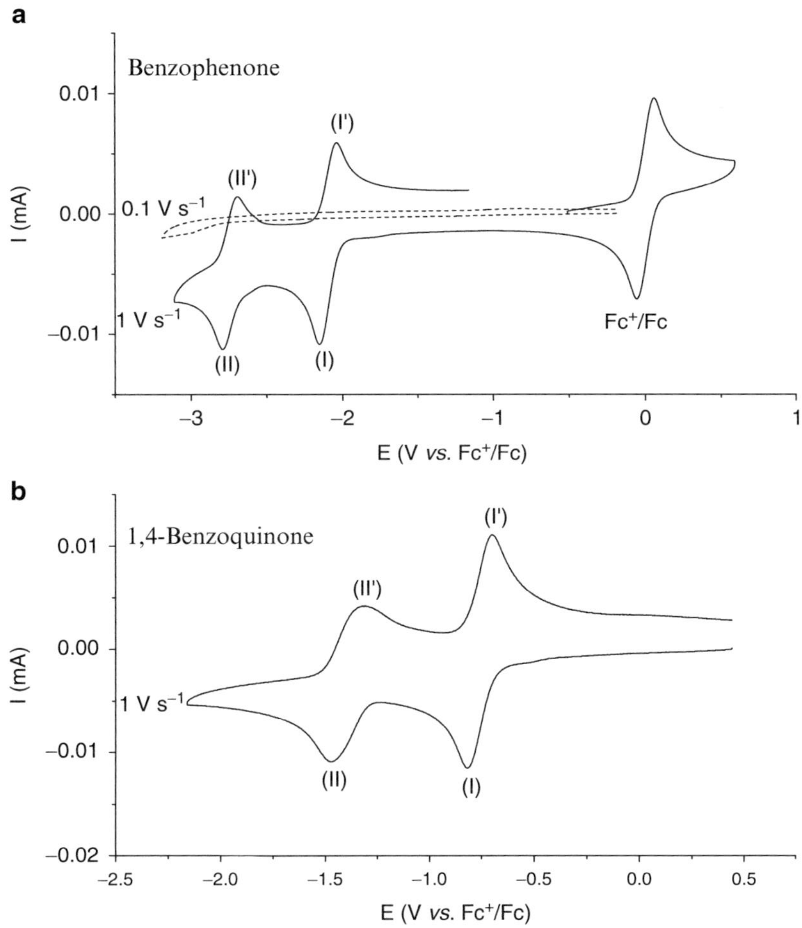

Fig. 14.16 Cyclic voltammograms obtained for reduction of 10 mM benzophenone (**a**) and 10 mM 1,4-benzoquinone in "dry" $[C_4mpyr][N(Tf)_2]$ (**b**) with a 1.0 mm diameter GC electrode at a scan rate of 1.0 V s^{-1} in a dry box under a N_2 atmosphere. In (**a**), the cyclic voltammogram obtained in the absence of benzophenone (*dashed lines*) is included for comparison. Adapted with permission from Zhao et al., *Electrochem Commun*, 2012, *16*, 14–18 [52], Copyright (2013), Elsevier

$[C_4mpyr][N(Tf)_2]$ under a nitrogen atmosphere in a dry box are shown in Fig. 14.16. In each case, two well-defined, reversible reduction processes are found that exhibit the mechanism summarized below:

B: 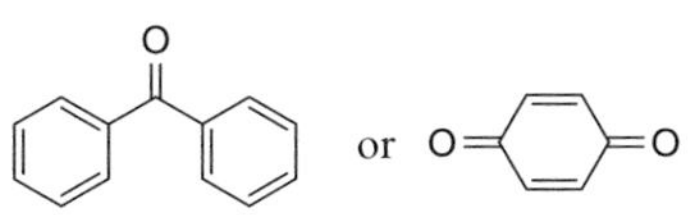

benzophenone 1,4-benzoquinone

$$B + e^- \rightleftharpoons B^{\cdot -} \qquad E_1^{0'} \tag{14.19}$$

$$B^{\cdot -} + e^- \rightleftharpoons B^{2-} \qquad E_2^{0'} \tag{14.20}$$

$$B^{2-} + B \rightleftharpoons BB^{2-} \tag{14.21}$$

$$BB^{2-} \rightleftharpoons B^{\cdot -} + B^{\cdot -} \tag{14.22}$$

where B is benzophenone or 1,4-benzoquinone.

Studies were also undertaken in "wet" $[C_4mpyr][N(Tf)_2]$ [56]. Cyclic voltammograms obtained from dissolved benzophenone and 1,4-benzoquinone under these conditions are shown in Fig. 14.17a. In the case of benzophenone, the second reduction step is now clearly irreversible in the presence of water (i.e., proton-sensitive radical anion intermediates), and at low scan rates partially

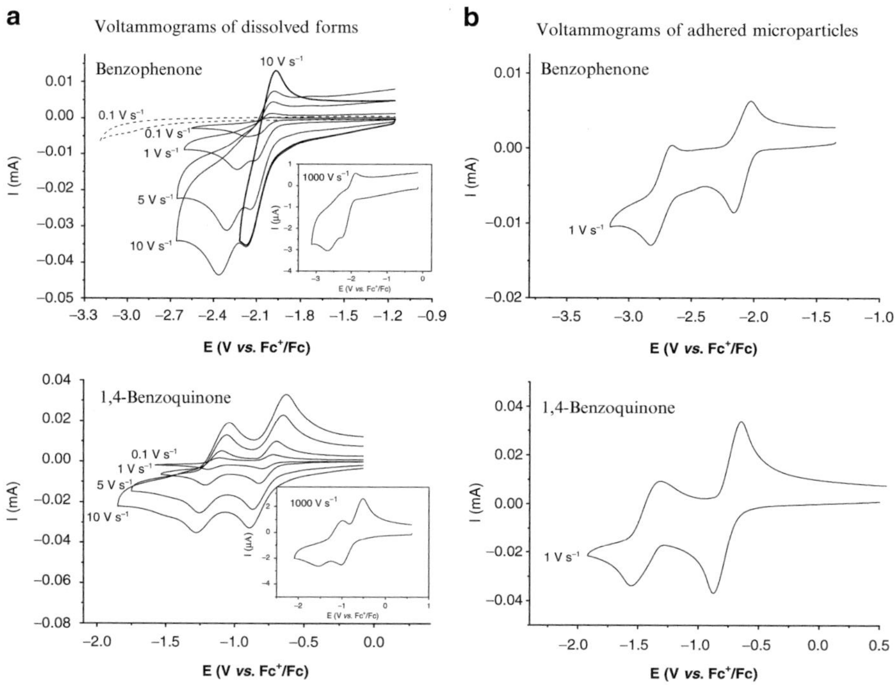

Fig. 14.17 Voltammograms of (**a**) 10 mM benzophenone and 10 mM 1,4-benzoquinone in "wet" (approx. 0.13 M water) $[C_4mpyr][N(Tf)_2]$ recorded with a 1.0 mm diameter GC electrode at designated scan rates under a N_2 atmosphere. The cyclic voltammogram obtained in the absence of benzophenone (*dashed lines*) is included for comparison. The inset figures show cyclic voltammograms for reduction of 100 mM benzophenone and 100 mM 1,4-benzoquinone in "wet" $[C_4mpyr][N(Tf)_2]$ recorded with a 33 μm diameter carbon fiber microelectrode at a scan rate of 1,000 V s^{-1} under a N_2 atmosphere; and (**b**) adhered microparticles of benzophenone and 1,4-benzoquinone on a 1.0 mm diameter GC electrode in contact with "wet" $[C_4mpyr][N(Tf)_2]$ under a N_2 atmosphere, scan rate = 1.0 V s^{-1}. Adapted with permission from Zhao et al., *Electrochem Commun*, 2012, *16*, 14–18 [52], Copyright (2013), Elsevier

overlaps with the first process (i.e., increased resolution is achieved at higher scan rates). The water effect is less obvious in the case of 1,4-benzoquinone, with the main difference being that the reversible potential for the second process is significantly more positive than in "dry" $[C_4mpyr][N(Tf)_2]$. While the use of elevated concentrations or employment of faster scan rates (or a combination of both approaches) may minimize the effect, it is clear that the presence of water has a dramatic influence on the voltammetry of these compounds in ionic liquids, consistent with reports in conventional aprotic solvents [53, 54].

The high local concentration of product achieved at the working electrode surface under adhered microparticles conditions [56] minimizes the effect of water, allowing voltammograms found in "dry" $[C_4mpyr][N(Tf)_2]$ to be obtained under "wet" conditions. Cyclic voltammograms obtained from adhered benzophenone and 1,4-benzoquinone solid in "wet" $[C_4mpyr][N(Tf)_2]$ are shown in Fig. 14.17b. By analogy with the voltammetry of adhered microparticles of benzophenone and 1,4-benzoquinone, the reaction scheme is summarized as follows:

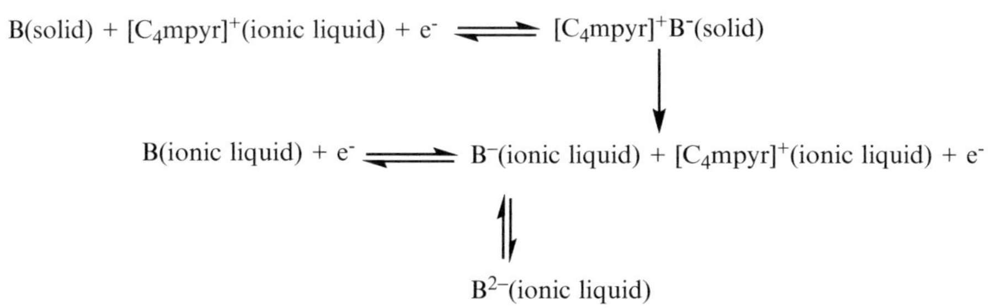

The reversible potentials for both compounds obtained under "wet" conditions ($[C_4mpyr][N(Tf)_2]$, $[H_2O] \approx 0.13$ M) are very similar to those obtained in extensively dried ionic liquid. For these compounds, it may be concluded that the voltammetry of adhered microparticles provides an efficient and convenient method for obtaining "water-free" reversible potentials for water-sensitive reduction processes in ionic liquids under bench top conditions, without the need to use extensively dried ionic liquids or operate under more cumbersome dry box or vacuum line conditions.

14.5 Conclusions

The history, theory, and application of the voltammetry of adhered microparticles in contact with both bulk and thin-layer forms of ionic liquids have been reviewed. In the short time since its introduction, the technique has enhanced the range of studies possible for a wide variety of compounds. For example, new insights have now become available when electroactive molecules undergo kinetically slow dissolution (e.g., for oxidation of Fc in an ionic liquid). Using this method, the formal potentials and voltammetric characteristics of Fc and its derivatives have been characterized in a range of ionic liquids. An important outcome of this application is that the reversible potential obtained from a voltammogram of adhered Fc can be used to provide a reference potential scale in ionic liquids. Furthermore, the technique can be employed for quantitative voltammetric studies on processes exhibiting electron-transfer with a coupled first-order homogeneous process (e.g., oxidation of *cis*-Mn in an ionic liquid), and in cases where reaction with the ionic liquid itself or impurities such as water in the ionic liquid occur (e.g., polyoxometalate reduction in DIMCARB or benzophenone reduction under "wet" conditions).

References

1. Scholz F, Nitschke L, Henrion G (1989) A new procedure for fast electrochemical analysis of solid materials. Naturwissenschaften 76(2):71–72. doi:10.1007/BF00396709
2. Scholz F, Meyer B (1998) Electroanalytical chemistry: a series of advances. Marcel Dekker, New York
3. Banks CE, Davies TJ, Evans RG, Hignett G, Wain AJ, Lawrence NS, Wadhawan JD, Marken F, Compton RG (2003) Electrochemistry of immobilised redox droplets: concepts and applications. PCCP 5(19):4053–4069. doi:10.1039/B307326M
4. Bond AM (2002) Broadening electrochemical horizons: principles and illustration of voltammetric and related techniques. Oxford University Press, Oxford
5. Zhang J, Bond AM (2005) Practical considerations associated with voltammetric studies in room temperature ionic liquids. Analyst (Cambridge, U K) 130(8):1132–1147. doi:10.1039/b504721h
6. Oldham KB (1998) Voltammetry at a three-phase junction. J Solid State Electrochem 2 (6):367–377. doi:10.1007/s100080050113
7. Scholz F, Nitschke L, Henrion G (1990) Abrasive stripping voltammetric analysis of tin–bismuth. Electroanalysis 2(1):85–87. doi:10.1002/elan.1140020116
8. Scholz F, Rabi F, Müller W-D (1992) The anodic dissolution of dental amalgams studied with abrasive stripping voltammetry. Electroanalysis 4(3):339–346. doi:10.1002/elan.1140040312
9. Dostal A, Kauschka G, Reddy SJ, Scholz F (1996) Lattice contractions and expansions accompanying the electrochemical conversions of Prussian blue and the reversible and irreversible insertion of rubidium and thallium ions. J Electroanal Chem 406(1–2):155–163. doi:10.1016/0022-0728(95)04427-2
10. Bond AM, Cooper JB, Marken F, Way DM (1995) Redox and electroinsertion processes associated with the voltammetry of microcrystalline forms of Dawson molybdate anion salts mechanically attached to graphite electrodes and immersed in aqueous electrolyte media. J Electroanal Chem 396(1–2):407–418. doi:10.1016/0022-0728(95)03863-C
11. Bond AM, Colton R, Mahon PJ, Tan WT (1997) Tetrabutylammonium cation expulsion versus perchlorate electrolyte anion uptake in the electrochemical oxidation of microcrystals of $[(C_4H_9)_4N][Cr(CO)_5I]$ mechanically attached to a gold electrode: a voltammetric and quartz crystal microbalance study. J Solid State Electrochem 1(1):53–61. doi:10.1007/s100080050022
12. Zhang J, Bond AM (2005) The role of dissolution in the voltammetry of microdroplets and microparticles adhered to electrode surfaces in contact with aqueous electrolytes or ionic liquids. J Electroanal Chem 574(2):299–309. doi:10.1016/j.jelechem.2004.08.011

13. Zhang J, Bond AM (2004) Voltammetric studies with adhered microparticles and the detection of a dependence of organometallic *cis*$^+$ → *trans*$^+$ first-order isomerization rate constants on the identity of the ionic liquid. J Phys Chem B 108(22):7363–7372. doi:10.1021/jp049294r
14. Welton T (1999) Room-temperature ionic liquids. Solvents for synthesis and catalysis. Chem Rev 99(8):2071–2084. doi:10.1021/cr980032t
15. Pârvulescu VI, Hardacre C (2007) Catalysis in ionic liquids. Chem Rev 107(6):2615–2665. doi:10.1021/cr050948h
16. Greaves TL, Drummond CJ (2007) Protic ionic liquids: properties and applications. Chem Rev 108(1):206–237. doi:10.1021/cr068040u
17. Martins MAP, Frizzo CP, Moreira DN, Zanatta N, Bonacorso HG (2008) Ionic liquids in heterocyclic synthesis. Chem Rev 108(6):2015–2050. doi:10.1021/cr078399y
18. Plaquevent J-C, Levillain J, Fdr G, Malhiac C, Gaumont A-C (2008) Ionic liquids: new targets and media for α-amino acid and peptide chemistry. Chem Rev 108(12):5035–5060. doi:10.1021/cr068218c
19. Barrosse-Antle LE, Bond AM, Compton RG, O'Mahony AM, Rogers EI, Silvester DS (2010) Voltammetry in room temperature ionic liquids: Comparisons and contrasts with conventional electrochemical solvents. Chem Asian J 5(2):202–230. doi:10.1002/asia.200900191
20. Hapiot P, Lagrost C (2008) Electrochemical reactivity in room-temperature ionic liquids. Chem Rev 108(7):2238–2264. doi:10.1021/cr0680686
21. Shi C, Anson FC (1998) A simple method for examining the electrochemistry of metalloporphyrins and other hydrophobic reactants in thin layers of organic solvents interposed between graphite electrodes and aqueous solutions. Anal Chem 70(15):3114–3118. doi:10.1021/ac980426k
22. Volkov AG, Deamer DW, Tanelian DL, Markin VS (1998) Liquid interfaces in chemistry and biology. Wiley, New York. doi:10.1021/ed075p1391
23. Zhang J, Bond AM (2003) Comparison of voltammetric data obtained for the *trans*-[Mn(CN)$(CO)_2\{P(OPh)_3\}(Ph_2PCH_2PPh_2)]^{0/+}$ process in $BMIMPF_6$ ionic liquid under microchemical and conventional conditions. Anal Chem 75(24):6938–6948. doi:10.1021/ac034921e
24. Zhang J, Bond AM (2003) Conditions required to achieve the apparent equivalence of adhered solid- and solution-phase voltammetry for ferrocene and other redox-active solids in ionic liquids. Anal Chem 75(11):2694–2702. doi:10.1021/ac026329f
25. Hultgren VM, Mariotti AWA, Bond AM, Wedd AG (2002) Reference potential calibration and voltammetry at macrodisk electrodes of metallocene derivatives in the ionic liquid [bmim] [PF_6]. Anal Chem 74(13):3151–3156. doi:10.1021/ac015729k
26. Bard AJ, Faulkner LR (2001) Electrochemical methods: fundamentals and applications, 2nd edn. Wiley, New York
27. Lee W-Y, Majda M, Brezesinski G, Wittek M, Möbius D (1999) Structure and lateral electron hopping in osmium-tris-4,7-diphenylphenanthroline perchlorate monolayers at the air/water interface. J Phys Chem B 103(33):6950–6956. doi:10.1021/jp991053+
28. Schröder U, Oldham KB, Myland JC, Mahon PJ, Scholz F (2000) Modelling of solid state voltammetry of immobilized microcrystals assuming an initiation of the electrochemical reaction at a three-phase junction. J Solid State Electrochem 4(6):314–324. doi:10.1007/s100080000130
29. Girault HH, Schiffrin DJ (1989) Electrochemistry of liquid–liquid interfaces. In: Bard AJ (ed) Electroanalytical chemistry, vol 15. Marcel Dekker, New York, pp 1–141
30. Stojanovic RS, Bond AM (1993) Examination of conditions under which the reduction of the cobaltocenium cation can be used as a standard voltammetric reference process in organic and aqueous solvents. Anal Chem 65(1):56–64. doi:10.1021/ac00049a012
31. Gritzner G, Kuta J (1984) Recommendations on reporting electrode potentials in nonaqueous solvents. Pure Appl Chem 56(4):461–466
32. Suarez PAZ, Selbach VM, Dullius JEL, Einloft S, Piatnicki CMS, Azambuja DS, de Souza RF, Dupont J (1997) Enlarged electrochemical window in dialkyl-imidazolium cation based room-temperature air and water-stable molten salts. Electrochim Acta 42(16):2533–2535. doi:10.1016/S0013-4686(96)00444-6

33. Brown RJC, Dyson PJ, Ellis DJ, Welton T (2001) 1-Butyl-3-methylimidazolium cobalt tetracarbonyl [Bmim][Co(CO)]: a catalytically active organometallic ionic liquid. Chem Commun 18:1862–1863. doi:10.1039/B104601M
34. Endres F, Schrodt C (2000) In situ STM studies on germanium tetraiodide electroreduction on Au(111) in the room temperature molten salt 1-butyl-3-methylimidazolium hexafluorophosphate. PCCP 2(24):5517–5520. doi:10.1039/B007897M
35. Schroder U, Wadhawan JD, Compton RG, Marken F, Suarez PAZ, Consorti CS, de Souza RF, Dupont J (2000) Water-induced accelerated ion diffusion: voltammetric studies in 1-methyl-3-[2,6-(*S*)-dimethylocten-2-yl]imidazolium tetrafluoroborate, 1-butyl-3-methylimidazolium tetrafluoroborate and hexafluorophosphate ionic liquids. New J Chem 24(12):1009–1015. doi:10.1039/B007172M
36. MacFarlane DR, Golding J, Forsyth S, Forsyth M, Deacon GB (2001) Low viscosity ionic liquids based on organic salts of the dicyanamide anion. Chem Commun 16:1430–1431. doi:10.1039/B103064G
37. Dickinson WME, Hendrickson SM, Masui H, Murray RW (1999) Hybrid redox polyether melts based on polyether-tailed counterions. J Am Chem Soc 121(4):613–616. doi:10.1021/ja983184e
38. Bonhôte P, Dias A-P, Papageorgiou N, Kalyanasundaram K, Grätzel M (1996) Hydrophobic, highly conductive ambient-temperature molten salts. Inorg Chem 35(5):1168–1178. doi:10.1021/ic951325x
39. Koch V, Dominey L, Nanjundiah C, Ondrechen M (1996) The intrinsic anodic stability of several anions comprising solvent-free ionic liquids. J Electrochem Soc 143(3):798–803. doi:10.1149/1.1836540
40. Zhang J, Bond AM, Schumann H, Suehring K (2005) Voltammetric studies on decaphenylferrocene, substituted decaphenylferrocenes, and their oxidized forms in dichloromethane and ionic liquids. Organometallics 24(9):2188–2196. doi:10.1021/om049117c
41. Zhang J, Bond AM, Belcher WJ, Wallace KJ, Steed JW (2003) Electrochemical studies on the modular podand 1,3,5-tris(3-((ferrocenylmethyl)amino)pyridiniumyl)-2,4,6-triethylbenzene hexafluorophosphate in conventional solvents and ionic liquids. J Phys Chem B 107 (24):5777–5786. doi:10.1021/jp027699a
42. Connelly NG, Hassard KA, Dunne BJ, Orpen AG, Raven SJ, Carriedo GA, Riera V (1988) The oxidative isomerisation of *cis*, *cis*-dicarbonylmanganese (I) derivatives; isolation of the resulting *trans*-dicarbonylmanganese (II) complexes and the X-ray structure of *cis*, *cis*-[MnBr $(CO)_2\{P\ (OPh)_3\}(Ph_2PCH_2PPh_2)$]. J Chem Soc Dalton Trans 6:1623–1629. doi:10.1039/DT9880001623
43. Bombin F, Carriedo GA, Miguel JA, Riera V (1981) Syntheses of cationic tri-and di-carbonyl complexes of manganese (I) with diphosphines from the appropriate neutral bromocarbonyls. J Chem Soc Dalton Trans 10:2049–2053. doi:10.1039/DT9810002049
44. Bond AM, Colton R (1997) Electrochemical studies of metal carbonyl compounds. Coord Chem Rev 166:161–180. doi:10.1016/S0010-8545(97)00022-2
45. Baffert C, Feldberg SW, Bond AM, Long D-L, Cronin L (2007) pH-Dependence of the aqueous electrochemistry of the two-electron reduced α-[$Mo_{18}O_{54}$ (SO_3)] sulfite Dawson-like polyoxometalate anion derived from its triethanolammonium salt. Dalton Trans 40:4599–4607. doi:10.1039/B705991D
46. Zhang J, Bond AM, Richardt PJS, Wedd AG (2004) Voltammetric reduction of α- and γ*-$[S_2W_{18}O_{62}]^{4-}$ and α-, β-, and γ-$[S_2W_{12}O_{40}]^{4-}$: isomeric dependence of reversible potentials of polyoxometalate anions using data obtained by novel dissolution and conventional solution-phase processes. Inorg Chem 43(26):8263–8271. doi:10.1021/ic049043x
47. Zhang J, Bhatt AI, Bond AM, Wedd AG, Scott JL, Strauss CR (2005) Voltammetric studies of polyoxometalate microparticles in contact with the reactive distillable ionic liquid DIMCARB. Electrochem Commun 7(12):1283–1290. doi:10.1016/j.elecom.2005.09.008

48. Keita B, Bouaziz D, Nadjo L (1988) Solvent effects on the redox potentials of potassium 12-tungstosilicate and 18-tungstodiphosphate. J Electrochem Soc 135(1):87–91. doi:10.1149/1.2095596
49. Gritzner G, Danksagmüller K, Gutmann V (1976) Outer-sphere coordination effects on the redox behaviour of the $Fe(CN)_6^{3-}/Fe(CN)_6^{4-}$ couple in non-aqueous solvents. J Electroanal Chem 72(2):177–185. doi:10.1016/S0022-0728(76)80166-0
50. Gritzner G, Danksagmüller K, Gutmann V (1978) Solvent effects on the redox potentials of tetraethylammonium hexacyanomanganate (III) and hexacyanoferrate (III). J Electroanal Chem 90(2):203–210. doi:10.1016/S0022-0728(78)80054-0
51. Zhang J, Bond AM, MacFarlane DR, Forsyth SA, Pringle JM, Mariotti AWA, Glowinski AF, Wedd AG (2005) Voltammetric studies on the reduction of polyoxometalate anions in ionic liquids. Inorg Chem 44(14):5123–5132. doi:10.1021/ic050032t
52. Zhao S-F, Lu J-X, Bond AM, Zhang J (2012) Remarkable sensitivity of the electrochemical reduction of benzophenone to proton availability in ionic liquids. Chem Eur J 18 (17):5290–5301. doi:10.1002/chem.201103365
53. Gupta N, Linschitz H (1997) Hydrogen-bonding and protonation effects in electrochemistry of quinones in aprotic solvents. J Am Chem Soc 119(27):6384–6391. doi:10.1021/ja970028j
54. Zhao S-F, Wang H, Lan Y-C, Liu X, Lu J-X, Zhang J (2012) Influences of the operative parameters and the nature of the substrate on the electrocarboxylation of benzophenones. J Electroanal Chem 664:105–110. doi:10.1016/j.jelechem.2011.11.001
55. Patai S, Rappoport Z (eds) (1988) The chemistry of the quinonoid compounds. Wiley, Chichester
56. Zhao S-F, Lu J-X, Bond AM, Zhang J (2012) Voltammetric studies in "wet" 1-butyl-1-methylpyrrolidinium bis(trifluoromethylsulfonyl)imide ionic liquid using electrodes with adhered microparticles. Electrochem Commun 16(1):14–18. doi:10.1016/j.elecom.2011.11.034

Chapter 15
Electrochemical Reaction of Organic Compounds in Ionic Liquids

Angel A.J. Torriero and Douglas R. MacFarlane

15.1 Introduction

In general, ILs are used principally as "inert" reaction media for classical organic chemical reactions [1]. However, it is becoming apparent that their rich physico-chemical properties (e.g., Lewis and/or Brønsted acidity/basicity, hydrogen-bonding ability, π–π interactions, etc.) influence chemical reactivity [2]. Considering that organic electrochemical reactions can be complex and are frequently dependent on experimental conditions (electrode material, solvent, electrolyte, etc.), the extrapolation of the understanding of many known and useful electro-chemical reactions from molecular solvent systems to an IL environment may not be straightforward. For example, although follow-up chemistry may, or may not, be the same in ILs, relative to traditional solvents, two common effects of ILs are evident: (a) mass transport rates (reported as diffusion coefficients) are smaller by a factor of between 10 and 100 in all IL media, due to the viscous nature of the media; and (b) the heterogeneous electron-transfer rate constants can be smaller by as much as two orders of magnitude for processes controlled by outer-sphere dynamics. This effect was attributed to the "special" solvent effect of ILs [3].

Hitherto, relatively few electrochemical studies have been published devoted to the study of the electrochemical behavior of organic compounds in IL media, relative to that found in conventional solvent media. This is also reflected in the limited number of recent review articles available [4–6]. Certainly, the effectiveness of the use of ILs in this area will be very much assisted by the availability of

A.A.J. Torriero (✉)
Centre for Chemistry and Biotechnology, School of Life and Environmental Sciences, Faculty of Science, Engineering and Built Environment, Deakin University, Melbourne Burwood Campus, 221 Burwood Highway, Burwood, VIC 3125, Australia
e-mail: angel.torriero@deakin.edu.au

D.R. MacFarlane
School of Chemistry, Monash University, Clayton, VIC 3800, Australia

A.A.J. Torriero (ed.), *Electrochemistry in Ionic Liquids*,
DOI 10.1007/978-3-319-15132-8_15

data on their influence on chemical behavior of the intermediate products of electrochemical processes.

We begin by discussing relevant works concerned with the application of ILs as solvent systems, with and without the addition of cosolvents, as well as the use of ILs as supporting electrolyte in conventional solvents. Particular attention is paid to the oxidation reaction of aromatic rings, amines, and sulfur-containing compounds. The reduction reactions of conjugated alkenes, haloalkanes, aromatic rings, and carbonyl- and nitro-containing compounds are also discussed.

Some of these examples are selected to illustrate the potential this field has to revolutionize industrial processes that involve oxidations or reductions by providing facile electrochemical, rather than chemical means of carrying out these processes. Ionic liquids offer great potential in this respect by providing a highly stable electrolyte medium of adjustable solvency properties over a wide range and from which volatile products can be readily removed by distillation. One of the particularly attractive features of electrosynthetic processes is that they are highly compatible with flow chemistry approaches to large-scale synthesis. Thus, these examples amply illustrate the breadth of practical electrosynthetic processes that ionic liquid solvents can support and therefore the huge range of possibilities in this field.

15.2 Oxidation Reactions

15.2.1 Oxidation of Aromatic Rings

Aromatic hydrocarbons are, in general, oxidized through loss of an electron from the highest occupied π-molecular orbitals generating a π-delocalised radical cation. The electron-transfer process is generally irreversible in acetonitrile under normal conditions because the radical cation reacts rapidly with nucleophile species present as impurities, including residual water [7, 8]. A second irreversible oxidation wave is seen at more positive potentials due to the formation of a transitory dication. Dichloromethane and mixtures of dichloromethane with fluorosulphonic acid or trifluoroacetic acid are better solvents in which to demonstrate the reversible one-electron oxidation of aromatic species [7, 8].

The electrochemical oxidation of substituted anthracenes (Scheme 15.1) to their respective radical cations was studied in $[C_2mim][N(Tf)_2]$, $[C_4mim][N(Tf)_2]$, $[C_8mim][N(Tf)_2]$, $[C_4mpyr][N(Tf)_2]$, $[N_{2,2,2,6}][N(Tf)_2]$, $[N_{1,8,8,8}][N(Tf)_2]$,

Scheme 15.1 Chemical structure of substituted anthracenes

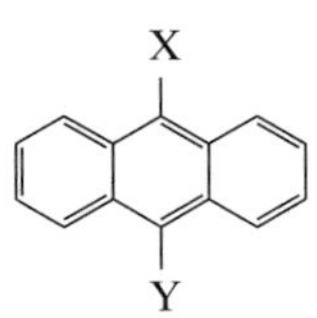

	X	Y
i	–H	–H
ii	–Cl	–H
iii	$–NH_2$	–H
iv	–Ph	–H
v	–Cl	–Cl
vi	–Ph	–Ph

$[C_4mim][BF_4]$, $[C_4mim][PF_6]$, $[C_4mim][NO_3]$, and $[C_4mim][CF_3SO_3]$ ILs [9, 10]. Compton and co-workers concluded that the direct radical–radical coupling (RRC) mechanism, with separate steps for dimerization and deprotonation (which may or may not be experimentally observable), [10] is the dominant process for the oxidation of anthracene (**i**), 9-chloroanthracene (**ii**), 9-aminoanthracene (**iii**), and 9-phenylanthracene (**iv**) in ILs (Eqs. 15.1–15.5). Under conditions of low nucleophilicity, the radical cation of the dimer (Eq. 15.5) may survive, as long as the concentration is kept at the millimolar level, and the dimer may then be obtained after back reduction to the neutral stage [8]. Furthermore, the heterogeneous rate constants for the oxidation of **iv** have been derived in the same range of ILs and found not to show any clear variation with either viscosity or static dielectric constant of the medium. However, their values are typically 2–3 orders of magnitude slower than in acetonitrile and do not follow the Marcus model for the dependence of the heterogeneous electron-transfer rate constant on solvodynamic radius that is expected for outer-sphere electron transfer [9].

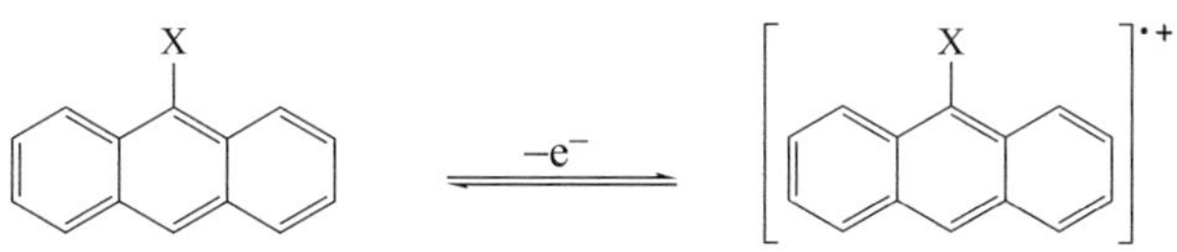

(15.1)

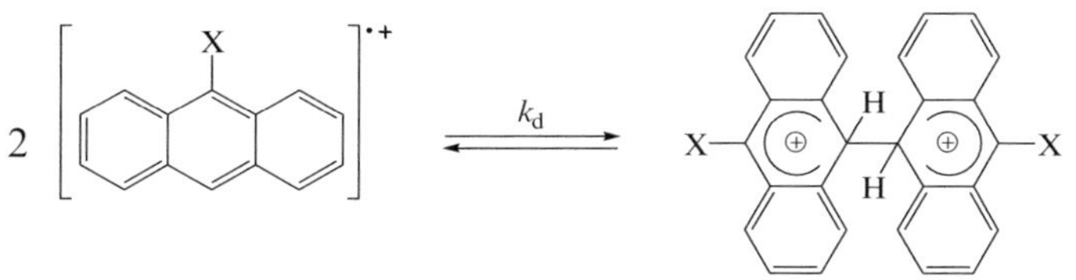

(15.2)

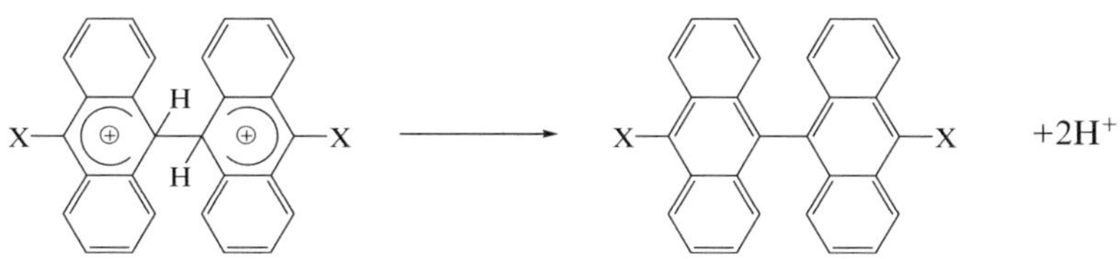

(15.3)

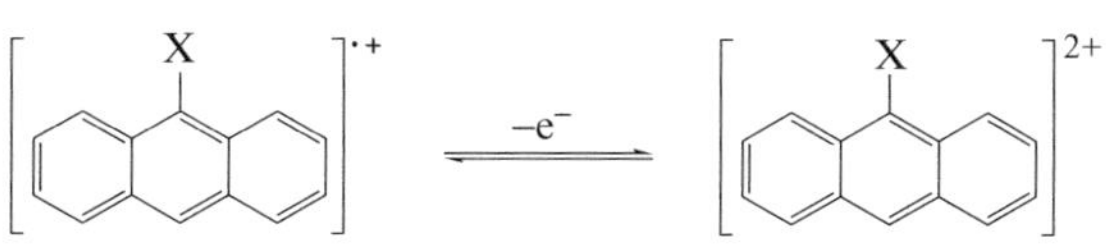

(15.4)

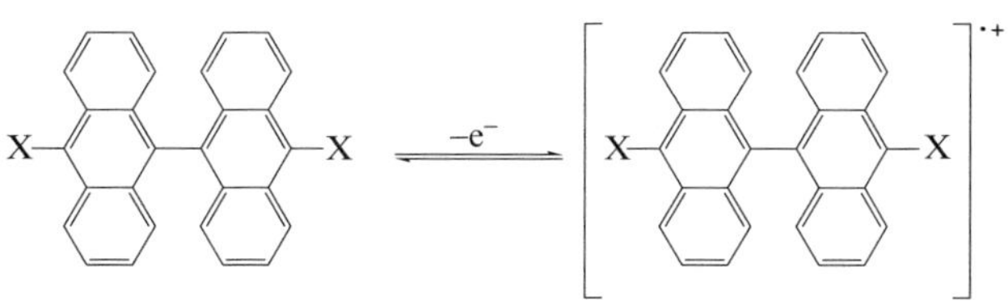

(15.5)

Disubstituted anthracenes, such as 9,10-dichloroanthracene (**v**) and 9,10-diphenylanthracene (**vi**), follow a simple electron-transfer (E) reaction mechanism in $[C_2mim][N(Tf)_2]$ with no homogeneous reactions [9], which is consistent with previous observations in organic solvent systems [11].

A similar experiment in $[C_2mim][N(Tf)_2]$ was conducted on a family of arenes, such as pyrene (**vii**), perylene (**viii**), chrysene (**ix**), 1,2-benzanthracene (**x**), and 2,3-benzanthracene (**xi**) [9]. For all these compounds, a radical–substrate coupling (RSC) mechanism (Eqs. 15.6–15.9) was observed, instead the radical–radical coupling (RRC) mechanism observed for anthracenes. Uniquely, **viii** was excellently modelled by a simple E reaction mechanism [9].

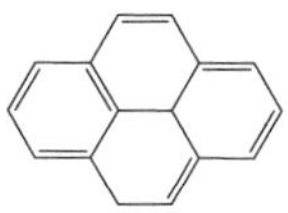

(vii)

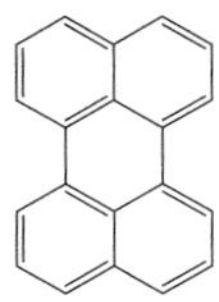

(viii)

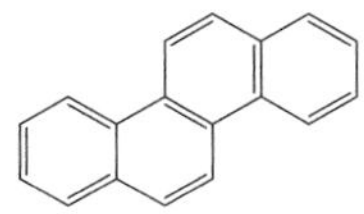

(ix)

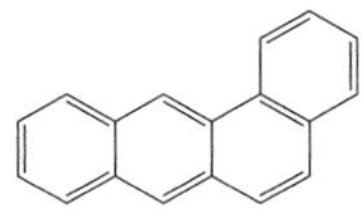

(x)

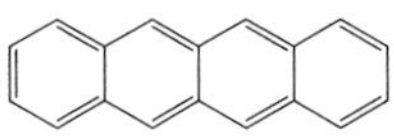

(xi)

$$Ar \rightleftharpoons Ar^{\bullet +} + e^- \quad (15.6)$$

$$Ar^{\bullet +} + Ar \xrightleftharpoons{k_d} Ar_2^{+} + H^{\bullet} \quad (15.7)$$

$$Ar^{\bullet +} \rightleftharpoons Ar^{2+} + e^- \quad (15.8)$$

$$Ar_2^{+} \rightleftharpoons Ar_2^{2+} + e^- \quad (15.9)$$

Oxidation of naphthalene (**xii**) and durene (**xiii**) was investigated in $[C_2mim][N(Tf)_2]$, $[C_4mim][N(Tf)_2]$, $[C_4mim][PF_6]$, and $[N_{1,8,8,8}][N(Tf)_2]$ ILs [10]. An irreversible electrochemical process was detected for **xii**. Similar results are commonly observed in molecular solvents, where the reactivity of the radical cation of a polyaromatic hydrocarbon is known to decrease with the size of the aromatic hydrocarbon [10]. Less delocalized cation radicals of alkylbenzenes, like the one obtained from **xiii**, are known to display a different reactivity due to the C–H acidity, and the deprotonation from the cation radical is the predominant type of decay (Eq. 11). The nature of the electrochemical mechanism was found to follow the same ECE-DISP1 mechanism (DISP = disproportionation) as that in acetonitrile (Eqs. 15.10–15.13) [10].

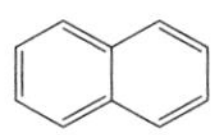

(xii)

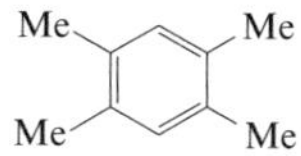

(xiii)

$$\mathrm{ArCH_3} \rightleftharpoons \mathrm{ArCH_3^{\bullet +}} + e^- \tag{15.10}$$

$$\mathrm{ArCH_3^{\bullet +}} \xrightleftharpoons{k_1} \mathrm{ArCH_2^{\bullet}} + \mathrm{H^+} \tag{15.11}$$

$$\mathrm{ArCH_2^{\bullet}} \rightleftharpoons \mathrm{ArCH_2^{+}} + e^- \tag{15.12}$$

$$\mathrm{ArCH_2^{\bullet}} + \mathrm{ArCH_3^{\bullet +}} \rightleftharpoons \mathrm{ArCH_2^{+}} + \mathrm{ArCH_3} \tag{15.13}$$

The oxidation of 1,2-dimethoxybenzene (veratrole, **xiv**) was studied in [C_2mim][N(Tf)$_2$] and [C_4mim][N(Tf)$_2$] ILs [10]. The mechanism involves the oxidative coupling of this aromatic diether with the formation of hexamethoxytriphenylene (**xv**), which is oxidized at a more negative potential than that of the starting monomer. The key intermediate during the electrochemical generation of **xv** is a dimer resulting from a first oxidative coupling of **xiv** that is rapidly followed by different electrochemical/chemical reactions leading to the final trimer. The dimerization rate constants measured in ILs were between 5 and 10 times smaller than those measured in acetonitrile, depending on the nature of ILs [10]. In the scan rate (v) range from 0.1 to 20 Vs^{-1}, the peak potential varies linearly as a function of log (v) with a slope close to 20 mV/log(v) for both ionic liquids. This shows that the first dimer is produced by an RRC mechanism similar to that described for anthracenes. The oxidation of **xiv** is only reversible at scan rates larger than 4,000 V s^{-1}, indicating that the radical cation of **xiv** has a lifetime in the range of several microseconds [10].

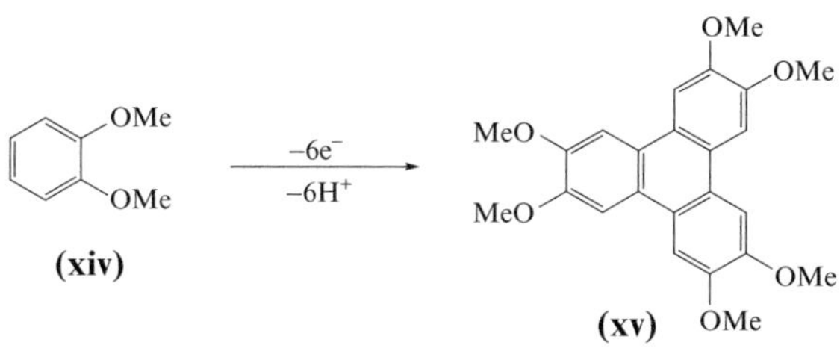

(15.14)

The oxidation of 4-methylanisole (**xvi**) was studied in methanol in the presence of [S_{222}][N(Tf)$_2$], and [C_4mim][N(Tf)$_2$] as the supporting electrolyte [12]. The oxidation of **xvi** starts at around 1.25 V vs. SCE. A transfer of 3.3 electrons per mol of **xvi** was observed when sodium perchlorate was used as the supporting electrolyte, which is smaller than the theoretical four-electron-transfer process expected for the oxidation of **xvi** to 4-methoxybenzaldehyde dimethyl acetal (**xvii**, Eq. 15.15). This difference was attributed to the presence of a side oligomerization reaction. When the ILs are present in the solution as the supporting electrolyte, a value of 3.7 electrons per mol of **xvi** was found, indicating the oligomerization reaction is more hindered under these experimental conditions, which could be a consequence of the intermediate stabilization by the IL anion [12]. The production of **xvii** in methanol containing ILs as the supporting electrolyte was performed with a yield of 45 %. Further improvement of around 10 % in this selectivity was observed with the further addition of 0.01 M KF [12].

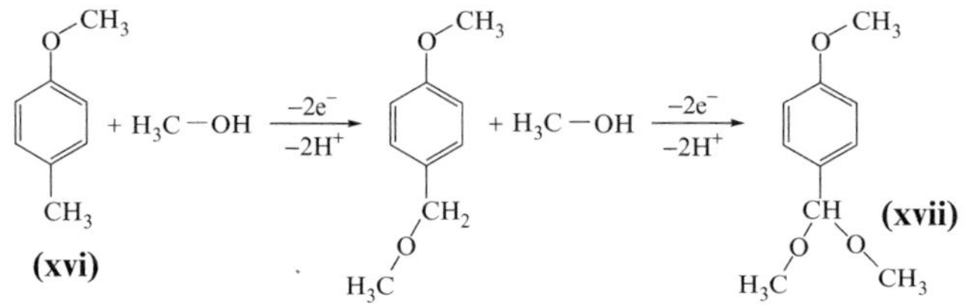

(15.15)

The electrochemical oxidation of hydroquinone (QH_2) has been studied in [C_2mim][$N(Tf)_2$], [C_4mim][BF_4], and [C_4mim][PF_6] [13, 14]. In dry ILs, the reduction follows an EE mechanism, consisting in the consecutive two-electron-transfer process (bottom horizontal row in Scheme 15.2) [14]. The voltammogram of QH_2 was also recorded in [C_4mim][BF_4] and [C_4mim][PF_6] in the presence of a proton acceptor (ammonia). This reaction proceeds with the one-electron oxidation of QH_2 to produce $QH_2^{\bullet-}$, which is deprotonated by ammonia to form $QH^{\bullet}$. This is further oxidized at the potential of the first wave to produce Q after a second deprotonation step (ECEC mechanism). The diffusion coefficient determined for QH_2 was 6.72×10^{-7} and 1.51×10^{-6} cm^2s^{-1} in [C_4mim][PF_6] and [C_4mim][BF_4], respectively [14].

Dopamine (**xviii**) was also studied in [C_2mim][$N(Tf)_2$] and [C_4mim][BF_4] ILs [13]. In this case, two chemically irreversible oxidation peaks were observed at 0.75 and 1.1 V vs. Pt wire. The first wave is attributed to the one-electron oxidation of dopamine, which results in the formation of a semiquinone (**xix**) (Eq. 15.16). The proton is not lost to the solution, but rather is probably taken up by the amine group. The second wave is assigned to a second one-electron oxidation, which results in a

Scheme 15.2 Square scheme showing all the possible pathways for the hydroquinone/ benzoquinone redox couple

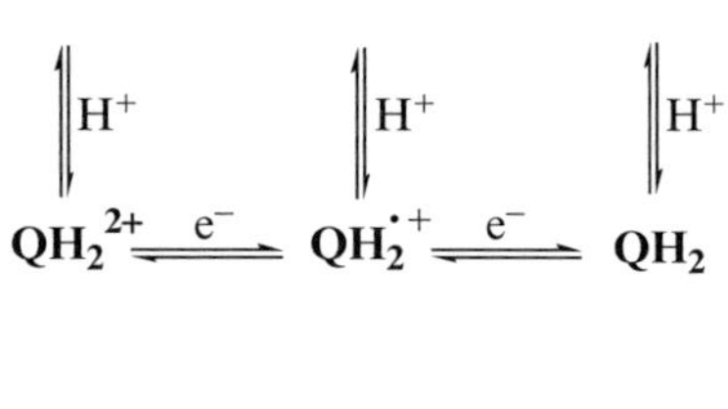

protonated quinone (Eq. 15.16). Both oxidation signals are at similar potentials to that of catechol [13]. In $[C_4mim][BF_4]$ a peak at less positive potential was observed for dopamine, similar to catechol, which is assigned to the adsorption of the neutral quinone species on the platinum electrode formed by the reaction of the removal of protons from the oxidized dopamine with the $[BF_4]$ anion or water molecules present in the IL as impurities [13].

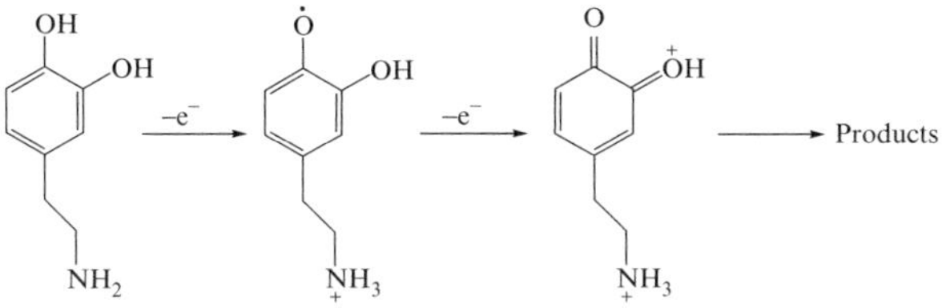

(xviii) **(xix)**

(15.16)

15.2.2 *Oxidation of Amines*

The oxidation of *N*,*N*-dimethyl-*p*-toluidine (**xx**) has been studied in $[C_4mpyr][N(Tf)_2]$ (Eqs. 15.17–15.21) [15]. The mechanism elucidated in this IL is consistent with that previously established in acetonitrile, and involves the deprotonation of the radical cation of **xx** by the starting molecule as a first and kinetic-determining step (Eq. 15.18), followed by a dimerization step (Eq. 15.19) [15]. This dimer is susceptible to two successive one-electron oxidations (Eqs. 15.20 and 15.21). In this case, the kinetic rate constant for the deprotonation step in IL was found to be similar to that measured in acetonitrile [15].

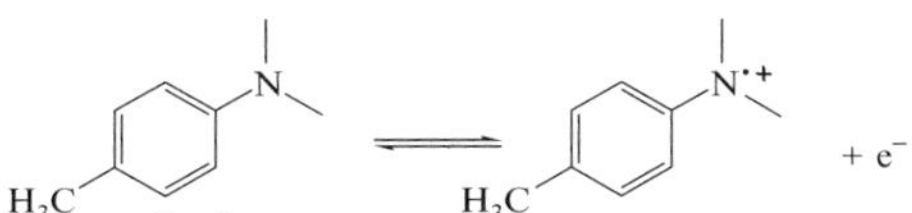

(xx)

(15.17)

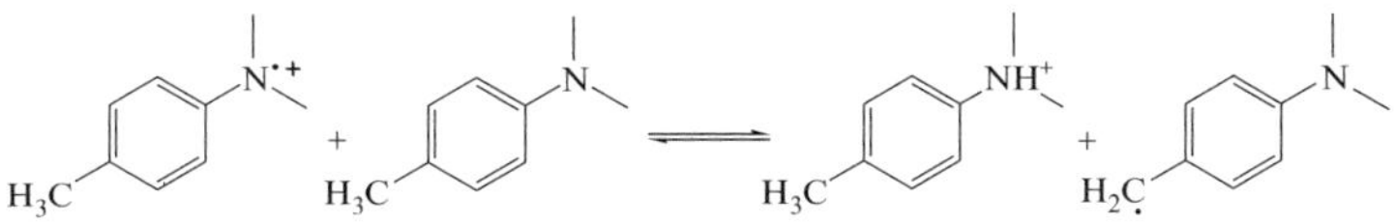

(15.18)

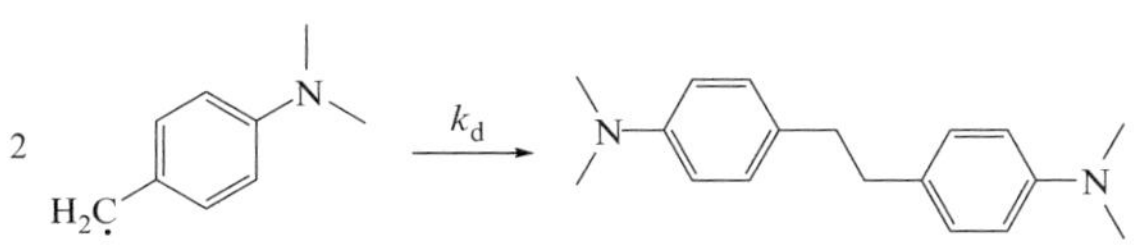

(15.19)

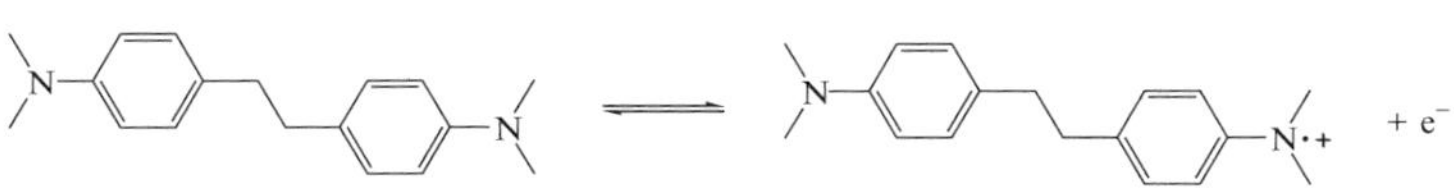

(15.20)

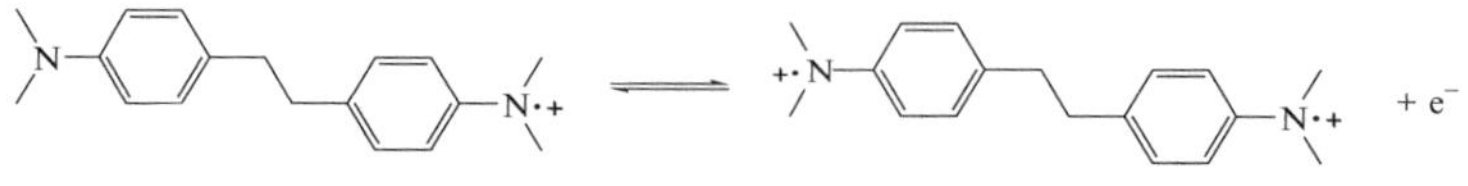

(15.21)

The electrochemical behavior of N,N,N′,N′-tetramethyl-*p*-phenylenediamine (**xxi**) and N,N,N′,N′-tetrabutyl-*p*-phenylenediamine (**xxii**), was studied in [C_2mim][N(Tf)$_2$], [C_4mim][N(Tf)$_2$], [C_8mim][N(Tf)$_2$], [C_{10}mim][N(Tf)$_2$], [C_4mpyr][N(Tf)$_2$], and [$N_{2,2,2,6}$][N(Tf)$_2$] ILs [16, 17]. In all these ILs, two well-defined reversible oxidation waves were detected, which can be assigned to the first and second one-electron oxidations of the phenylenediamines (Eq. 15.22).

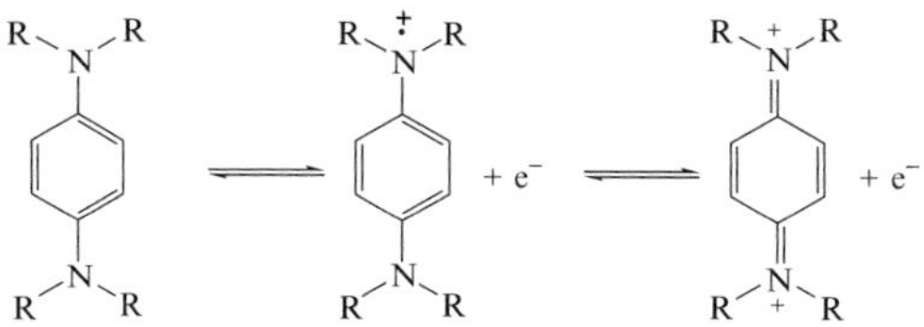

R = CH_3 (**xxi**)
= *n*-C_4H_9 (**xxii**)

(15.22)

The main effect of changing the cationic component of the ionic liquid was found to be its effect on the solvent viscosity, as the diffusion coefficient (D) of each species was found to be inversely proportional to the viscosity across the series of ionic liquids, in accordance with Stoke's equation [16, 17]. The only deviation from this relationship arose for the case when [$P_{6,6,6,14}$][N(Tf)$_2$] was used as the solvent [17]. Here, though relatively normal voltammetry was observed for the oxidation and immediate re-reduction of **xxi** (first electron transfer in Eq. 15.22), sweeping the potential at a more positive value to promote the formation of the dication, resulted in an anomalous wave shape. This behavior was rationalized in terms of a dimerization reaction between the dication and neutral **xxi** (Eq. 15.23), with the further dimer oxidation at a potential more positive than that for the first oxidation of the monomer (Eq. 15.24) [17].

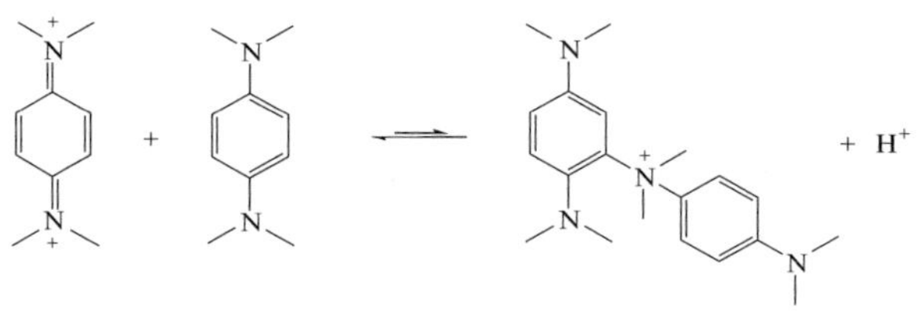

(15.23)

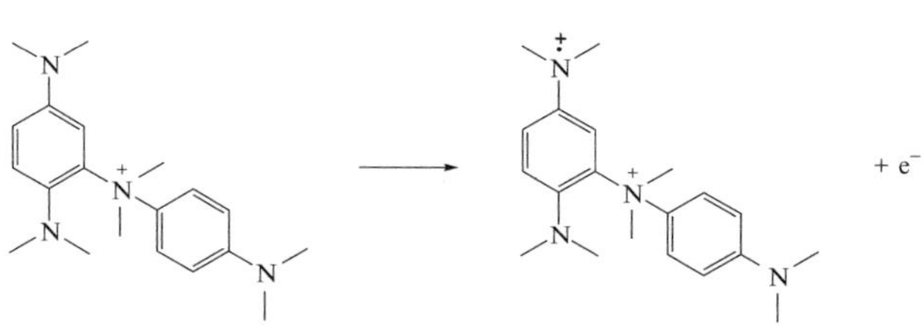

(15.24)

The electrochemical behavior of tris(4-bromophenyl)amine (**xxiii**) and 2,2,6,6-tetramethylpiperidine-1-oxyl (TEMPO, **xxiv**) was comparatively studied in [C_4mim][CF_3SO_3], [C_4mim][BF_4], and [C_4mim][PF_6] [18]. TEMPO showed a well-defined electrochemical reversibility with an anodic to cathodic peak current ratio equal to unity in all of the ILs studied. In contrast, a current ratio greater than unity was observed for **xxiii** at all the scan rates employed (0.01–1 V s^{-1}). Controlled potential bulk electrolysis of **xxiii** in the different ILs yielded a dark-blue colored solution of its monocation, which is stable under nitrogen. Diffusion coefficients of the former **xxiii** monocation were determined in all three ILs by chronoamperometry and found to be at least one order of magnitude less than the calculated value for **xxiii**. This effect, in conjunction with the deviation of the anodic to cathodic peak current ratio from unity was ascribed to ion-pairing interaction between the **xxiii** radical cation and the anion of the ILs [18]. These observations emphasize the potential for solute–solvent interactions in ionic liquids despite the ionic nature of the medium [19].

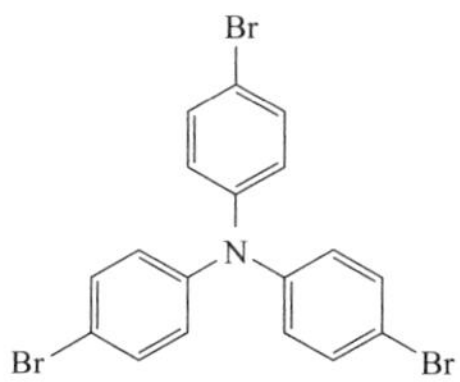

(xxiii)

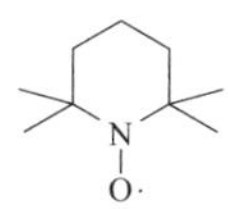

(xxiv)

15.2.3 Oxidation of Sulfur-Containing Compounds

Organic π-donors based on tetrathiafulvalene (TTF) and its derivatives have been the subject of numerous studies in organic media, due to their ability to form charge-transfer compounds and their wide range of potential applicability as, for instance, molecular conductors, semiconducting films, electroactive polymers (including low-band gap materials), molecular switches, and superconductors [8].

The electrochemical oxidation of TTF was studied in [C_2mim][BF_4] [20]. The two-electron oxidation of TTF was reversible and proceeded through two consecutive one-electron steps, which is consistent with the sequential oxidation of TTF to TTF^+ and TTF^{2+}. During slow scans, a dark orange precipitate was observed on the electrode surface. If the electrode was not cleaned between experiments, a decrease in the oxidation wave of TTF^+ to TTF^{2+} was observed [20]. Similar precipitation phenomena have been reported in chloroaluminate melts in which TTF^+ and TTF^{2+} are also produced through two consecutive one-electron oxidation of TTF [21].

15.3 Reduction Reactions

15.3.1 Reduction of Conjugated Alkenes

In general, aliphatic and nonconjugated alkene bonds are not reducible under normal electrochemical conditions. Meanwhile, alkene conjugation lowers the energy level of the lowest π*-orbital, so they become now reducible in an accessible

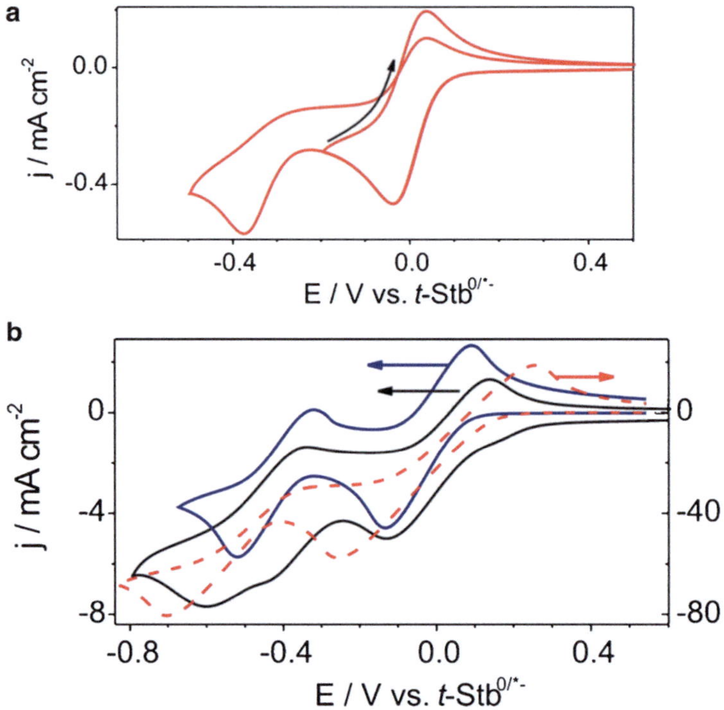

Fig. 15.1 Cyclic voltammograms obtained with a GC electrode at $\nu = 0.10\ \mathrm{V\ s^{-1}}$ for reduction of (**a**) 1 mM **xxv** in CH_3CN (0.1 M $[Bu_4N][PF_6]$) (**b**) (*dashed red lines*) 170 mM **xxv** in CH_3CN (0.1 M $[Bu_4N][PF_6]$), (*blue lines*) 150 mM **xxv** in $[C_4mpyr][N(Tf)_2]$, and (*black lines*) 220 mM **xxv** in $[(N_{111})(N_{112})BH_2][N(Tf)_2]$. Reproduced from [22] with permission from ACS

potential range. For example, Fig. 15.1a shows the cyclic voltammetry for the electrochemical reduction of *trans*-stilbene (**xxv**) in dry acetonitrile containing 0.1 M $[Bu_4N][PF_6]$ as the supporting electrolyte. Under this condition, it is possible to observe the fast and reversible one-electron reduction of the alkene bond to yield a radical anion (Eq. 15.25) with a midpoint potential of -2.67 V vs. $Fc^{0/+}$ [22]. At even more negative potential (≈ -3.1 V vs. $Fc^{0/+}$), the radical anion is further reduced in a one-electron irreversible process to produce a highly basic dianion (Eq. 15.26), which undergoes rapid protonation to form *1,2*-diphenylethane (Eq. 15.27) [22–27].

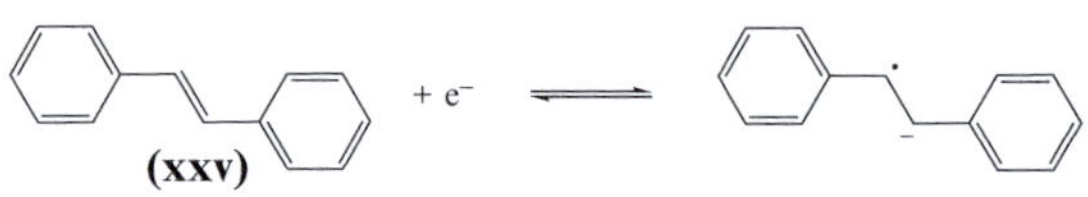

(15.25)

(15.26)

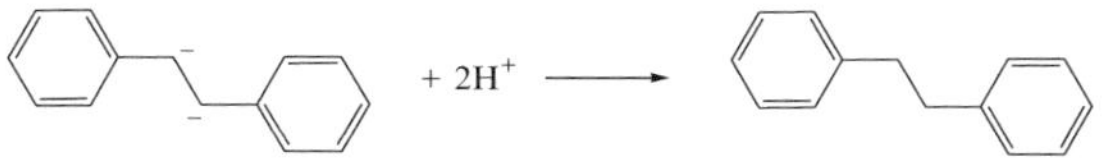

(15.27)

Convincing evidence for generation of the stable trans-stilbene dianion on the time scale of cyclic voltammetry at room temperature is acquired when the electroreduction is carried out in $[(N_{111})(N_{112})BH_2][N(Tf)_2]$, and $[C_4mpyr][N(Tf)_2]$ ionic liquids at atypically high concentrations >0.1 M of **xxv** (Fig. 15.1b) [22]. In the case of $[C_4mpyr][N(Tf)_2]$, almost complete chemical reversibility is observed for the second reduction process, while partial reversibility is attained with $[(N_{111})(N_{112})BH_2][N(Tf)_2]$. Experiments performed in the ILs at concentrations of **xxv** below 15 mM provided chemically irreversible voltammograms that resemble those obtained in acetonitrile. The origin of the chemical irreversibility was attributed to abstraction of the protons from trace amounts of water or other impurities present in the ionic liquids [22]. The origin of the peak at ca. −0.6 V appearing prior to the reversible $\mathbf{xxv}^{-/2-}$ process in $[(N_{111})(N_{112})BH_2][N(Tf)_2]$ (Fig. 15.1b) was attributed to partial protonation of the dianions by adventitious proton source. The separation in the reversible midpoint potential (E_m) values for $\mathbf{xxv}^{0/-}$ and $\mathbf{xxv}^{-/2-}$ is 0.37 V in $[C_4mpyr][N(Tf)_2]$ and 0.44 V in $[(N_{111})(N_{112})BH_2][N(Tf)_2]$ [22].

The reductive coupling of 2-cyclohexen-1-one (**xxvi**) has been investigated extensively in aqueous, organic, and a mixture of these solvents, because this reaction is able

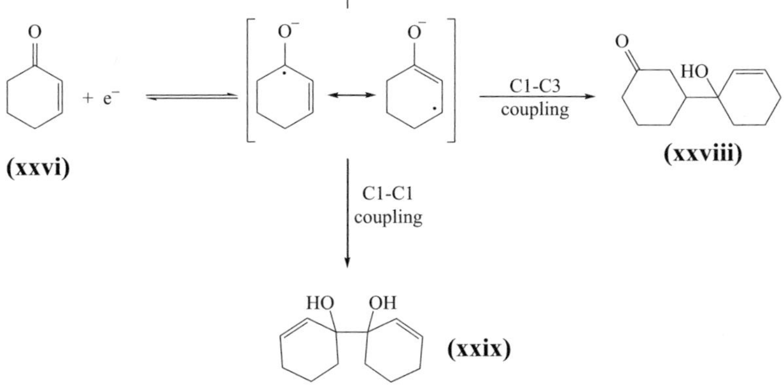

Scheme 15.3 Mechanism and possible products obtained from a reductive coupling of **xxvi**

to afford different products, such as 3-(3-oxocyclohexyl)cyclohexan-1-one (**xxvii**), 3-(1-hydroxycyclohex-2-enyl)cyclohexanone (**xxviii**), and 1-(1-hydroxycyclohex-2-enyl)cyclohex-2-enol (**xxix**), depending on the reaction conditions (Scheme 15.3) [7, 28, 29]. The reaction starts by addition of an electron to the lowest unoccupied molecular orbital, LUMO, of the carbonyl moiety present in **xxvi**, forming a radical anion intermediate able to stabilize the charge by resonance.

This reaction was performed in an electrolytic cell at a constant current of 50 mA, containing [C_4mim][BF_4] as the solvent system and using a platinum plate as the cathode material [28]. At room temperature, the reductive coupling product **xxvii** was obtained in low yield (ca. 11 %). The problem of high viscosity of the [C_4mim][BF_4], which limited the mass transport and the reaction rate, was overcome by heating the solution up to 50 °C and by adding water as a cosolvent (20 % v/v). Under these conditions, **xxvii** was produced at a maximum yield of 96 % [28]. This was explained by considering that the [C_4mim] cation is able to form ion pairs and/or hydrogen bonds with the intermediate radical anion, which produces a strong steric hindrance at the carbon C1 of **xxvi** for a C1–C3 or a C1–C1 coupling to proceed. Therefore, the C3–C3 coupling reaction is promoted to form the product **xxvii**.

15.3.2 Reduction of Halogenated Organic Compounds

The electrochemical reduction of halogenated organic compounds (RX) in organic solvents is well known [7, 8]. The accepted general electrochemical mechanism starts with the transfer of one electron from the working electrode, followed by a fast elimination of the halide anion to form a radical intermediate $R^{\bullet}$. This radical can either be reduced further to the carbanion R^{-}, which generally occurs at potentials more positive with respect to the reduction of the parent molecule RX, or it can couple to produce a dimeric product as a consequence of an RRC process. Therefore, the mechanism and the structure of the obtained products depend on the starting material and conditions selected for the electrochemical reduction.

Electrolytic reduction of benzoyl chloride (**xxx**) in [C_4mim][BF_4] has been carried out at 50 °C and using silver and magnesium as the cathode and anode material, respectively [30]. Silver as a cathode electrode fulfills a double role by catalytically activating the C–Cl bond and favoring the radical coupling between two acyl radicals. When a potential of −1.2 V vs. Ag wire QRE is applied, **xxx** is reduced by an irreversible one-electron-transfer process to an acyl radical (Eq. 15.28). This is followed by a fast chemical reaction, where two acyl radicals couple to produce benzil (**xxxi**, Eq. 15.29) [30]. The yield of **xxxi** ($\approx$51 % under these working conditions) declined $\leq$25 % when the applied potential was more cathodic than −1.2 V, as this potential is negative enough to further reduce benzil in a two-electron process to yield 2-hydroxy-1,2-diphenylvinyl benzoate (Eq. 15.30) [31].

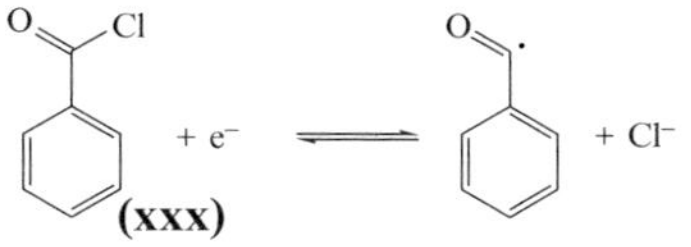

(15.28)

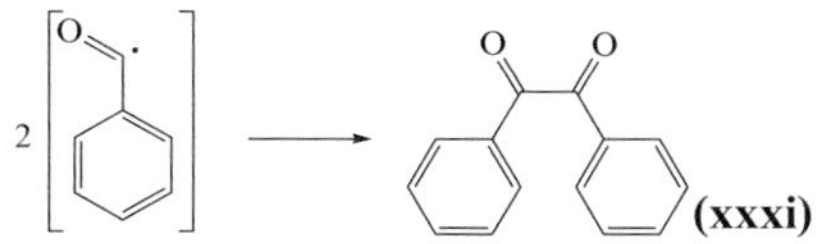

(15.29)

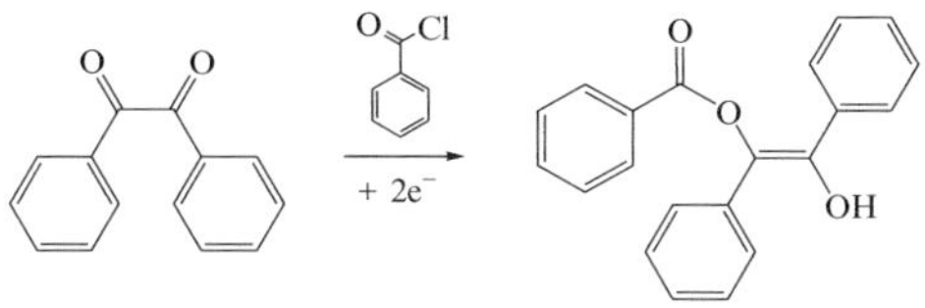

(15.30)

The electrolytic reduction of α-chloroethylbenzene (**xxxii**) was performed in *N*, *N*-diethyl-*N*-methyl-*N*-(2-methoxyethyl)ammonium bis(trifluoromethanesulfonyl) amide, $[N_{2,2,1,m}][N(Tf)_2]$ at 40 °C in the presence of carbon dioxide atmosphere (5 MPa) [32]. For this experiment, a platinum plate was used as the working electrode and a magnesium rod as the sacrificial anode. A platinum wire was used as the QRE. Under these conditions, 2-phenylpropanoic acid was produced with a yield of 65 % following the mechanism depicted in Eqs. 15.31 and 15.32. Even though the reaction mechanism was influenced by the solvation properties of the ionic liquid, it was found that the reaction efficiency was largely improved by the presence of magnesium as the anode electrode. It was suggested that the magnesium cation plays an important role in this reaction by producing an insoluble magnesium carboxylate form of the product in reaction 15.32, which prevents further reaction [32].

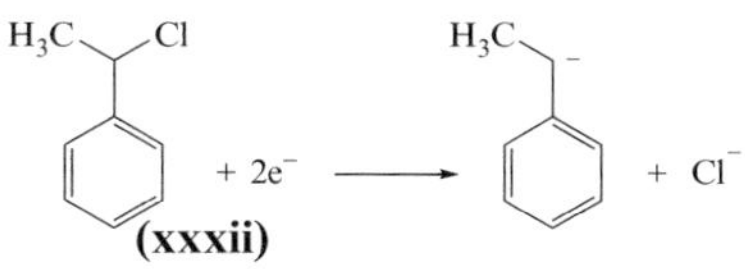

(15.31)

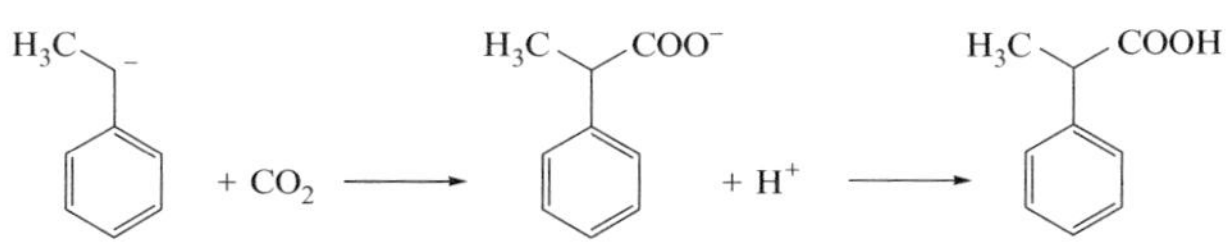

(15.32)

The electroreduction of 9-chloroanthracene (**ii**, Scheme 15.1) was investigated in $[C_4mim][N(Tf)_2]$, $[N_{1,1,1,4}][N(Tf)_2]$, and $[N_{2,2,2,4}][N(Tf)_2]$ [33]. In all of these ILs the voltammogram exhibits two successive one-electron processes. The first one was observed at around −2.2 V vs. $Fc^{0/+}$, showing characteristics for a chemically irreversible process. Meanwhile, the second wave (−2.4 V vs. $Fc^{0/+}$) is chemically reversible and displays all the characteristics associated with the anthracene|anthracene radical anion system. Upon increasing the scan rates, this second process decreases to the point of vanishing as the first wave simultaneously becomes reversible. Therefore, the second cathodic process was assigned to the reduction of the product formed during the first reduction process, and it is consistent with the well-known mechanism described for the electrochemical reduction of aryl halides [33]:

(15.33)

(15.34)

Radical anthracene is converted into anthracene via a series of fast reactions, which do not consume electrons, certainly involving H-atom transfer from the ionic liquid to the radical. Then, anthracene is reversibly reduced at the second wave into its anion radical [33].

The electrocatalytic reduction of diethyl-2,7-dibromooctanedioate (**xxxiii**) and 2,3-dibromo-1,2,3,4-tetrahydronaphthalene (**xxxiv**) were performed in [C_2mim][BF_4] and [C_4mim][BF_4] using cobalt(II)salen (**xxxv**) as the mediator [34]. Figure 15.2, curve a, shows the reversible voltammogram related to the one-electron reduction of **xxxv** (Eq. 15.35) in [C_4mim][BF_4] with a midpoint potential of -1.19 V vs. SCE [34]. A similar reversible process was also observed

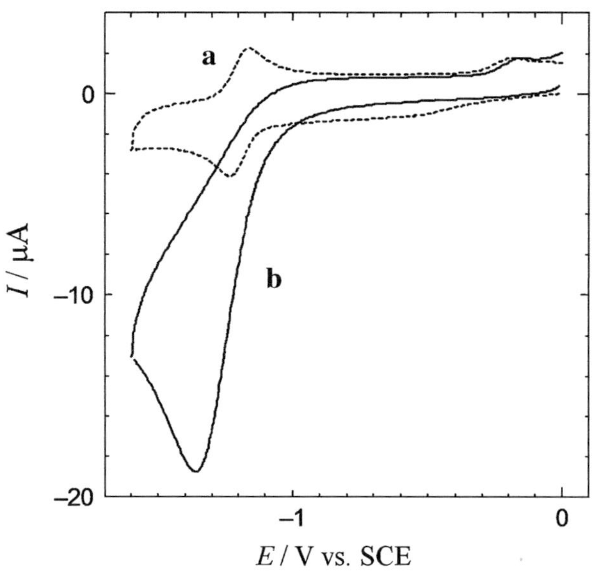

Fig. 15.2 Cyclic voltammograms of (a) Co(II)salen (3 mM) and (b) Co(II)salen (3 mM) in the presence of **xxxiii** (5 mM) at a glassy carbon electrode ($d = 3$ mm) in [C_4mim][BF_4]. Scan rate: 100 mV/s. Reproduced from ref. [34] with permission from Springer

at −1.14 V vs. SCE in [C_2mim][BF_4]. Given that the irreversible peak potentials for the direct reduction of compounds **xxxiii** and **xxxiv** at the glassy carbon electrode surface in [C_2mim][BF_4] and [C_4mim][BF_4] are ca. 1 V more cathodic with respect to the reduction potential of **xxxv**, the increase in the cathodic current on voltammogram b (Fig. 15.2) is attributed to the catalytic regeneration of **xxxv** resulting from the reaction of **xxxv**$^-$ with the brominated substrates (for example, Fig. 15.2 shows the experimental data related to the reaction of **xxxv**$^-$ with **xxxiii**) producing the debrominated product **xxxvi** with a yield of 51 % (Eqs. 15.36–15.38) [34]. This product differs from that obtained in DMF, where the compound **xxxvii** is obtained as a consequence of an intramolecular cyclization reaction (Eq. 15.39) [35]. Similarly, **xxxv**$^-$ reacts with **xxxiv** in a four-electron-two-proton reaction to form 1,4-dihydronaphthalene as the final product with a yield of around 70 % (Eqs. 15.40–15.42) [34]. The proton source has not been identified, but it is likely to be related either to the acidic imidazolium proton on position C2 or to the presence of water molecules in the respective ionic liquids.

(15.35)

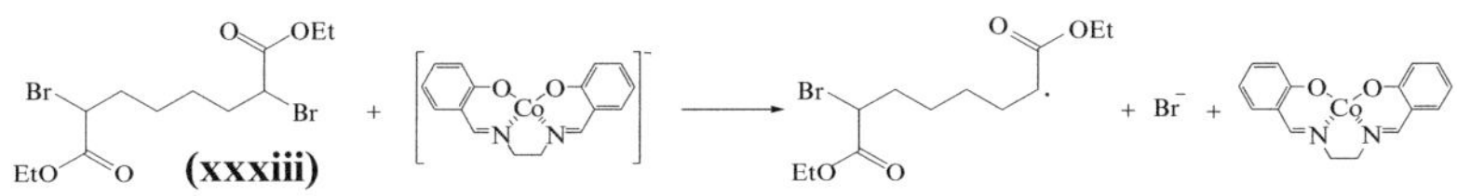

(15.36)

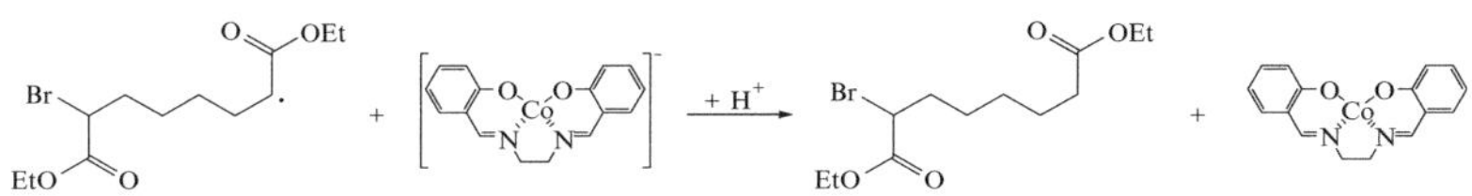

(15.37)

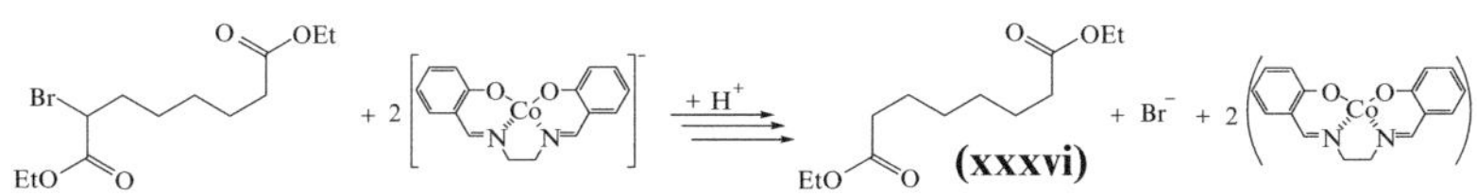

(15.38)

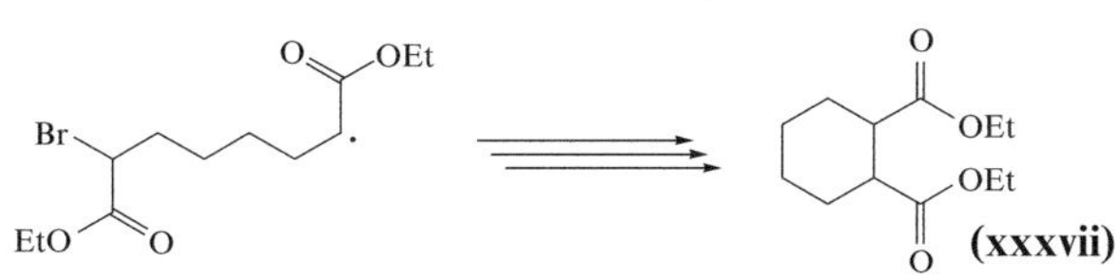

(15.39)

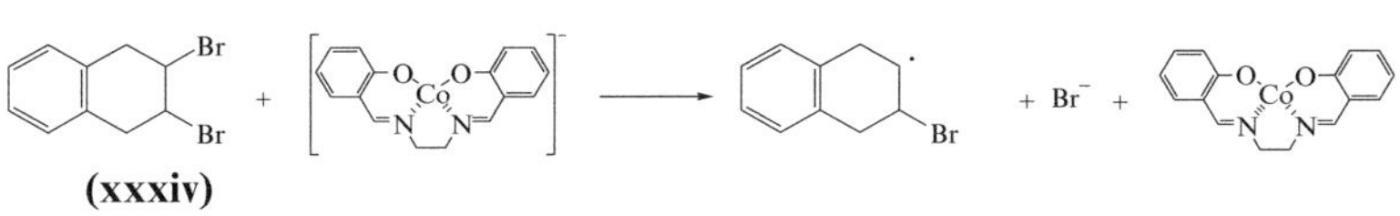

(15.40)

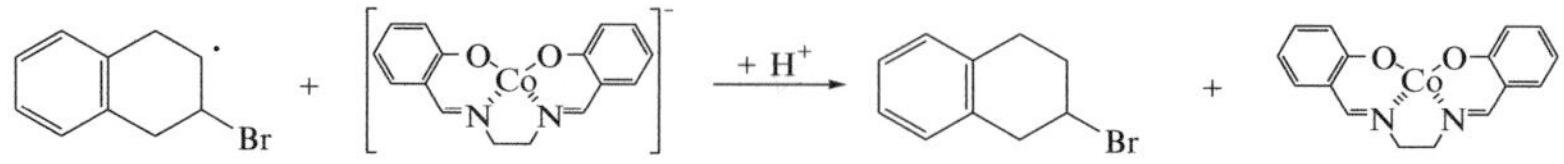 (15.41)

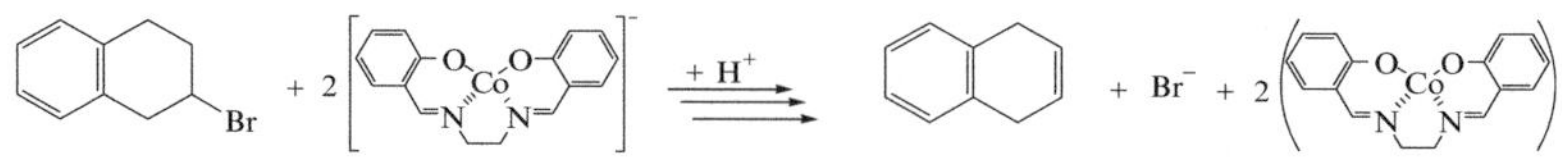 (15.42)

Very few papers have been devoted to the direct electroreduction of halogenated alkanes in ILs. Reduction potentials for the direct electroreduction of trichloroacetic acid, dichloroacetic acid, dibromoethane, and benzyl chloride was reported as −1.91, −1.82, −1.83, and −2.05 V vs. SCE, respectively, in $[C_4mim][PF_6]$ on a glassy carbon electrode [36]. More detailed investigations of the direct electroreduction of a series of Freon compounds (F113 (**xxxviii**), F113B2 (**xxxix**), F114B2 (**xL**), and F13B1 (**xLi**)) were reported in $[C_4mim][N(Tf)_2]$, $[C_4mpyr][N(Tf)_2]$, $[C_4mim][PF_6]$, and $[P_{6,6,6,14}][N(Tf)_2]$ ILs [37]. Irreversible but well-defined voltammograms were obtained using a glassy carbon, platinum, and silver as working electrodes. Importantly, the known ability of silver working electrodes to act as an electrocatalyst for Freon electroreduction in conventional solvent media is maintained in the IL media [37]. Unfortunately, the measured reduction potentials were reported using a Pt wire as a pseudo-reference electrode. Therefore, no clear effect of IL media on the peak potential for the reduction of Freon compounds was reported.

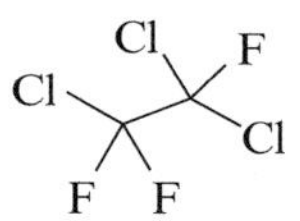

(xxxviii)

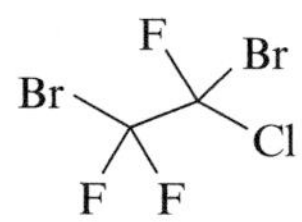

(xxxiv)

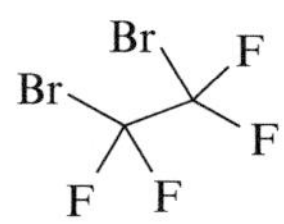

(xL)

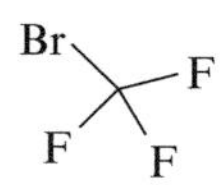

(xLi)

15.3.3 Reduction of Aromatic Rings

The electrochemical reduction of 1,4-benzoquinone (Q) was investigated on a microelectrode (10 μm diameter Au) in $[C_2mim][N(Tf)_2]$ [38]. Two reduction peaks were observed (Fig. 15.3), corresponding to the reduction of BQ to the radical anion, $Q^{\bullet -}$, at a midpoint potential of approximately −0.45 V vs. Ag wire, followed by reduction to the dianion, Q^{2-} at more negative potentials ($E_m = -0.83$ V vs. Ag wire), with the whole process being described as an EE mechanism (top horizontal row in Scheme 15.2). The voltammetric behavior changes from almost steady state at slow scan rates, to be more transient at faster scan rates, with the back peak becoming more prominent as the scan rate increases. In this system, the redox couples are separated by ca. 0.4 V, and it is a case where the second electron transfer takes place with greater thermodynamic difficulty than the first one. Therefore, comproportionation reactions are favored to occur, where the electrogenerated Q^{2-} reacts with the initial Q species to form two molecules of

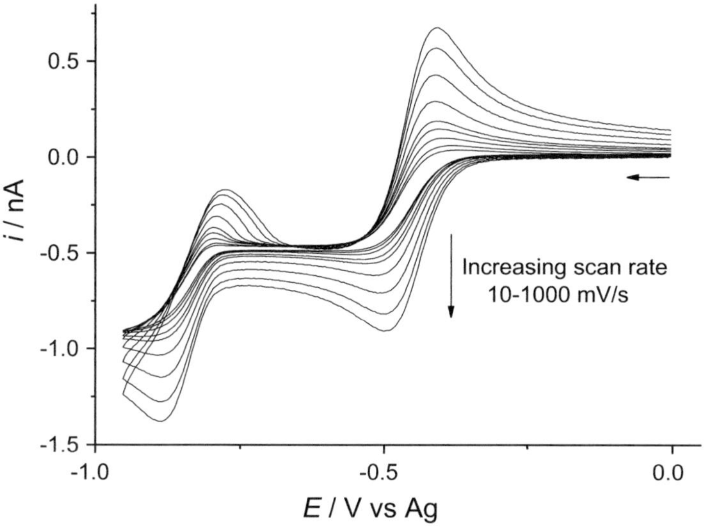

Fig. 15.3 Cyclic voltammetry for the reduction of 5 mM BQ on a 10 μm diameter Au microelectrode in $[C_2mim][N(Tf)_2]$ at scan rates of 10, 20, 40, 70, 100, 200, 400, 700, and 1,000 mV s^{-1} in the potential range 0.00 to −0.95 V. Reproduced from ref. [38] with permission from Elsevier

the intermediate $Q^{\bullet-}$ radical anion, as in Eq. 15.43. The diffusion coefficients determined for Q, $Q^{\bullet-}$, and Q^{2-} species were 4.73×10^{-7}, 1.73×10^{-7}, and 1.55×10^{-7} $cm^2\ s^{-1}$, respectively [38]. The same mechanism was also postulated for the reduction of Q in dry $[C_4mim][BF_4]$, and $[C_4mim][PF_6]$ [14]. The diffusion coefficient of Q was 5.34×10^{-7} and 1.37×10^{-6} $cm^2\ s^{-1}$ in $[C_4mim][PF_6]$ and $[C_4mim][BF_4]$, respectively [14]. The reduction of Q in $[C_4mim][BF_4]$, and $[C_4mim][PF_6]$ was also studied in the presence of a propanoic acid, used in this experiment as a proton donor. Under these conditions, each electron-transfer step is followed by a proton-transfer reaction (ECEC mechanism, Scheme 15.2).

$$Q + Q^{2-} \rightleftharpoons 2Q^{\bullet -} \tag{15.43}$$

The reduction of tetrachloro-1,4-benzoquinone (chloranil, **xLii**), 2-methyl-1,4-benzoquinone (toluquinone, **xLiii**), and 9,10-antraquinone (**xLiv**) in $[C_4mim][BF_4]$ follow the same EE mechanism as the nonsubstituted Q [39]. The experimental potentials vs. $Fc^{0/+}$ obtained for the first (E_m1) and second (E_m2) electron transfer are [39]:

	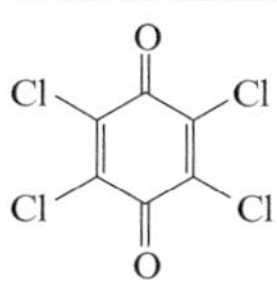	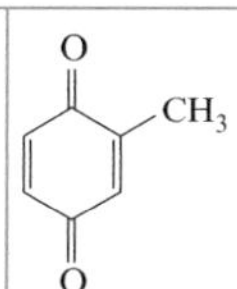	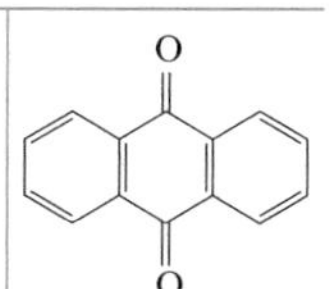
	(xLii)	**(xLiii)**	**(xLiv)**
$\boldsymbol{E_m 1}/V$ vs. $Fc^{0/+}$	−0.28	−0.77	−1.14
$\boldsymbol{E_m 2}/V$ vs. $Fc^{0/+}$	−0.78	−0.95	−1.48

The reduction of guanine (**xLv**) was studied in a number of imidazolium, ammonium, pyrrolidinium, and phosphonium-based ILs. However, only in $[N_{2,2,2,6}][N(Tf)_2]$ and $[P_{6,6,6,14}][FAP]$, ILs was a one-electron, diffusion-controlled process observed [40]. The E_m for this process was −1.8 and −1.6 V vs. $Fc^{0/+}$ in $[N_{2,2,2,6}][N(Tf)_2]$ and $[P_{6,6,6,14,}][FAP]$, respectively, and this was related to the formation of an unstable radical anion, which was proposed to undergo a dimerization reaction after a proton abstraction from the water molecules present in the IL (Eq. 15.44) [40]. The dimer structure is dependent on the positions of maximum unpaired electron density in the intermediate molecule. Therefore, it is reasonable to assume that the 8-position is the most likely site to form the corresponding dimer. This dimer is oxidized at potentials around 0.3 V vs. $Fc^{0/+}$ to regenerate the parent guanine, oxidizable around 1.3 V. The dimerization rate depends on the nature of the IL used, being 6×10^3 and 8.5×10^4 mol^{-1} L s^{-1} in $[N_{2,2,2,6}][N(Tf)_2]$ and $[P_{6,6,6,14}][FAP]$, respectively [40].

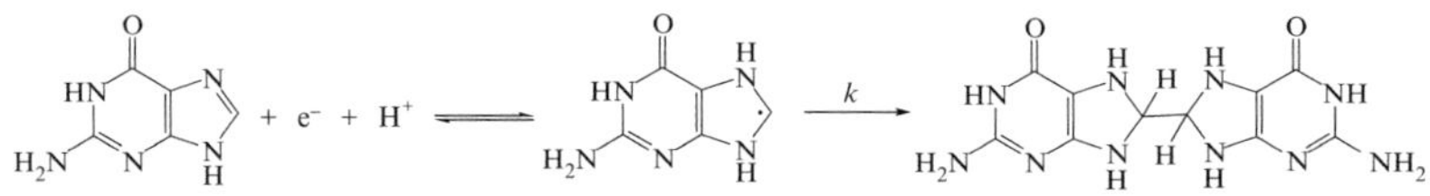

(**xLv**)

(15.44)

Adenine was electrochemically reduced at −1.70 V vs. $Fc^{0/+}$ in $[N_{2,2,2,6}][N(Tf)_2]$, where a similar behavior to that reported for **xLv** was observed. This is in a noticeable difference to their previously described behavior, which is, either two successive two-electron steps or a single four-electron step described in aqueous media on a dropping mercury electrode [41, 42].

15.3.4 Reduction of Carbonyl Compounds

The electrochemical reduction of benzophenone (**xLvi**) was studied in five different ionic liquids [43]. In dry (<20 ppm H_2O) and aprotic ILs, such as $[C_4mpyr][N(Tf)_2]$ and $[C_4mpip][N(Tf)_2]$, two reversible and well-resolved one-electron processes were observed (Fig. 15.4). These processes are related to the addition of an electron to the LUMO of the carbonyl moiety present in **xLvi** to form a radical anion (Eq. 15.45), in which the new charge is delocalized over the π-system conferring stability to this redox process. The second reversible one-electron process, observed at 0.664 V more cathodic with respect to the parent reduction, is related to the formation of a dianion reduction product (Eq. 15.46). The magnitude of this second process was smaller than the first one. This was attributed to the presence of a following comproportionation reaction (Eq. 15.47) [43].

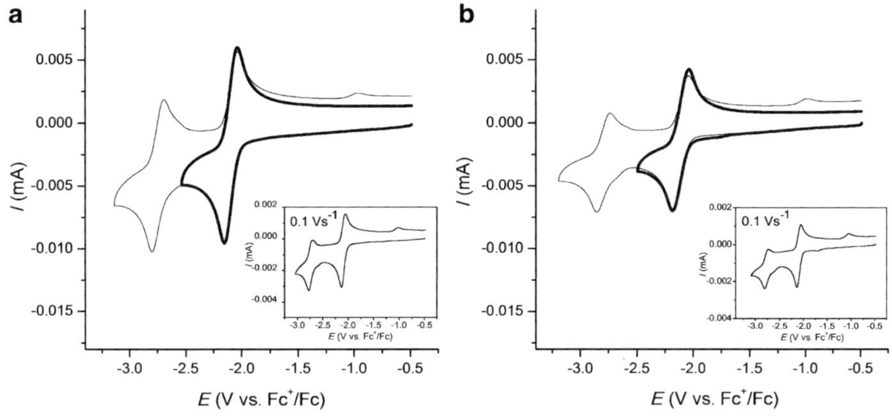

Fig. 15.4 Cyclic voltammograms for the reduction of 10 mM benzophenone under a N_2 atmosphere with a glassy carbon electrode ($d = 1.0$ mm) in (**a**) dry [C_4mpyr][N(Tf)$_2$] and (**b**) dry [C_4mpip][N(Tf)$_2$] at a scan rate of 1.0 V s^{-1} (*insets*: scan rate = 0.1 V s^{-1}). Reproduced from ref. [43] with permission from Wiley

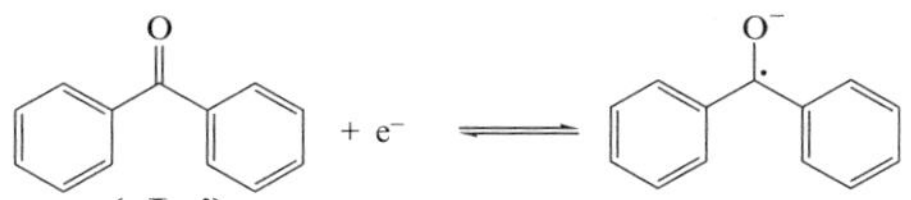

(xLvi)

(15.45)

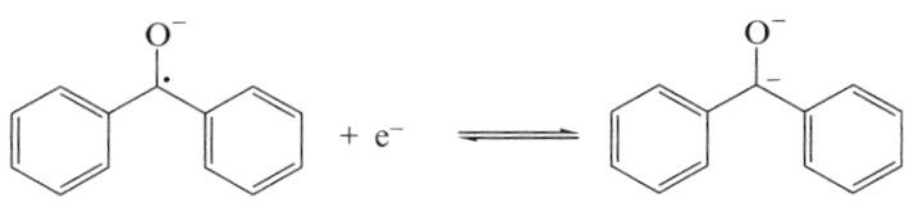

(15.46)

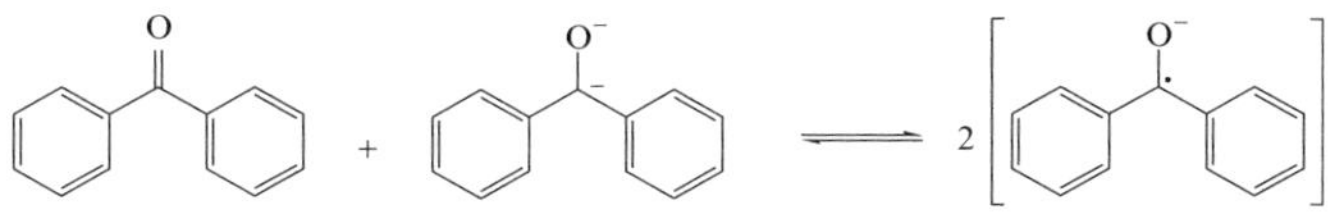

(15.47)

Upon addition of water, the second process became chemically irreversible. The potential of this second process shifts anodically as a function of the water content, until it merges into a single proton-coupled two-electron-reduction process when about 0.6 M water was present [43]. Meanwhile, in dry (<20 ppm H_2O) imidazolium-based ILs ([C_2mim][N(Tf)$_2$], [C_4mim][PF_6], and [C_4mim][BF_4]), the first reduction process was again reversible, whilst the second reduction process became chemically irreversible, even at scan rates as high as 100 Vs^{-1}, owing to the availability of the relatively acidic proton at the C2 position in the imidazolium ring.

The reduction of **xLvi**, 3-methylbenzophenone (**xLvii**), 4,4′-dimethylbenzophenone (**xLviii**), and 4,4′-dimethoxybenzophenone (**xLix**) was investigated in [C_4mim][N(Tf)$_2$] and [C_4mpyr][N(Tf)$_2$] [44]. In this last IL, the reduction of the substituted benzophenones occurs via two consecutive one-electron processes leading initially to the radical anion and subsequently to the dianion species at around 0.62 V more cathodic than the parent reduction. The radical anion generated is highly stable at all measured timescales while the voltammetry of the dianion species is reversible at scan rates of $\geq$10.0 V s^{-1}. In contrast, the reductive electrochemistry of substituted benzophenones in $[C_4mim][N(Tf)_2]$ depends on the timescale of the measurements. At fast scan rates, the initial reduction (radical anion generation) is partially reversible, whereas the second reduction process (dianion generation) is irreversible and shifted anodically, which is a consequence of the protonation of the electrogenerated dianion. However, at low scan rates ($\leq$0.1 V s^{-1}), reduction occurs via a two-electron process with the characteristic of an ECE- or DISP1-type mechanism, where the protonation of the radical anion is the rate-determining step via a slow proton-transfer reaction from the $[C_4mim]$ cation [44].

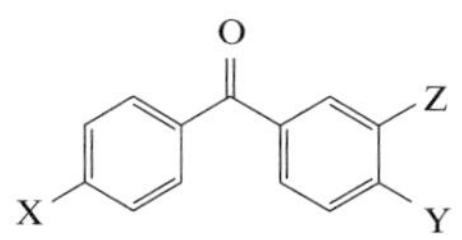

	X	Y	Z
xLvi	-H	-H	-H
xLvii	-H	-H	$-CH_3$
xLviii	$-CH_3$	$-CH_3$	-H
xLix	$-OCH_3$	$-OCH_3$	-H
L	-H	-Cl	-H

Consistently, with this observation, the reduction of 4-chlorobenzophenone (**L**) displayed different patterns in $[C_4mim][N(Tf)_2]$, $[N_{1,1,1,4}][N(Tf)_2]$, and $[N_{2,2,2,4}]$ $[N(Tf)_2]$, than those observed in acetonitrile (Fig. 15.5) [33]. In the ILs, two one-electron reduction waves are observed, as in organic media, but the first wave is partially reversible, whereas the second wave is irreversible (in the scan-rate range of 0.05–50 V s^{-1}). Moreover, the reversibility of the first reduction wave decreases when the initial concentration of **L** is increased, indicating that the rate-determining step is now a bimolecular reaction. Simultaneously, upon increasing the initial concentration of **L**, the second irreversible wave tends to disappear [33]. For comparison, the electrochemical reduction of **Li**, was performed in the same ILs, which displayed the same general behaviors observed for the reduction of **L** [33]. From this similarity, Lagrost and colleagues concluded that the first monoelectronic reduction of **L** corresponds to the production of its radical anion (Eq. 15.48), and the second wave to the reduction of this radical anion into the dianion (Eq. 15.49). The negative charge in the radical anion obtained as a product of Eq. 15.48 is more localized on the oxygen atom of the carbonyl group, probably because a specific ion-pair association between the radical anion and the cation of the ionic liquid that stabilizes the unpaired electron, with a consequent decrease in the C–Cl cleavage rate. As a consequence of this effect, the dimerization between two radical anions that prevails over the cleavage reaction was favored, leading to a different mechanism with respect to that observed in organic media [33].

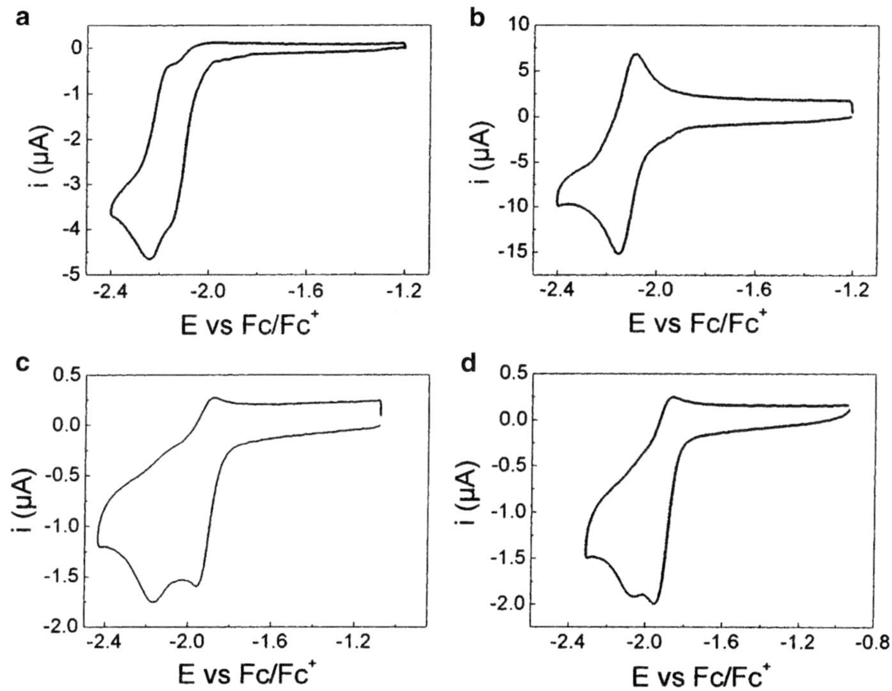

Fig. 15.5 Cyclic voltammograms of 2 mM **L** in acetonitrile at (**a**) 0.2 and (**b**) 5 V s^{-1} and 10 mM **L** in (**c**) $[N_{1,1,1,4}][N(Tf)_2]$ and (**d**) $[C_4mim][N(Tf)_2]$ at 0.2 V s^{-1} on a glassy carbon electrode. Reproduced from ref. [33] with permission from The American Chemical Society

(15.48)

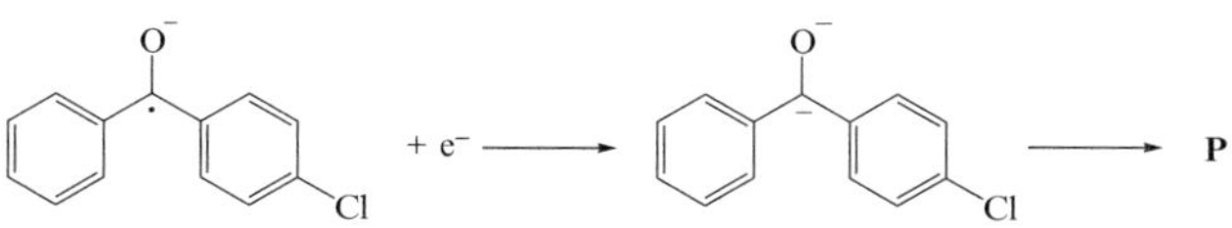

(15.49)

The reduction of benzaldehyde (**Li**) was investigated in $[C_4mim][N(Tf)_2]$ and $[C_4mpyr][N(Tf)_2]$ [45, 46]. Unlike those discussed previously, the voltammograms exhibit two reversible one-electron reduction processes in these two ILs. The first process is assigned to the primary reduction of **Li** to the radical anion, with an E_m of −1.75 V vs. $Fc^{0/+}$ (Eq. 15.50). This radical anion is involved in two competitive pathways: (1) a dimerization reaction, leading to the corresponding pinacol (Eq. 15.51) and (2) a second electron transfer (E_m of −2.33 V vs. $Fc^{0/+}$) that reduces the radical anion to the dianionic species, which can lead to the formation of alcohol or alcoholate product (Eq. 15.50) [45, 46]. The apparent second-order rate constant for the dimerization (k_d) was reported to be 1.0×10^4 and 1.4×10^4 L mol^{-1} s^{-1} in

[C_4mim][$N(Tf)_2$] and [C_4mpyr][$N(Tf)_2$], respectively. Meanwhile, the pseudo-first-order rate constant, k_1, for the alcohol/alcoholate formation was 1×10^3 and $1.6 \times 10^3\ s^{-1}$ in [C_4mim][$N(Tf)_2$] and [C_4mpyr][$N(Tf)_2$], respectively [45, 46]. However, it should be noticed that determination of the second-order rate constants may be somewhat imprecise because of the occurrence of two follow-up chemical reactions.

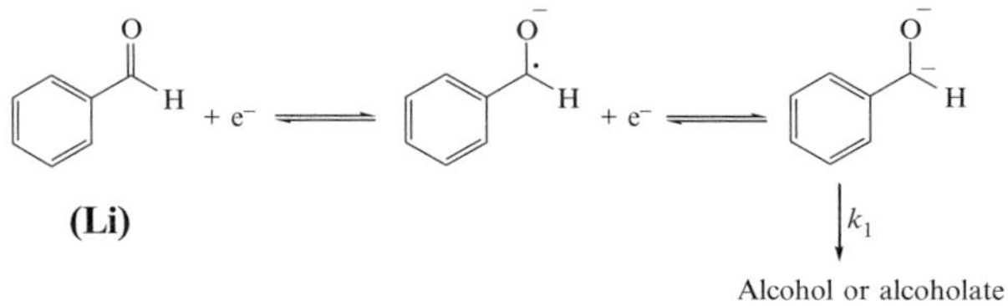

(15.50)

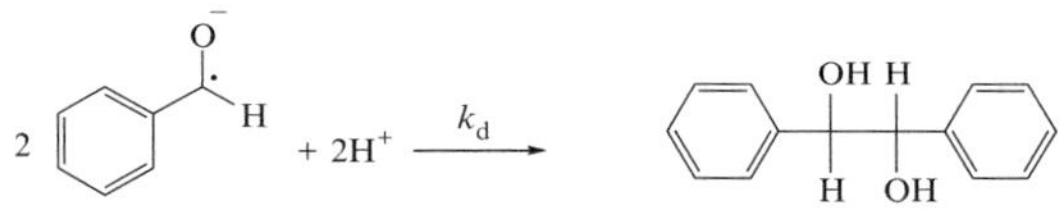

(15.51)

Acetophenone (**Lii**) was investigated in [C_4mim][$N(Tf)_2$], [$N_{1,1,1,4}$][$N(Tf)_2$], and [$N_{2,2,2,4}$][$N(Tf)_2$] ILs [47]. In contrast to the results obtained in molecular solvents, preparative scale electrochemical reduction showed the exclusive formation of the corresponding pinacol as a mixture of two diastereoisomers, D-L mixture, and *meso* compound, and not the corresponding alcohol. This change in the nature of the reactivity was explained by the existence of strong ion-pair associations between the charged radical anion and the cation of the ILs. The dimerization rate between the two acetophenone radical anions is a function of the nature of the IL cation, and decreases in the order [C_4mim]$^+$ > [$N_{2,2,2,4}$]$^+$ > [$N_{1,1,1,4}$]$^+$, indicating that associations are stronger in imidazolium-based ILs than in the quaternary ammonium-based ILs [47].

The electrocatalytic carboxylation of acetophenone (**Lii**) 4′-methylacetophenone (**Liii**) and 4′-methoxyacetophenone (**Liv**) by reaction with carbon dioxide in [C_4mim] [$N(Tf)_2$] was investigated [48]. The voltammogram showed an irreversible two-electron process at −1.7 V vs. Ag wire QRE, attributed to the reduction of acetophenone to a dianion product, followed by its fast reaction with carbon dioxide to yield the carboxylated product. Then, methyliodide was added in excess as an alkylation agent, and the mixture was stirred at 55 °C for 6 h. Using platinum as the cathode material, 2.25 mA cm^{-2} of current density, and a reaction temperature of 50 °C, the following reaction and yields were obtained [48]:

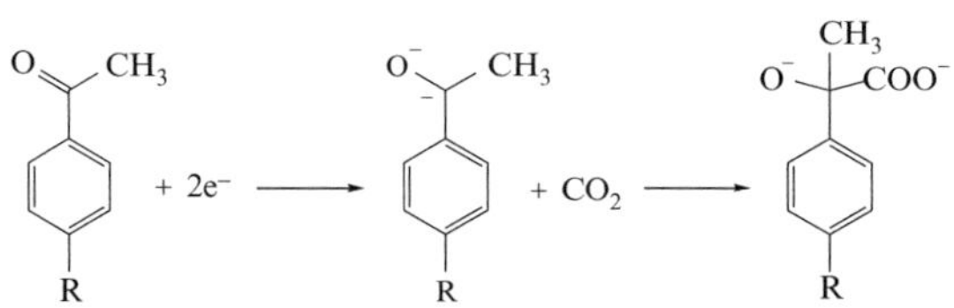

(15.52)

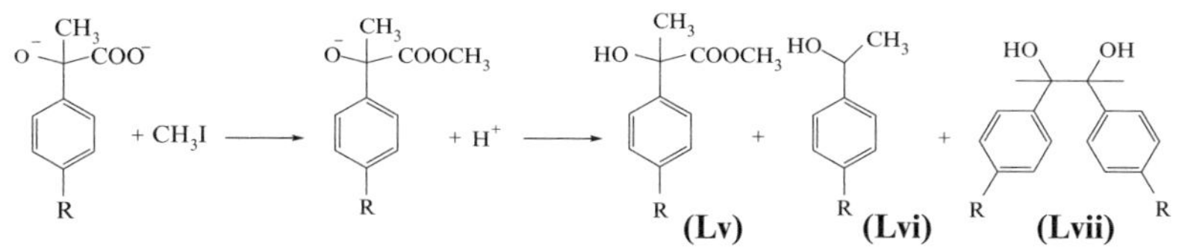

(15.53)

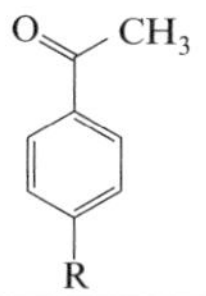		R	Yield of **Lv** (%)	Yield of **Lvi** (%)
	Lii	–H	62	20
	Liii	$-CH_3$	55	31
	Liv	$-OCH_3$	56	28

15.3.5 Reduction of Nitro-Containing Compounds

Nitrobenzene was studied in $[C_4dmim][N(Tf)_2]$ on a 25 μm diameter gold microdisk electrode at a scan rate of 0.1 V s^{-1} (Fig. 15.6) [49]. The voltammetric wave at −1.2 V vs. Ag wire QRE (peak 1) corresponds to the reversible one-electron reduction of nitrobenzene to the radical anion, which is commonly known in aprotic solvents [8]. A second peak was observed at −1.85 V vs. Ag wire (peak 2), which is related with a two-electron chemically irreversible process. This result is in contrast with conventional aprotic solvents [8], where a four-electron reduction with the consequent formation of phenylhydroxylamine is achieved. Compton et al. suggested that the dianion, formed by further reduction of the radical anion, is unstable in this environment and most likely undergoes rapid

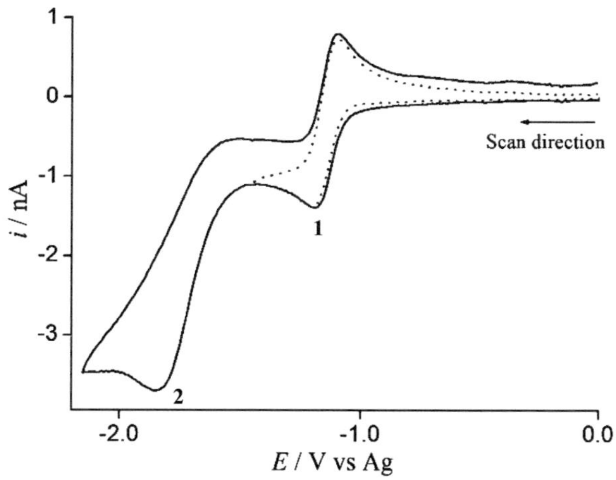

Fig. 15.6 Cyclic voltammograms on a gold microelectrode (diameter = 25 μm) for the reduction of 8 mM nitrobenzene in $[C_4dmim][N(Tf)_2]$ at a scan rate of 0.1 V s^{-1}. (*Solid lines*) reversed at −2.2 V, (*dashed lines*) reversed at −1.5 V. Reproduced from ref. [49] with permission from Elsevier

Table 15.1 Electrochemical data for nitro derivatives in different IL media

Compound	IL	E_m (V)[a]	D (10^{-7} cm^2 s^{-1})	RE	Ref
Nitrobenzene	$[C_4dmim][N(Tf)_2]$	−1.20	2.0	Ag wire	[49]
	$[C_4mim][BF_4]$	−1.42	–	Ag|Ag$^+$	[51]
	$[C_4mpyr]$ $[N(Tf)_2]$	−1.81	3.4	Ag|Ag$^+$	[50]
4-Nitrotoluene	$[C_4mim][N(Tf)_2]$	−1.39	4.5	Fc|Fc$^+$	[3]
	$[N_{2,2,2,4}][N(Tf)_2]$	−1.54	2.2	Fc|Fc$^+$	[3]
2-Nitro-*m*-xylene	$[C_4mim][N(Tf)_2]$	−1.56	3.3	Fc|Fc$^+$	[3]
	$[N_{2,2,2,4}][N(Tf)_2]$	−1.73	1.8	Fc|Fc$^+$	[3]
2-Nitromesitylene	$[C_4mim][N(Tf)_2]$	−1.57	4.0	Fc|Fc$^+$	[3]
	$[N_{2,2,2,4}][N(Tf)_2]$	−1.74	1.0	Fc|Fc$^+$	[3]
Pentamethylnitrobenzene	$[C_4mim][N(Tf)_2]$	−1.60	0.6	Fc|Fc$^+$	[3]
	$[N_{2,2,2,4}][N(Tf)_2]$	−1.76	0.3	Fc|Fc$^+$	[3]
1,3,5-Tri-*tert*-butyl-2-nitrobenzene	$[C_4mim][N(Tf)_2]$	−1.66	0.5	Fc|Fc$^+$	[3]
	$[N_{2,2,2,4}][N(Tf)_2]$	−1.84	0.1	Fc|Fc$^+$	[3]
2-Methyl-2-nitropropane	$[C_4mim][N(Tf)_2]$	−1.86	3.1	Fc|Fc$^+$	[3]
	$[N_{2,2,2,4}][N(Tf)_2]$	−2.04	1.0	Fc|Fc$^+$	[3]
1,4-Dinitrobenzene	$[C_4mim][N(Tf)_2]$	−1.03[b]	–	Ag|Ag$^+$	[51]
1,2-Dinitrobenzene	$[C_4mim][N(Tf)_2]$	−1.11[b]	–	Ag|Ag$^+$	[51]

[a]Formal potential related to the first reduction peak
[b]Potential related with a single two-electron process

protonation after which the loss of a water molecule yields nitrosobenzene. However, the occurrence of a small anodic peak at −0.39 V vs. Ag wire in Fig. 15.6 might suggest that a dimerization process to yield azoxybenzene could also be applicable [49].

The electrochemical reduction of a series of aromatic and aliphatic nitro-containing compounds was also carried out under similar conditions, showing nearly similar behavior to that of nitrobenzene [3]. Table 15.1 summarizes some of the electrochemical parameters obtained in different IL media. From this table, it is possible to see that the E_m for the aromatic molecules became more negative with the number of methyl groups as expected when introducing donating groups. It is worth mentioning that this comparison is only valid when the E_m is compared using both the same ionic liquid and the same reference system.

The reduction of 1-bromo-4-nitrobenzene (**Lviii**) in $[C_4mpyr][N(Tf)_2]$ proceeds via a DISP1 type mechanism, were **Lviii** is reduced by one electron to form a radical anion (Eq. 15.54) [50]. This step is followed by a debromination step, which also involves the fast scavenge of a proton from the solvent system to yield nitrobenzene (Eq. 15.55). Due to the aprotic nature of this ionic liquid, the protonation step should involve water (represented as AH in Eq. 15.55) as the proton donor molecules, which may be present in the IL system as impurities. Finally, the product of this reaction is then reduced further to its radical anion through a disproportionation reaction (Eq. 15.56).

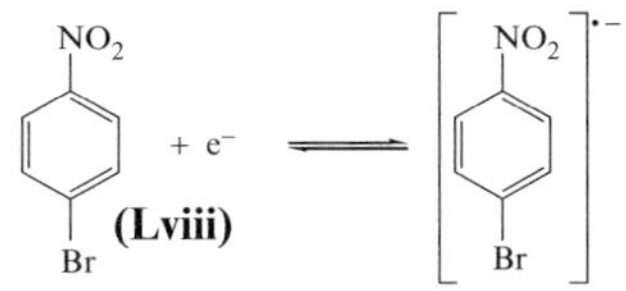

(15.54)

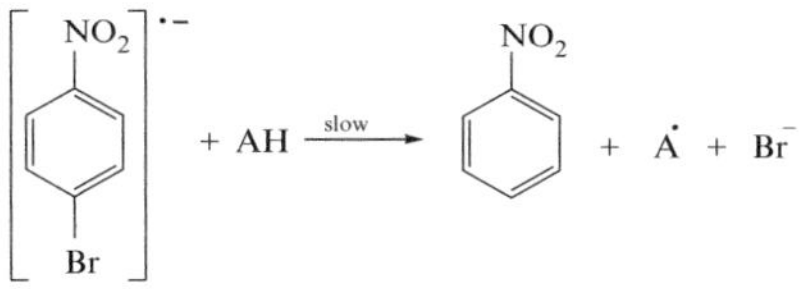

(15.55)

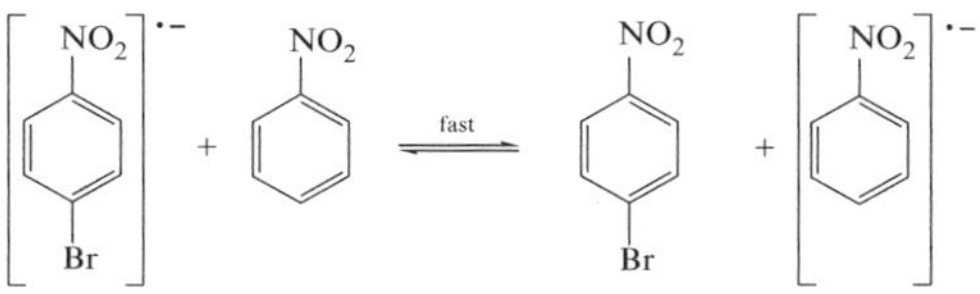

(15.56)

It is important to mention that even when the one-electron mechanism to form the stable radical anion (Eq. 15.54) is also observed in a range of organic solvents (e.g., acetonitrile, dimethyl sulfoxide, and *N*,*N*-dimethylformamide), the debromination step only occurs in organic solvents after the absorption of light by the 1-bromo-4-nitrobenzene radical anion (photochemical reaction) [50, 52–54].

The reduction of 4-nitrophenol (**Lix**) in $[C_4dmim][N(Tf)_2]$ is more complex than that previously mentioned for nitrobenzene [49]. The radical anion formed after the first electron transfer is rapidly protonated by an acidic parent molecule, resulting in further chemical reactions, following the scheme proposed by Amatore et al. (Eqs. 15.57–15.60) [55]. Three reductive voltammetric waves were observed; the first two were irreversible (with the first wave appearing as a shoulder on the larger second wave), and the third wave was chemically reversible [49]. Two additional anodic waves were observed on the oxidative potential sweep scan, and were attributed to oxidations of product species formed during the first reductive potential sweep scan. The theoretical simulation of the experimental voltammograms allowed the following conclusions [49]: (a) the first and second waves arise from the initial reduction of a parent molecule, and generation of more parent molecules later in the reaction scheme (Eqs. 15.57–15.60), (b) the third reversible wave is most likely the one-electron reduction of 4-nitrophenolate, which is a product resulting from the primary reaction scheme (Eq. 15.58). This reduction process was found to be kinetically controlled by the equilibrium and rate constants of the reactions depicted in Eqs. 15.58 and 15.59.

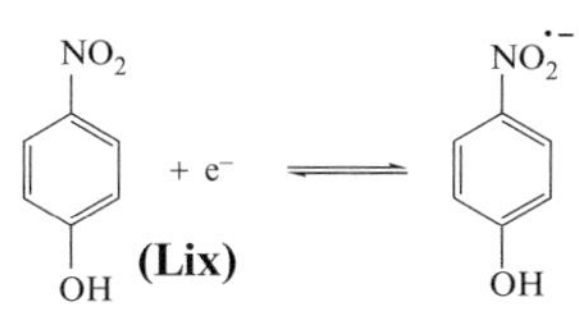

(15.57)

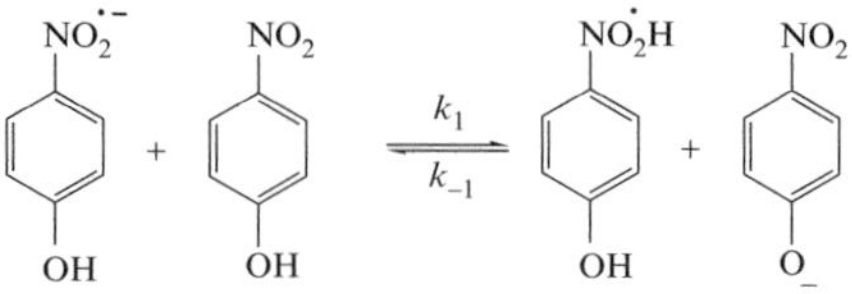

(15.58)

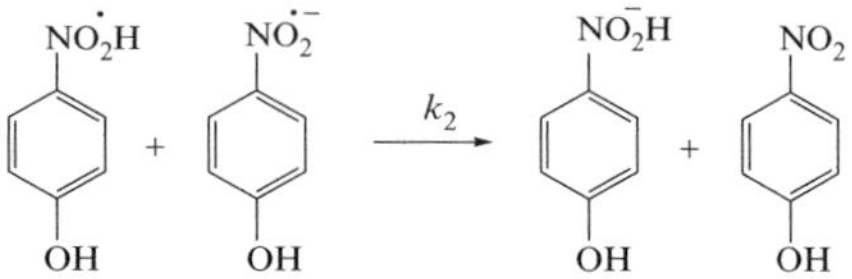

(15.59)

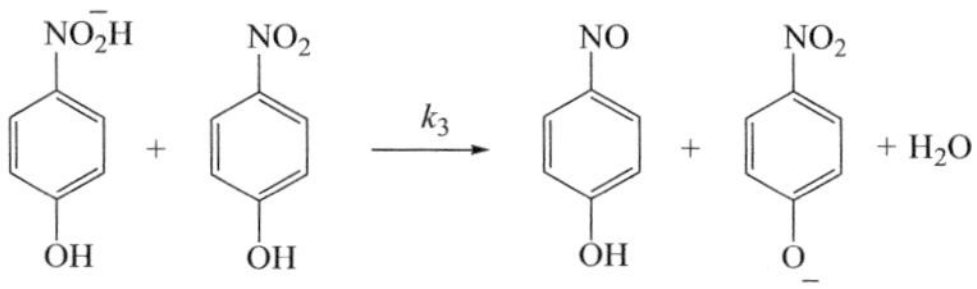

(15.60)

15.4 Conclusions

In this chapter, we have attempted to provide an up-to-date perspective on electrochemical processes that can be carried out in ionic liquids both from a fundamental as well as from a practical point of view. Ionic liquids offer a powerful combination of being good supporting electrolytes in which to carry out electrochemical processes coupled with varied solvency. The possibility of using them in both small and large-scale industrial chemistry processes has many appealing aspects. Such a process would likely be more atom-efficient, involve (in some cases) a more "Green" solvent, and operate at an overall lower total energy cost than any traditional chemical oxidation or reduction (other than, perhaps, air oxidation!). Many more studies of synthetically important processes are needed, along with further investigation of practical approaches to the separation of products and of increasing yields, and the interested researcher is strongly encouraged in that direction.

References

1. Wasserscheid P, Welton T (eds) (2008) Ionic liquids in synthesis, vol 1, 2nd edn. Wiley-VCH, Weinheim
2. Earle MJ, Katdare SP, Seddon KR (2004) Paradigm confirmed: the first Use of ionic liquids to dramatically influence the outcome of chemical reactions. Org Lett 6(5):707–710. doi:10.1021/ol036310e
3. Lagrost C, Preda L, Volanschi E, Hapiot P (2005) Heterogeneous electron-transfer kinetics of nitro compounds in room-temperature ionic liquids. J Electroanal Chem 585(1):1–7

4. Fuchigami T, Inagi S (2011) Selective electrochemical fluorination of organic molecules and macromolecules in ionic liquids. Chem Commun (Camb) 47(37):10211–10223. doi:10.1039/c1cc12414e
5. Feroci M, Orsini M, Rossi L, Inesi A (2012) The double role of ionic liquids in electroorganic synthesis: green solvents and precursors of N-heterocyclic carbenes. Curr Org Synth 9:40–52
6. Wagner M, Kvarnström C, Ivaska A (2012) Synthesis of organic electroactive materials in ionic liquids. In: Wei D (ed) Electrochemical nanofabrication: principles and applications. CRC press, Boca Raton
7. Grimshaw J (2000) Electrochemical reactions and mechanisms in organic chemistry. Elsevier, Amsterdam
8. Lund H, Hammerich O (eds) (1991) Organic electrochemistry. Marcel Dekker, New York
9. Belding SR, Rees NV, Aldous L, Hardacre C, Compton RG (2008) Behavior of the heterogeneous electron-transfer rate constants of arenes and substituted anthracenes in room-temperature ionic liquids. J Phys Chem C 112(5):1650–1657
10. Lagrost C, Carrie D, Vaultier M, Hapiot P (2003) Reactivities of some electrogenerated organic cation radicals in room-temperature ionic liquids: toward an alternative to volatile organic solvents? J Phys Chem A 107(5):745–752
11. Clegg AD, Rees NV, Klymenko OV, Coles BA, Compton RG (2004) Marcus theory of outer-sphere heterogeneous electron transfer reactions: dependence of the standard electrochemical rate constant on the hydrodynamic radius from high precision measurements of the oxidation of anthracene and its derivatives in nonaqueous solvents using the high-speed channel electrode. J Am Chem Soc 126(19):6185–6192
12. Bystron T, Bouzek K (2013) Ionic liquids as potential supporting electrolytes for the anodic oxidation of 4-methylanisole. J Electrochem Soc 160(8):G117–G123. doi:10.1149/2.041308jes
13. Barnes EO, Mahony AM, Aldous L, Hardacre C, Compton RG (2010) The electrochemical oxidation of catechol and dopamine on platinum in 1-ethyl-3-methylimidazolium bis(trifluoromethylsulfonyl)imide ([C2mim][NTf2]) and 1-butyl-3-methylimidazolium tetrafluoroborate [C4mim][BF4]: adsorption effects in ionic liquid voltammetry. J Electroanal Chem 646 (1–2):11–17
14. Bhat MA (2012) Mechanistic, kinetic and electroanalytical aspects of quinone–hydroquinone redox system in N-alkylimidazolium based room temperature ionic liquids. Electrochim Acta 81:275–282. doi:10.1016/j.electacta.2012.07.059
15. Evans RG, Compton RG (2006) A kinetic study of the reaction between N, N-dimethyl-p-toluidine and its electrogenerated radical cation in a room temperature ionic liquid. ChemPhysChem 7(2):488–496
16. Evans RG, Klymenko OV, Hardacre C, Seddon KR, Compton RG (2003) Oxidation of N, N, N′, N′-tetraalkyl-para-phenylenediamines in a series of room temperature ionic liquids incorporating the bis(trifluoromethylsulfonyl)imide anion. J Electroanal Chem 556:179–188
17. Evans RG, Klymenko OV, Price PD, Davies SG, Hardacre C, Compton RG (2005) A comparative electrochemical study of diffusion in room temperature ionic liquid solvents versus acetonitrile. ChemPhysChem 6(3):526–533
18. Herath AC, Becker JY (2010) Electrochemical study of tris(4-bromophenyl)amine and 2,2,6,6-tetramethylpiperidine-1-oxyl in room-temperature ionic liquids. Electrochim Acta 55 (27):8319–8324
19. MacFarlane DR, Pringle JM, Johansson KM, Forsyth SA, Forsyth M (2006) Lewis base ionic liquids. Chem Commun 18:1905–1917 (Cambridge, England)
20. Fuller J, Carlin RT, Osteryoung RA (1997) The room temperature ionic liquid 1-ethyl-3-methylimidazolium tetrafluoroborate: electrochemical couples and physical properties. J Electrochem Soc 144(11):3881–3886
21. Carter MT, Osteryoung RA (1994) Heterogeneous and homogeneous electron transfer reactions of tetrathiafulvalene in ambient temperature chloroaluminate molten salts. J Electrochem Soc 141(7):1713–1720

22. Abdul-Rahim O, Simonov AN, Ruther T, Boas JF, Torriero AAJ, Collins DJ, Perlmutter P, Bond AM (2013) The observation of dianions generated by electrochemical reduction of trans-stilbenes in ionic liquids at room temperature. Anal Chem 85:6113–6120. doi:10.1021/ac400915z
23. Wawzonek S, Blaha EW, Behkey R, Runner ME (1955) Polarographic studies in acetonitrile and dimethylformamide: II. Behavior of aromatic olefins and hydrocarbons. J Electrochem Soc 102(5):235–242. doi:10.1149/1.2430036
24. Grodzka PG, Elving PJ (1963) Polarographic reduction of the phenyl-substituted ethenes: II. Electrochemical kinetic parameters and mechanism in dimethylformamide. J Electrochem Soc 110(3):231–236. doi:10.1149/1.2425716
25. Funt BL, Gray DG (1970) A study of some primary processes in electropolymerization by cyclic voltammetry of phenyl-substituted ethylenes. J Electrochem Soc 117(8):1020–1024. doi:10.1149/1.2407711
26. Troll T, Baizer MM (1974) Cyclic voltammetry of phenyl-substituted ethylenes in DMF and HMPA. Electrochim Acta 19(12):951–953. doi:10.1016/0013-4686(74)85049-8
27. Kojima H, Bard AJ (1975) An A.C. technique for determining the rates of rapid electrode reactions of aromatic compounds in aprotic media. J Electroanal Chem Interfacial Electrochem 63(2):117–129. doi:10.1016/S0022-0728(75)80285-3
28. Jones A, Kronenwetter H, Manchanayakage R (2012) Electrochemical reductive coupling of 2-cyclohexen-1-one in a mixture of ionic liquid and water. Electrochem Commun 25:8–10. doi:10.1016/j.elecom.2012.09.020
29. Bastida RM, Brillas E, Costa JM (1987) Electrohydrodimerization of 4,4-dimethyl-2-cyclohexen-1-one in dimethylformamide on a mercury electrode. J Electroanal Chem Interfacial Electrochem 227(1–2):67–75. doi:10.1016/0022-0728(87)80066-9
30. Feng Q, Huang K, Liu S, Wang H, Yan W (2012) Electrochemical reduction of benzoyl chloride to benzil in ionic liquid BMIMBF4. J Phys Org Chem 25(6):506–510. doi:10.1002/poc.1946
31. Cheek GT, Horine PA (1984) Electrochemical reduction of benzoyl chloride and benzoyl fluoride. J Electrochem Soc 131(8):1796–1801. doi:10.1149/1.2115963
32. Hiejima Y, Hayashi M, Uda A, Oya S, Kondo H, Senboku H, Takahashi K (2010) Electrochemical carboxylation of α-chloroethylbenzene in ionic liquids compressed with carbon dioxide. Phys Chem Chem Phys 12(8):1953–1957. doi:10.1039/b920413j
33. Lagrost C, Gmouh S, Vaultier M, Hapiot P (2004) Specific effects of room temperature ionic liquids on cleavage reactivity: example of the carbon-halogen bond breaking in aromatic radical anions. J Phys Chem A 108(29):6175–6182. doi:10.1021/jp049017k
34. Shen Y, Inagi S, Atobe M, Fuchigami T (2013) Electrocatalytic debromination of open-chain and cyclic dibromides in ionic liquids with cobalt(II)salen complex as mediator. Res Chem Intermed 39(1):89–99. doi:10.1007/s11164-012-0634-6
35. Tokuda M, Hayashi A, Suginome H (1991) Synthesis of dimethyl 1,2-cycloalkanedicarboxylates by electrochemical cyclization of dimethyl α,α′-dibromoalkanedioates using a copper anode. Bull Chem Soc Jpn 64(8):2590–2592. doi:10.1246/bcsj.64.2590
36. Gaillon L, Bedioui F (2004) Voltammetric analysis of the catalytic reactivity of electrogenerated CoI-salen with organohalogenated derivatives in an ionic liquid at room temperature. J Mol Catal A Chem 214(1):91–94
37. Doherty AP, Koshechko V, Titov V, Mishura A (2007) Freon electrochemistry in room-temperature ionic liquids. J Electroanal Chem 602(1):91–95
38. Wang Y, Rogers EI, Belding SR, Compton RG (2010) The electrochemical reduction of 1,4-benzoquinone in 1-ethyl-3-methylimidazolium bis(trifluoromethane-sulfonyl)-imide, [C_2mim][NTf_2]: a voltammetric study of the comproportionation between benzoquinone and the benzoquinone dianion. J Electroanal Chem 648(2):134–142. doi:10.1016/j.jelechem.2010.07.016
39. Nikitina VA, Nazmutdinov RR, Tsirlina GA (2011) Quinones electrochemistry in room-temperature ionic liquids. J Phys Chem B 115(4):668–677. doi:10.1021/jp1095807

40. Boldrin Zanoni MV, Rogers EI, Hardacre C, Compton RG (2010) The electrochemical reduction of the purines guanine and adenine at platinum electrodes in several room temperature ionic liquids. Anal Chim Acta 659(1–2):115–121
41. Janik B, Elving PJ (1968) Polarographic behavior of nucleosides and nucleotides of purines, pyrimidines, pyridines, and flavines. Chem Rev 68(3):295–319. doi:10.1021/cr60253a003
42. Santhanam KSV, Elving PJ (1974) Redox pattern for purine and 6-substituted purines in nonaqueous media. Free radical behavior. J Am Chem Soc 96(6):1653–1660. doi:10.1021/ja00813a002
43. Zhao S-F, Lu J-X, Bond AM, Zhang J (2012) Remarkable sensitivity of the electrochemical reduction of benzophenone to proton availability in ionic liquids. Chemistry 18(17):5290–5301. doi:10.1002/chem.201103365
44. O'Toole S, Pentlavalli S, Doherty AP (2007) Behavior of electrogenerated bases in room-temperature ionic liquids. J Phys Chem B 111(31):9281–9287
45. Brooks CA, Doherty AP (2005) Electrogenerated radical anions in room-temperature ionic liquids. J Phys Chem B 109(13):6276–6279
46. Doherty AP, Brooks CA (2004) Electrosynthesis in room-temperature ionic liquids: benzaldehyde reduction. Electrochim Acta 49(22–23):3821–3826
47. Lagrost C, Hapiot P, Vaultier M (2005) The influence of room-temperature ionic liquids on the stereoselectivity and kinetics of the electrochemical pinacol coupling of acetophenone. Green Chem 7(6):468–474
48. Feng Q, Huang K, Liu S, Yu J, Liu F (2011) Electrocatalytic carboxylation of aromatic ketones with carbon dioxide in ionic liquid 1-butyl-3-methylimidazoliumtetrafluoborate to alpha-hydroxy-carboxylic acid methyl ester. Electrochim Acta 56(14):5137–5141. doi:10.1016/j.electacta.2011.03.061
49. Silvester DS, Wain AJ, Aldous L, Hardacre C, Compton RG (2006) Electrochemical reduction of nitrobenzene and 4-nitrophenol in the room temperature ionic liquid [C4dmim][N(Tf)2]. J Electroanal Chem 596(2):131–140
50. Ernst S, Ward KR, Norman SE, Hardacre C, Compton RG (2013) Changed reactivity of the 1-bromo-4-nitrobenzene radical anion in a room temperature ionic liquid. Phys Chem Chem Phys 15(17):6382–6389. doi:10.1039/c3cp51004b
51. Fry AJ (2003) Strong ion-pairing effects in a room temperature ionic liquid. J Electroanal Chem 546:35–39
52. Compton RG, Dryfe RAW, Fisher AC (1994) Photoelectrochemical reduction of p-halonitrobenzenes. J Chem Soc Perkin Trans 2(7):1581–1587. doi:10.1039/p29940001581
53. Lawless JG, Hawley MD (1969) Mechanistic studies of the decomposition of halonitrobenzene anion radicals. J Electroanal Chem Interfacial Electrochem 21(2):365–375. doi:10.1016/S0022-0728(69)80104-X
54. Cooper JA, Compton RG (1998) Photoelectrochemical reduction of p-bromonitrobenzene: mechanistic discrimination via channel microband array voltammetry. Electroanalysis 10(17):1182–1187. doi:10.1002/(sici)1521-4109(199811)10:17<1182::aid-elan1182>3.0.co;2-k
55. Amatore C, Capobianco G, Farnia G, Sandona G, Saveant JM, Severin MG, Vianello E (1985) Kinetics and mechanism of self-protonation reactions in organic electrochemical processes. J Am Chem Soc 107(7):1815–1824. doi:10.1021/ja00293a003

Chapter 16
Electrode Reactions of Tris(2,2′-Bipyridine) Complexes of Some Transition Metals in Ionic Liquids

Yasushi Katayama

16.1 Introduction

ILs are interesting media for studying electrochemistry of various charged species since strong electrostatic interaction of the charged species with the ions composing ILs is expected to affect such electrochemical parameters as redox potentials, diffusion coefficients, and heterogeneous rate constants of the species. In case of conventional solutions, in contrast, the dielectric interaction of the charged species with the neutral polar molecules of the solvents has been considered as the major factors influencing thermodynamic and kinetic properties [1]. Electrostatic interaction of the charged species with the ions dissolved in the solutions has been also considered by means of activity coefficients, which are given by classical Debye-Hückel theory. However, Debye-Hückel theory is able to be used only at low ionic strength. Thus, ILs are regarded as the limiting case of concentrated solutions since ILs are composed of only ions (high ionic strength) without neutral polar molecules (little dielectric interaction).

Electrochemical methods are useful for determining the thermodynamic and kinetic parameters of the species in ILs when the species are able to be reduced or oxidized on an electrode. Electrochemical reactions of various metal species have been studied in a variety of ILs. However, there are not so many reports about the electrode kinetics of the species. Redox reactions of tris(2,2′-bipyridine) complexes of various metals, $[M(bpy)_3]^{n+}$, have been investigated in aqueous and organic solutions [2]. These complexes are able to take a variety of electric charges from −1 to +3. In case of $[M(bpy)_3]^{3+}$ and $[M(bpy)_3]^{2+}$, the oxidation states of the centered metal are II and III, respectively, since all of bpy ligands are regarded as neutral. $[M(bpy)_3]^{2+}$ is able to be further reduced by donating electrons to bpy ligands.

Y. Katayama (✉)
Department of Applied Chemistry, Faculty of Science and Technology, Keio University, 3-14-1 Hiyoshi, Kohoku-ku, Yokohama, Kanagawa 223-8522, Japan
e-mail: katayama@applc.keio.ac.jp

A.A.J. Torriero (ed.), *Electrochemistry in Ionic Liquids*,
DOI 10.1007/978-3-319-15132-8_16

However, these low-oxidation-state complexes are sometimes unstable when the donor ability of solvent molecules or anions of electrolytes is strong. Thus, it is necessary to choose the reaction media having low donor ability in order to stabilize the low-oxidation-state complexes.

The ILs composed of $[N(Tf)_2]^-$ have low donor ability due to its low charge density. In addition, it is possible to obtain the ILs having low melting points by combining various organic cations with $[N(Tf)_2]^-$. $[BETI]^-$ is similar to $[N(Tf)_2]^-$ while the viscosity of $[BETI]^-$-based ILs is higher than that of $[N(Tf)_2]^-$-based ones. Therefore, these amide-based ILs are expected as the suitable media for investigating the redox reactions of tris(bpy) complexes. In addition, the charge density of these complexes is considered proportional to the electric charge on them since the sizes of these complexes are able to be regarded as constant regardless of the centered metals and the oxidation states. In this chapter, the effects of the electrostatic interaction between the charged species and the ions composing the ILs are discussed through investigating the electrode reactions of tris(bpy) complexes of some transition metals.

16.2 Some Tips on Electrochemical Measurements in ILs

Since ILs are defined as the liquids solely composed of ions without any solvent, contamination of water, and other neutral molecules must be avoided in order to study the phenomenal characteristic of ILs. One of the major impurities is water, which easily comes into ILs from the atmosphere. In case of the amide-type ILs, it is possible to eliminate water from the ILs by heating under vacuum. The dried ILs should be handled under dry atmosphere in a glovebox, for example. The reagents to be added to ILs should be also dried before use unless they are shipped under inert and dry atmosphere.

Electrochemical measurements should be also conduced under dry atmosphere. It may be possible to use ordinary electrochemical cells open to the atmosphere in the glovebox by drawing the cables from electrochemical instruments although extension of the cables often reduces the quality of electric signals. In addition, it is sometimes difficult to control the temperature of the cells in the limited space of the glovebox. Therefore, it is better to use airtight electrochemical cells. Figure 16.1 shows the illustration of an airtight cell designed for electrochemical measurements in ILs. The body is made of such fluorinated polymers as PFA (tetrafluoroethylene-perfluoroalkylvinylether copolymer) and FEP (tetrafluoroethylene-hexafluoropropylene copolymer), which have high chemical stability and moderate hardness for machining. The electrolyte volume required for electrochemical measurements is a few mL in order to reduce the amount of expensive ILs. It has glass windows in order to observe the inside of cell. The windows are held on the body by the fixtures made of stainless steel and bolts. It is also possible to place a plate electrode instead of the glass window. The electrodes are fabricated with copper rods, of which the diameters are fit to the vacuum fittings. The disc of such desired electrode material

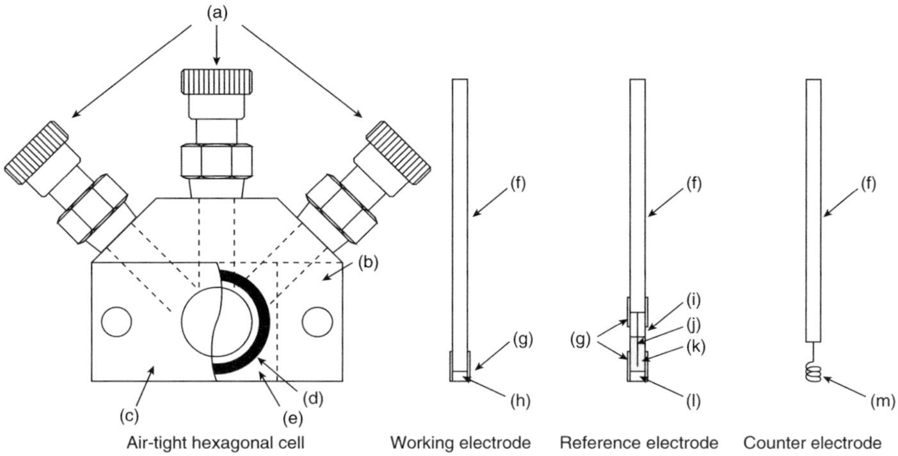

Fig. 16.1 Schematic illustration of an airtight hexagonal cell and electrodes designed for electrochemical measurements in ILs. (**a**) Vacuum fittings, (**b**) PFA body, (**c**) Fixture (stainless steel), (**d**) O-ring, (**e**) Cover glass, (**f**) Cu rod, (**g**) Heat shrink tube, (**h**) Pt disc, (**i**) Glass tube, (**j**) Ag wire, (**k**) Inner IL containing Ag(I), (**l**) Vycol (porous) glass, and (**m**) Pt wire

as platinum and glassy carbon is attached to the end of working electrode with silver paste and heat shrink tube. The counter electrode is composed of a metal wire connected to the copper rod. The reference electrode is composed of a metal wire with an inner IL, which is separated with a porous glass like Vycol [3].

Choice of reference electrodes is one of the most important points in electrochemical measurements in ILs. The reference electrodes are required to show stable electrode potentials, which are usually determined by an equilibrium between reversible redox couples. The redox reaction between silver and silver cation, Ag/Ag(I), is often used as the redox couple for reference electrode in conventional nonaqueous electrolytes. The reference electrode based on Ag/Ag(I) has been also used in various ILs. However, the potentials of Ag/Ag(I) reference electrodes are different in different ILs since the Gibbs energy for formation of Ag(I) depends on the ions composing the ILs. Therefore, it is necessary to calibrate the potentials of reference electrodes against a common standard redox potential. A redox couple of ferrocenium (Fc^+) and ferrocene (Fc) is often used for this purpose although its redox potential is considered slightly dependent on ILs. Platinum or silver electrodes immersed in ILs are sometimes used as quasi-reference electrodes. The potentials of these quasi-reference electrodes may seem to be stable in the ILs without any redox species. However, their potentials are unstable and unreliable since they are not determined by any redox equilibrium. Thus, use of quasi-reference electrodes should be avoided even when the potentials are calibrated by Fc^+/Fc couple.

Common electrochemical techniques are able to be used also in ILs. However, it is important to notice low conductivity of most of the ILs often affects the accuracy and reliability of electrochemical data. Cyclic voltammetry is one of the most popular methods, which will give much useful information on electrode reactions with simple operation. It is possible to estimate such electrochemical parameters as half-wave potential, diffusion coefficient, and rate constant with the equations found in textbooks. However, the ohmic drop due to low conductivity of ILs often affects peak potentials and linearity of potential sweep rates. Thus, cyclic voltammetry is not recommended to estimate these electrochemical parameters unless the effects of the ohmic drop are eliminated properly by such techniques as ultramicro electrode and IR compensators. Chronopotentiometry is also influenced by the ohmic drop. In contrast, the effects of the ohmic drop are expected to be insignificant in chronoamperometry when enough potential changes are applied.

16.3 Redox Reactions of the Tris(bpy) Complexes of Some Transition Metals

Figure 16.2 shows the cyclic voltammograms of tris(bpy) complexes of iron [4], ruthenium [5], nickel [6], manganese, and chromium [7] in an amide-type ionic liquid, 1-butyl-1-methylpyrrolidinium bis(trifluoromethylsulfonyl)amide ($[C_4mPyr][N(Tf)_2]$). In case of Fe, Ru, Mn, and Cr, there are four pairs of cathodic and anodic current peaks, which are attributed to the following reactions in order from positive to negative potentials.

$$[M(bpy)_3]^{3+} + e^- \leftrightarrows [M(bpy)_3]^{2+} \tag{16.1}$$

$$[M(bpy)_3]^{2+} + e^- \leftrightarrows [M(bpy)_3]^{+} \tag{16.2}$$

$$[M(bpy)_3]^{+} + e^- \leftrightarrows M(bpy)_3 \tag{16.3}$$

$$M(bpy)_3 + e^- \leftrightarrows [M(bpy)_3]^{-} \tag{16.4}$$

In case of Ni, $[Ni(bpy)_3]^{2+}$ is considered to be reduced to $Ni(bpy)_3$ [8, 9].

$$[Ni(bpy)_3]^{2+} + 2e^- \leftrightarrows Ni(bpy)_3 \tag{16.5}$$

Owing to the wide electrochemical potential window and low donor ability of $[N(Tf)_2]^-$, a variety of oxidation states of these complexes are able to be observed electrochemically. The redox potentials of $[M(bpy)_3]^{3+/2+}$ are listed in Table 16.1 together with those reported in acetonitrile. The potentials in $[C_4mPyr][N(Tf)_2]$ except for $[Ni(bpy)_3]^{3+/2+}$ are slightly higher than those in acetonitrile solutions, indicating the donor ability of $[N(Tf)_2]^-$ is weaker than that of acetonitrile. The donor number of $[C_4mPyr][N(Tf)_2]$ has been estimated to be ~7 [10], which is smaller than that of acetonitrile, 14.1 [2].

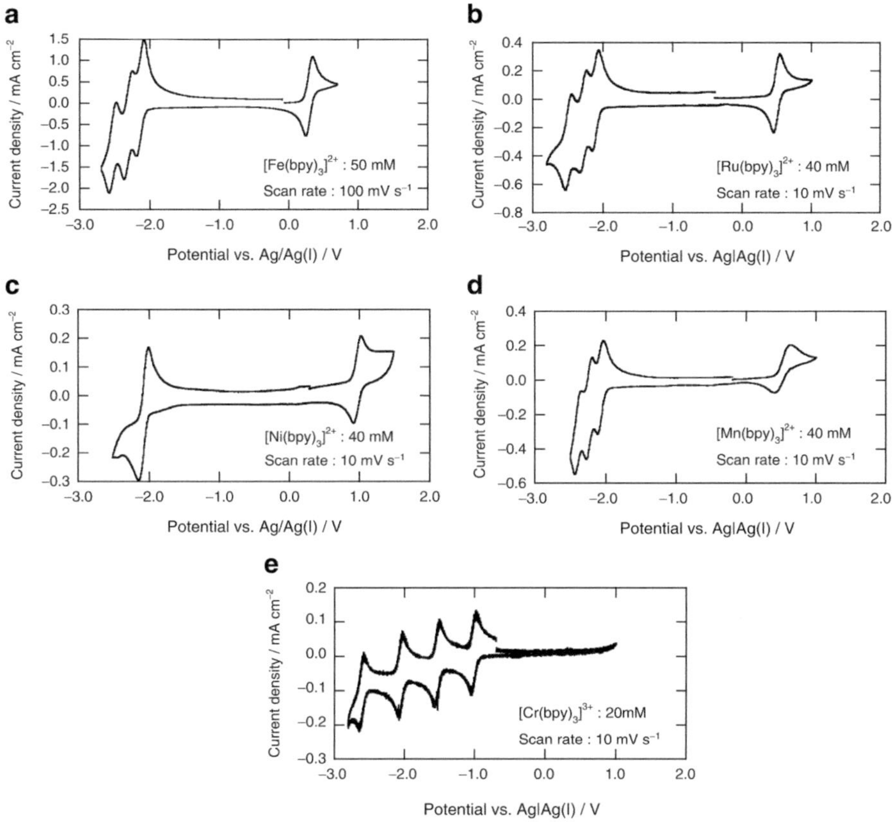

Fig. 16.2 Cyclic voltammograms of a Pt electrode in $[C_4mPyr][N(Tf)_2]$ containing (**a**) 50 mM $[Fe(bpy)_3]^{2+}$, (**b**) 40 mM $[Ru(bpy)_3]^{2+}$, (**c**) 40 mM $[Ni(bpy)_3]^{2+}$, (**d**) 40 mM $[Mn(bpy)_3]^{2+}$, and (**e**) 20 mM $[Cr(bpy)_3]^{3+}$ at 25 °C

Table 16.1 Potentials of $[M(bpy)_3]^{3+/2+}$ (M = Fe, Ru, Ni, Mn, and Cr) vs. Fc|Fc$^+$ in $[C_4mPyr][N(Tf)_2]$ and acetonitrile solutions at 25 °C

Redox couple	$[C_4mPyr][N(Tf)_2]$	Acetonitrile solution
$[Fe(bpy)_3]^{3+/2+}$	0.71 V [4]	0.67 V [11] (0.1 M $LiClO_4$)
$[Ru(bpy)_3]^{3+/2+}$	0.93 V [5]	0.89 V [12] (0.05 M TEAP[a])
$[Ni(bpy)_3]^{3+/2+}$	1.37 V [6]	1.39 V [13] (0.1 M TBAP[b])
$[Mn(bpy)_3]^{3+/2+}$	0.94 V [7]	0.93 V [14] (0.1 M TEAP[a])
$[Cr(bpy)_3]^{3+/2+}$	−0.63 V [7]	−0.60 V [15] (0.05 M TEAP[a])

[a]TEAP = tetraethylammonium perchlorate
[b]TBAP = tetrabutylammonium perchlorate

16.4 Diffusion of Tris(bpy) Complexes in Some ILs

Diffusion coefficients of $[M(bpy)_3]^{3+}$ and $[M(bpy)_3]^{2+}$ have been determined by chronoamperometry at various temperatures. Figure 16.3 shows the chronoamperogram and the Cottrell plots of a Pt electrode in $[C_4mPyr][N(Tf)_2]$ containing both 20 mM $[Ru(bpy)_3]^{3+}$ and 20 mM $[Ru(bpy)_3]^{2+}$, for example. The Pt electrode showed the equilibrium potential (0.49 V vs. Ag|Ag(I)) under open circuit. The potential was stepped anodically and cathodically to various potentials in order to find the step potentials giving the identical current decay curves, which were attained when the surface concentration of $[Ru(bpy)_3]^{3+}$ (for cathodic potential step) or $[Ru(bpy)_3]^{2+}$ (for anodic potential step) was assumed to be zero. In Fig. 16.3a, the current decay at 0.7 V overlapped with that at 0.8 V for anodic potential step, indicating the boundary conditions for Cottrell equation were fulfilled. In the same way, 0.2 V was chosen as the step potential for cathodic potential step. Figure 16.3b shows the Cottrell plots for the step potentials of 0.8 and 0.2 V. The diffusion coefficient, D, is calculated from the dependence of the current density, j, on the time, t, using the Cottrell equation.

$$j = \frac{nFD^{1/2}C}{\pi^{1/2}t^{1/2}} \tag{16.6}$$

where n is the number of electrons involved in the electrode reaction, F is Faraday constant, and C is the bulk concentration of the electroactive species.

Table 16.2 listed the diffusion coefficients of $[M(bpy)_3]^{3+}$ and $[M(bpy)_3]^{2+}$ ($M = Fe$, Ni, and Ru) by chronoamperometry in some ILs, as described above. In case of $[C_4mPyr][N(Tf)_2]$, the diffusion coefficients of $[M(bpy)_3]^{3+}$ are close to

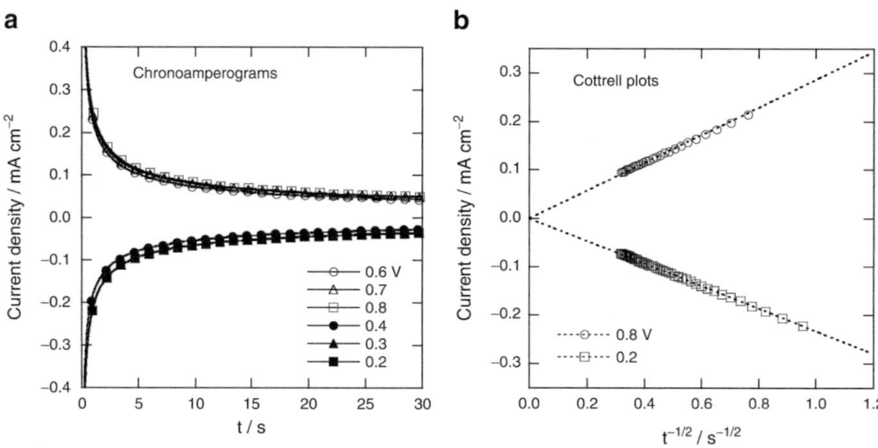

Fig. 16.3 (**a**) Chronoamperograms and (**b**) Cottrell plots of a Pt electrode in $[C_4mPyr][N(Tf)_2]$ containing 20 mM $[Ru(bpy)_3]^{3+}$ and 20 mM $[Ru(bpy)_3]^{2+}$ at 25 °C. The electrode potential was stepped from the equilibrium potential to the various potentials indicated in the figures

Table 16.2 Diffusion coefficients (D) of $[M(bpy)_3]^{3+}$ and $[M(bpy)_3]^{2+}$ in some ILs. The activation energies for the diffusion coefficients ($E_a(D)$) and the viscosity ($E_a(\eta)$), the ratio of the diffusion coefficient of $[M(bpy)_3]^{3+}$ to that of $[M(bpy)_3]^{2+}$ (D_{3+}/D_{2+}) are also listed

IL	Species	$10^8\, D/cm^2\, s^{-1}$	$E_a(D)$/ kJ mol^{-1}	η/mPa s	$E_a(\eta)$/ kJ mol^{-1}	D_{3+}/D_{2+}	References
$[C_4mPyr][N(Tf)_2]$	$[Ru(bpy)_3]^{3+}$	4.6	26	82	25	0.64	[5]
$[C_4mPyr][N(Tf)_2]$	$[Ru(bpy)_3]^{2+}$	7.3	23				[5]
$[C_2mim][N(Tf)_2]$	$[Ru(bpy)_3]^{3+}$	9.5	25	36	26	0.65	[5]
$[C_2mim][N(Tf)_2]$	$[Ru(bpy)_3]^{2+}$	15	25				[5]
$[C_4mPyr]$[BETI]	$[Ru(bpy)_3]^{3+}$	2.9	35	210	39	0.84	[5]
$[C_4mPyr]$[BETI]	$[Ru(bpy)_3]^{2+}$	3.5	37				[5]
$[C_2mim][BF_4]$	$[Ru(bpy)_3]^{3+}$	3.3	25	37	30	0.42	[16]
$[C_2mim][BF_4]$	$[Ru(bpy)_3]^{2+}$	7.8	23				[16]
$[C_4mPyr][N(Tf)_2]$	$[Ni(bpy)_3]^{3+}$	4.2	28	–	25	0.60	[6]
$[C_4mPyr][N(Tf)_2]$	$[Ni(bpy)_3]^{2+}$	7.0	22				[6]
$[C_4mPyr][N(Tf)_2]$	$[Fe(bpy)_3]^{3+}$	4.9	26	80		0.66	[17]
$[C_4mPyr][N(Tf)_2]$	$[Fe(bpy)_3]^{2+}$	7.5	26				[17]
$[C_2mim][N(Tf)_2]$	$[Fe(bpy)_3]^{3+}$	11	27	44	24	0.65	[17]
$[C_2mim][N(Tf)_2]$	$[Fe(bpy)_3]^{2+}$	17	26				[17]
AlC13-BPCl	$[Ru(bpy)_3]^{2+}$	26		24			[18]
Acetonitrile	$[Ru(bpy)_3]^{2+}$	9.7×10^2					[19]
DMF	$[Fe(bpy)_3]^{2+}$	360		0.80			[20]

one another regardless of the centered metal. Similarly, there's no significant difference in the diffusion coefficients of $[M(bpy)_3]^{2+}$. The diffusion coefficient of a spherical species with the radius of a (often called Stokes radius) is generally related to the viscosity coefficient of the diffusing medium, η, by Stokes-Einstein relation.

$$D = \frac{kT}{c\pi\eta a} \quad (16.7)$$

where k is the Boltzmann constant, T is the absolute temperature, and c is the constant, which is 4 or 6 at the limiting case for "slip" or "stick" conditions, respectively. Since the size of the bpy ligand is enough larger than that of the metal ion, the size of $[M(bpy)_3]^{3+}$ or $[M(bpy)_3]^{2+}$ is able to be regarded as constant and independent of the metal. Thus, the diffusion coefficients of $[M(bpy)_3]^{3+}$ or $[M(bpy)_3]^{2+}$ are close to one another in the same IL, indicating the diffusion coefficient of a species is basically determined by the size of the species and the viscosity of the diffusing medium even in ILs. On the other hand, the diffusion coefficients of $[M(bpy)_3]^{3+}$ are always smaller than those of $[M(bpy)_3]^{2+}$ for the same metal in the same IL even though the size of $[M(bpy)_3]^{3+}$ is also able to be regarded as the same as that of $[M(bpy)_3]^{2+}$. This discrepancy in the diffusion coefficients of $[M(bpy)_3]^{3+}$ and $[M(bpy)_3]^{2+}$ is considered attributed to electrostatic interaction between the charged species and the ions composing the IL. Stokes-Einstein relation is derived from the balance between the thermodynamic force by the chemical potential (concentration) gradient and the viscous drag from the viscosity of the diffusing medium. In case of ILs, the electrostatic interaction should be taken into account in addition to the viscous drag, as illustrated in Fig. 16.4. This "electrostatic drag" is expected to increase with an increase in the charge density of the charged species or the ions having the opposite charge to the species. Since $[M(bpy)_3]^{3+}$ and $[M(bpy)_3]^{2+}$ are able to be regarded as the same in size as mentioned above, the charge density of $[M(bpy)_3]^{3+}$ is expected to be higher than that of $[M(bpy)_3]^{2+}$. Thus, the electrostatic drag for $[M(bpy)_3]^{3+}$ is considered stronger than that for $[M(bpy)_3]^{2+}$. The ratio of the diffusion coefficient of $[M(bpy)_3]^{3+}$ to that of $[M(bpy)_3]^{2+}$, D_{3+}/D_{2+}, is expected to be used as an index for measuring the extent of electrostatic drag for $[M(bpy)_3]^{3+/2+}$. The D_{3+}/D_{2+} approaches to unity when the electrostatic drag is very weak. Inversely, the D_{3+}/D_{2+} decreases with an increase in the electrostatic drag. The D_{3+}/D_{2+} values for $[M(bpy)_3]^{3+/2+}$ are also listed in Table 16.2. The D_{3+}/D_{2+} values for $[M(bpy)_3]^{3+/2+}$ are close to one another in the ILs having the same anion, indicating the electrostatic drag for the cationic species is determined by the charge density of the anion of IL. In addition, the D_{3+}/D_{2+} in the $[N(Tf)_2]^-$-based IL is larger than that in the $[BF_4]^-$-based IL and smaller than that in the $[BETI]^-$-based IL since the sizes of these anions are in the order of

Fig. 16.4 Balance among the thermodynamic force, viscous drag, and electrostatic drag for diffusion of a charge species in ILs

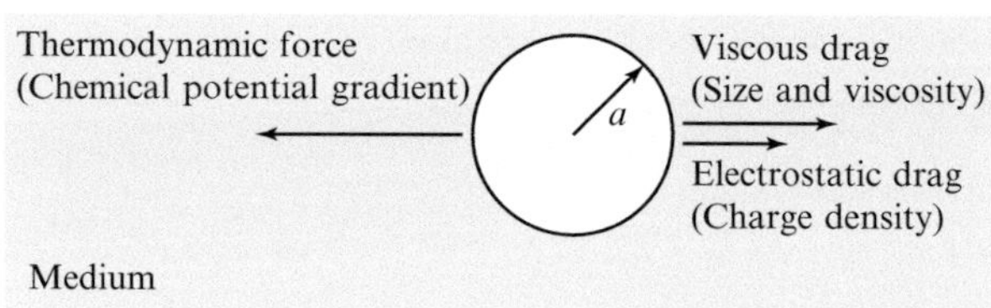

$[BF_4]^- < [N(Tf)_2]^- < [BETI]^-$ [16]. Thus, the D_{3+}/D_{2+} is verified to reflect the charge density of the anion of IL.

Since the diffusion coefficient in ILs is strongly affected by the electrostatic drag related to the charge densities of the diffusing species and the ions composing ILs, Stokes-Einstein relation is regarded as invalid. However, the viscosity of ILs is considered still a dominant factor for determining the diffusion coefficient of $[M(bpy)_3]^{3+}$ or $[M(bpy)_3]^{2+}$ since the activation energies for the diffusion coefficients are close to those for the viscosity coefficients of ILs. For example, the diffusion coefficient of $[M(bpy)_3]^{3+}$ or $[M(bpy)_3]^{2+}$ in $[C_4mPyr][N(Tf)_2]$ is smaller than that in $[C_2mim][N(Tf)_2]$ since the viscosity coefficient of $[C_4mPyr][N(Tf)_2]$ is higher than that of $[C_2mim][N(Tf)_2]$ at the same temperature. On the other hand, the diffusion coefficient of $[Ru(bpy)_3]^{3+}$ or $[Ru(bpy)_3]^{2+}$ in $[C_2mim][N(Tf)_2]$ is larger than that in $[C_2mim][BF_4]$ although the viscosity coefficient of $[C_2mim][N(Tf)_2]$ is close to that of $[C_2mim][BF_4]$. This irregularity is able to be explained by the electrostatic drag. The D_{3+}/D_{2+} for $[Ru(bpy)_3]^{3+/2+}$ in $[C_2mim][N(Tf)_2]$ is larger than that in $[C_2mim][BF_4]$, as listed in Table 16.2, indicating the diffusion becomes slower with an increase in the electrostatic drag in the ILs having the similar viscosity coefficients.

The invalidity of Stokes-Einstein relation in ILs implies the caution is needed to use the conventional relations about the transport properties of electrolytes. Walden's rule is one of these relations, which are often used even in ILs. Walden's rule is formulated as follows.

$$\Lambda_m \eta = \text{const.} \tag{16.8}$$

where Λ_m is the molar conductivity of electrolyte. Λ_m is the sum of the molar conductivity of ions in the electrolyte. When the electrolyte contains a single cation and anion, of which the molar conductivities are given by λ_+ and λ_-, respectively, Λ_m is represented by:

$$\Lambda_m = \lambda_+ + \lambda_- \tag{16.9}$$

On the other hand, λ_+ and λ_- are expressed with the diffusion coefficients of the cation and anion, D_+ and D_-, respectively, using Nernst-Einstein relation.

$$\Lambda_m = \frac{z_+^2 D_+ F^2}{RT} + \frac{z_-^2 D_- F^2}{RT} = \left(z_+^2 D_+ + z_-^2 D_-\right) \frac{F^2}{RT} \tag{16.10}$$

If the diffusion coefficients are assumed to be given by Stokes-Einstein relation, Λ_m is written as follows.

$$\Lambda_m = \left(\frac{z_+^2}{a_+} + \frac{z_-^2}{a_-}\right) \frac{e^2 N_A}{c\pi\eta} \tag{16.11}$$

where a_+ and a_- are the Stokes radius of the cation and anion, respectively, e is the fundamental electricity, and N_A is Avogadro number. Thus, Walden's rule is alternatively given by the following representation.

$$\eta \Lambda_{\mathrm{m}} = \left(\frac{z_+^2}{a_+} + \frac{z_-^2}{a_-} \right) \frac{e^2 N_{\mathrm{A}}}{c\pi} \tag{16.12}$$

For $z_+ = |z_-| = 1$, the value of the right hand side is constant when a_+ and a_- are independent of the media with different viscosity coefficients. Accordingly, Walden's rule is based on Stokes-Einstein relation, and is not applicable to the ILs, in which Stokes-Einstein relation is invalid due to strong electrostatic interaction. Although the relation similar to Stokes-Einstein relation for ILs has not been devised yet, the apparent Stokes radius tends to increase with an increase in electrostatic interaction. Thus, $\Lambda_{\mathrm{m}}\eta$ is expected to decrease with an increase in the charge densities of ions of ILs.

16.5 Rate Constants of Tris(bpy) Complexes in Some ILs

The redox reactions of $[\mathrm{M(bpy)_3}]^{3+/2+}$ are classified into an outer-sphere electron-transfer reaction, in which the coordination environment of the centered metal is unchanged before and after the electron transfer. The rate for such an outer-sphere electron-transfer reaction is often fast as compared with the inner-sphere electron-transfer reactions accompanied with formation and dissociation of chemical bonds. The rate constant, k, for the outer-sphere electron-transfer reaction is given by Marcus theory [2, 21].

$$k = K_{\mathrm{P}} \nu_{\mathrm{n}} \kappa_{\mathrm{el}} \exp\left(-\frac{\Delta^{\ddagger} G}{RT} \right) \tag{16.13}$$

where K_{P} is the precursor equilibrium constant, ν_{n} is the nuclear frequency factor, κ_{el} is the electron transmission coefficient, and $\Delta^{\ddagger}G$ is the activation Gibbs energy. When the electrode potential is equal to the formal potential, where the concentrations of an oxidized and reduced forms, C_{O} and C_{R}, are the same, $\Delta^{\ddagger}G$ is virtually expressed as follows.

$$\Delta^{\ddagger} G = \frac{E_{\lambda}}{4} \tag{16.14}$$

where E_{λ} is the reorganization energy. The rate constant at the formal potential is called the standard rate constant, k^0. The exchange current density, j_0, is given by the following equation.

$$j_0 = nFk^0 C \tag{16.15}$$

where $C = C_{\mathrm{O}} = C_{\mathrm{R}}$. The linear approximation of Butler-Volmer equation under small overpotentials gives the charge transfer resistance, R_{ct}, for unit area as follows.

$$R_{ct} = \frac{RT}{Fj_0} \tag{16.16}$$

Thus, the standard rate constant is able to be calculated by measuring the charge transfer resistance.

The charge transfer resistance for an electrode reaction is often determined by AC impedance method. For an electrolyte containing both oxidized and reduced forms at the same concentration, the electrode immersed in the electrolyte show the reversible potential, which is corresponding to the formal potential. Under this condition, the impedance at the interface between the electrode and the electrolyte is able to be obtained by applying a sine wave with a small amplitude (<10 mV) to the electrode potential and changing the frequency. The impedance is often analyzed by the least square fitting using an equivalent circuit. A typical equivalent circuit for the electrode reaction, in which the oxidized and reduced forms are both soluble in the electrolyte, is represented by Fig. 16.5, which is called Randles-type equivalent circuit. R_Ω is the electrolyte resistance, C_d is the double layer capacitance, R_{ct} is the charge transfer resistance, and Z_W is the Warburg resistance. C_d is sometimes replaced with a constant phase element (CPE), which is considered suitable for rough and/or inhomogeneous surface. Figure 16.6 shows the typical results of AC impedance measurement of a Pt electrode in $[C_4mPyr][N(Tf)_2]$ containing both 20 mM $[Ru(bpy)_3]^{3+}$ and 20 mM $[Ru(bpy)_3]^{2+}$ at 25 °C. A shield for electromagnetic waves, called Faraday cage, is often necessary for the impedance measurement in order to obtain better results. The real and imaginary parts of the impedance at various frequencies are plotted on a two-dimensional plane, as called Nyquist plots. The solid curve shows the locus calculated from the parameters determined by the least square fitting. In this way, the charge transfer resistances and standard rate constants were obtained for some ionic liquids containing $[M(bpy)_3]^{3+}$ and $[M(bpy)_3]^{2+}$ at various temperatures. The standard rate constants at 25 °C and the activation energies are listed in Table 16.3.

The rate constants for $[Fe(bpy)_3]^{3+/2+}$ and $[Ru(bpy)_3]^{3+/2+}$ in $[C_4mPyr][N(Tf)_2]$ are 9.0 and 4.5×10^{-4} cm s^{-1}, which are much smaller than those in dimethylformamide (DMF) [20, 22]. In addition, the rate constant of $[Ru(bpy)_3]^{3+/2+}$ decreases with an increase in the viscosity of the ILs, in the order of $[C_2mim][N(Tf)_2]$, $[C_4mPyr][N(Tf)_2]$, and $[C_4mPyr][BETI]$. Similarly, the rate constant of $[Fe(bpy)_3]^{3+/2+}$ in $[C_4mPyr][N(Tf)_2]$ is also smaller than that in $[C_2mim][N(Tf)_2]$.

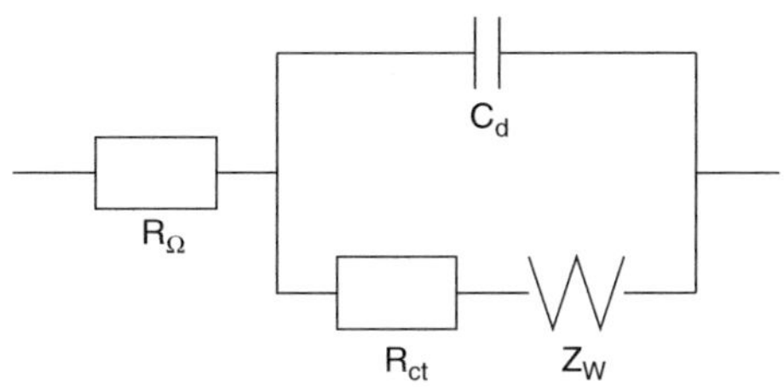

Fig. 16.5 Randles-type equivalent circuit for the interface between the electrode and the electrolyte

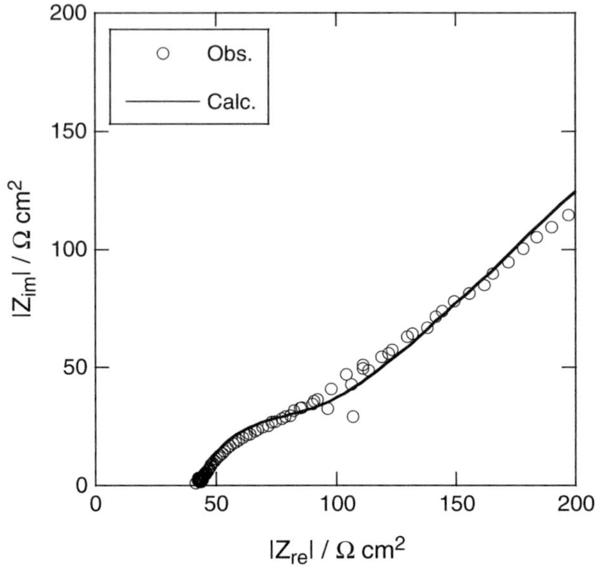

Fig. 16.6 Nyquist plots of a Pt electrode in $[C_4mPyr][N(Tf)_2]$ containing both 20 mM $[Ru(bpy)_3]^{3+}$ and 20 mM $[Ru(bpy)_3]^{2+}$ at 25 °C. Amplitude: 10 mV-rms, Frequency: 100 kHz–1 Hz

Table 16.3 The rate constant (k^0), the activation energy of k^0 ($E_a(k^0)$), and the activation energy of viscosity coefficient ($E_a(\eta)$) for $[M(bpy)_3]^{3+/2+}$ (M = Fe and Ru) in some ionic liquids at 25 °C

Ionic liquid	Redox couple	$10^4\,k^0$/cm s^{-1}	η/mPa s	$E_a(k^0)$/ kJ mol^{-1}	$E_a(\eta)$/ kJ mol^{-1}	References
$[C_2mim][N(Tf)_2]$	$[Ru(bpy)_3]^{3+/2+}$	27 ± 1	36	37	26	[5]
$[C_4mPyr][N(Tf)_2]$	$[Ru(bpy)_3]^{3+/2+}$	4.5 ± 0.4	82	37	25	[5]
$[C_4mPyr][BETI]$	$[Ru(bpy)_3]^{3+/2+}$	1.2 ± 0.2	210	53	39	[5]
$[C_2mim][BF_4]$	$[Ru(bpy)_3]^{3+/2+}$	0.67 ± 0.02	37	46	30	[6]
$[C_4mPyr][N(Tf)_2]$	$[Fe(bpy)_3]^{3+/2+}$	9.0 ± 0.5	80	24	25	[4]
$[C_4mPyr][N(Tf)_2]$	$[Ni(bpy)_3]^{3+/2+}$	0.083 ± 0.001	–	67	25	[16]
$[C_2mim][N(Tf)_2]$	$[Fe(bpy)_3]^{3+/2+}$	22	44	42	24	[17]
$[C_4mim][N(Tf)_2]$	$[Fe(bpy)_3]^{3+/2+}$	36		23		[24]
DMF	$[Ru(bpy)_3]^{3+/2+}$	$(9.5 \pm 3.9) \times 10^2$	0.80			[22]
DMF	$[Fe(bpy)_3]^{3+/2+}$	1.1×10^4	0.80			[20]

The rate constant generally depends on the viscosity of the medium since the ν_n in Eq. (16.13) is known to be inversely proportional to the viscosity. Thus, the apparent activation energy for the rate constant is expected to be expressed as follows.

$$k^0 \propto \exp\left(-\frac{E_a(\eta) + E_\lambda/4}{RT}\right) \tag{16.17}$$

where $E_a(\eta)$ is the activation energy for the viscosity. Thus, the difference in the activation energy for the apparent rate constant and that for the viscosity is considered to correspond to the reorganization energy. The reorganization energy is divided into inner and outer components. The inner component of reorganization energy reflects the bonds between the centered metal and the ligands. The outer component of reorganization energy corresponds to the interaction of the complex with the molecules or ions in the medium. In case of $[Ru(bpy)_3]^{3+/2+}$ and $[Fe(bpy)_3]^{3+/2+}$, the reorganization energies calculated from the difference in the activation energies for the apparent rate constants and the viscosity of ILs are ranging from 0 to 18 kJ mol^{-1}, suggesting the rate constant is mainly affected by the viscosity (dynamics of ILs) rather than the reorganization energy of the complexes. However, the rate constant for $[Ru(bpy)_3]^{3+/2+}$ in $[C_2mim][BF_4]$ is much smaller than that in $[C_2mim][N(Tf)_2]$ although the viscosity of $[C_2mim][BF_4]$ is close to that of $[C_2mim][N(Tf)_2]$. Since the electrostatic interaction of $[Ru(bpy)_3]^{3+/2+}$ with $[C_2mim][BF_4]$ is considered stronger than that in $[C_2mim][N(Tf)_2]$, as expected from D_{3+}/D_{2+}, the rate constant may be also influenced by the electrostatic interaction in addition to the viscosity.

On the other hand, the rate constant of $[Ni(bpy)_3]^{3+/2+}$ is significantly smaller than those of $[Ru(bpy)_3]^{3+/2+}$ and $[Fe(bpy)_3]^{3+/2+}$ in $[C_4mPyr][N(Tf)_2]$. Since there is no difference in the diffusion coefficients of these complexes, the slow kinetics for $[Ni(bpy)_3]^{3+/2+}$ is caused by the difference in the intrinsic properties among these complexes [23]. The redox reactions of $[Ru(bpy)_3]^{3+/2+}$ and $[Fe(bpy)_3]^{3+/2+}$ take place essentially in the same electron configurations, $(\pi d)^5/(\pi d)^6$, which are the bonding orbitals of the complexes. In case of $[Ni(bpy)_3]^{3+/2+}$, the electron configurations for $[Ni(bpy)_3]^{3+}$ and $[Ni(bpy)_3]^{2+}$ are represented by $(\pi d)^6(\sigma^*d)^1$ and $(\pi d)^6(\sigma^*d)^2$, respectively. Since σ*d orbital is the antibonding orbital, the occupation of an electron in the antibonding orbital leads to weaken the bond strength between the centered metal and the ligands, resulting in the significant change of the inner component of reorganization energy. The difference in the activation energies for the apparent rate constant of $[Ni(bpy)_3]^{3+/2+}$ and the viscosity is very large compared with those of $[Ru(bpy)_3]^{3+/2+}$ and $[Fe(bpy)_3]^{3+/2+}$, indicating the influence of the reorganization energy on the reaction kinetics is also observed in the ILs.

16.6 Temperature Dependence of the Redox Potentials of $[M(bpy)_3]^{n+1/n}$

The electrode potential generally depends on the temperature. The temperature dependence of the standard electrode potential of the complexes, of which the coordination environment does not change before and after electron transfer, has been investigated in aqueous and organic electrolytes [25]. The temperature dependence of the standard electrode potential, dE°/dT, of a single electrode reaction is able to be related to the reaction entropy, $\Delta_{rc}S^\circ$, assuming the temperature dependence of the reaction enthalpy is negligible.

$$\frac{\mathrm{d}E^\circ}{\mathrm{d}T} = \frac{\Delta_{rc}S^\circ}{nF} \tag{16.18}$$

For the redox reaction between the oxidized and reduced forms with the same radius of r, it is known that $\Delta_{rc}S^\circ$ is expected to be estimated by the following equation based on Born equation for solvation Gibbs energy [25].

$$\Delta_{rc}S^\circ = \frac{e^2 N_A}{2\varepsilon T}\left(\frac{\mathrm{d}\ln\varepsilon}{\mathrm{d}T}\right)\frac{z_O^2 - z_R^2}{r} \tag{16.19}$$

where ε is the permittivity of medium, z_O and z_R are the charges of the oxidized and reduced forms, respectively. This equation is applicable for dielectric media.

The temperature dependence of the formal potential of $[M(bpy)_3]^{n+1/n}$, $\mathrm{d}E^{\circ'}/\mathrm{d}T$, has been evaluated in some ILs [26, 27]. The $\mathrm{d}E^{\circ'}/\mathrm{d}T$ of a redox couple (O and R) is able to be measured by constructing a non-isothermal cell, represented by as follows, for example.

$$\mathrm{Pt}(T_1)|\mathrm{O},\ \mathrm{R}(T_1)||\mathrm{O},\ \mathrm{R}(T_2)|\mathrm{Pt}(T_2) \tag{16.20}$$

When the concentration of O is identical to that of R, each electrode shows the formal potential at each temperature. Thus, the $\mathrm{d}E^{\circ'}/\mathrm{d}T$ is obtained simply by the following equation.

$$\frac{\mathrm{d}E^{\circ'}}{\mathrm{d}T} = \frac{E^{\circ'}(T_2) - E^{\circ'}(T_1)}{T_2 - T_1} = \frac{\Delta_{rc}S^{\circ'}}{nF} \tag{16.21}$$

Note that $\Delta_{rc}S^{\circ'}$ is considered different from $\Delta_{rc}S^\circ$ since the formal potential contains the activity coefficient terms. Table 16.4 lists the $\mathrm{d}E^{\circ'}/\mathrm{d}T$ and $\Delta_{rc}S^{\circ'}$ values for Fc^+/Fc couple in some ILs with the non-isothermal cell. The $\mathrm{d}E^{\circ'}/\mathrm{d}T$ values are found to be small regardless of the ILs, indicating Fc^+/Fc couple is a good potential standard for these ILs with a small temperature dependence of potential.

Table 16.4 $dE^{\circ'}/dT$ and $\Delta_{rc}S^{\circ'}$ for Fc^+/Fc couple in some ILs [27]

IL	$dE^{\circ'}/dT$/mV K^{-1}	$\Delta_{rc}S^{\circ'}$/J mol K^{-1}
$[C_4mPyr][N(Tf)_2]$	0.10 ± 0.01	10 ± 1
$[C_3mPip][N(Tf)_2]$	0.10 ± 0.01	10 ± 1
$[C_2mim][N(Tf)_2]$	0.09 ± 0.01	9 ± 1
$[C_4mPyr][BETI]$	0.09 ± 0.01	9 ± 1
$[C_2mim][BF_4]$	0.10 ± 0.01	10 ± 1

Table 16.5 $dE^{\circ'}/dT$ and $\Delta_{rc}S^{\circ'}$ for $[Fe(bpy)_3]^{n+1/n}$ and $[Ru(bpy)_3]^{n+1/n}$ in some ILs [27]

Ionic liquids	Redox couple	$dE^{\circ'}/dT$/mV K^{-1}	$\Delta_{rc}S^{\circ'}$/J mol^{-1} K^{-1}
$[C_4mPyr][N(Tf)_2]$	$[Ru(bpy)_3]^{3+/2+}$	0.51 ± 0.02	49 ± 2
	$[Fe(bpy)_3]^{3+/2+}$	0.45 ± 0.02	43 ± 2
	$[Ru(bpy)_3]^{2+/+}$	0.26 ± 0.02	25 ± 2
	$[Fe(bpy)_3]^{2+/+}$	0.29 ± 0.02	28 ± 2
	$[Ru(bpy)_3]^{+/0}$	0.08 ± 0.02	8 ± 2
	$[Fe(bpy)_3]^{+/0}$	0.08 ± 0.02	8 ± 2
	$[Ru(bpy)_3]^{0/-}$	-0.13 ± 0.02	-13 ± 2
	$[Fe(bpy)_3]^{0/-}$	-0.14 ± 0.02	-14 ± 2
$[C_2mim][N(Tf)_2]$	$[Fe(bpy)_3]^{3+/2+}$	0.50 ± 0.02	48 ± 2
$[C_3mPip][N(Tf)_2]$	$[Fe(bpy)_3]^{3+/2+}$	0.49 ± 0.02	47 ± 2
	$[Fe(bpy)_3]^{2+/+}$	0.25 ± 0.02	24 ± 2
	$[Fe(bpy)_3]^{+/0}$	0.11 ± 0.01	11 ± 1
	$[Fe(bpy)_3]^{0/-}$	-0.05 ± 0.01	-5 ± 1
$[C_4mPyr][BETI]$	$[Fe(bpy)_3]^{3+/2+}$	0.33 ± 0.02	32 ± 2
	$[Fe(bpy)_3]^{2+/+}$	0.15 ± 0.01	14 ± 1
	$[Fe(bpy)_3]^{+/0}$	0.05 ± 0.01	5 ± 1
	$[Fe(bpy)_3]^{0/-}$	-0.15 ± 0.01	-14 ± 1
$[C_2mim][BF_4]$	$[Fe(bpy)_3]^{3+/2+}$	0.48 ± 0.02	46 ± 2

However, it is sometimes troublesome to prepare the ILs containing both oxidized and reduced forms at the same concentration. The $dE^{\circ'}/dT$ is able to be measured using an isothermal cell with a reference electrode, of which the temperature dependence of potential is known. The temperature dependence of a Ag/Ag (I) reference electrode has been examined by measuring the potential of the following isothermal cell at different temperatures [27].

$$\mathrm{Ag}\big|0.1\,\mathrm{M\ \ Ag(I)\ \ in\ \ [C_4mPyr]\big[N(Tf)_2\big]\big|\big|Fc^+,\ \ Fc\ \ (in\ some\ ILs)\big|Pt} \qquad (16.22)$$

The temperature dependence of the potential of Ag|Ag(I) reference electrode was found to be in the range from 0.54 to 0.57 mV K^{-1}. Using this Ag|Ag(I) reference electrode, the $dE^{\circ'}/dT$ values for $[M(bpy)_3]^{n+1/n}$ were determined by measuring the half-wave potentials of $[M(bpy)_3]^{n+1/n}$ couples with differential pulse voltammetry

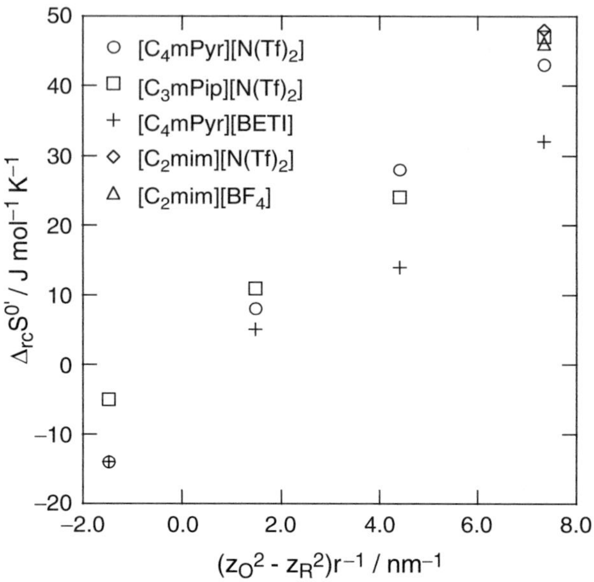

Fig. 16.7 Dependence of $\Delta_{\rm rc}S^{\circ\prime}$ for $[{\rm Fe(bpy)_3}]^{n+1/n}$ on $\left(z_{\rm O}^2 - z_{\rm R}^2\right)/r$ in some ionic liquids

at different temperatures, as listed in Table 16.5 together with the $\Delta_{\rm rc}S^{\circ\prime}$ values. The $\Delta_{\rm rc}S^{\circ\prime}$ value for $[{\rm Fe(bpy)_3}]^{n+1/n}$ is linearly dependent on the $\left(z_{\rm O}^2 - z_{\rm R}^2\right)/r$, as shown in Fig. 16.7, suggesting the $\Delta_{\rm rc}S^{\circ\prime}$ value is closely related to the electrostatic interaction, as known in conventional solutions. In [C$_4$mPyr][N(Tf)$_2$], the $\Delta_{\rm rc}S^{\circ\prime}$ values for $[{\rm Ru(bpy)_3}]^{n+1/n}$ are close to those for $[{\rm Fe(bpy)_3}]^{n+1/n}$, verifying the charge density of the redox couples is a major factor for determining the $\Delta_{\rm rc}S^{\circ\prime}$ value. In addition, the $\Delta_{\rm rc}S^{\circ\prime}$ values also depend on the charge density of the ions composing the ILs. The $\Delta_{\rm rc}S^{\circ\prime}$ values for the positively charged redox couples in the ILs composed of $[{\rm N(Tf)_2}]^-$ are close to one another. However, they are slightly larger than those in [C$_4$mPyr][BETI], probably because the charge density of $[{\rm BETI}]^-$ is lower than that of $[{\rm N(Tf)_2}]^-$. On the other hand, although [C$_3$mPip][N(Tf)$_2$] and [C$_4$mPyr][N(Tf)$_2$] have the same anion, the absolute value of $\Delta_{\rm rc}S^{\circ\prime}$ for $[{\rm Fe(bpy)_3}]^{0/-}$ in [C$_3$mPip][N(Tf)$_2$] is smaller than that in [C$_4$mPyr][N(Tf)$_2$] since the charge density of $[{\rm C_3mPip}]^+$ is expected to be lower than that of $[{\rm C_4mPyr}]^+$. Therefore, the entropy term (including the activity coefficient terms) of the redox reaction is dependent on the difference in the electrostatic energies of the oxidized and reduced form. The electrostatic energy is the function of the charge densities of both redox species and the ions composing the IL, suggesting the absolute value of the entropy term increases with an increase in the charge densities of the redox species and/or the ions of the IL. However, it is suggested that such factors as polarization and steric shape of ions also affect the entropy term since the $\Delta_{\rm rc}S^{\circ\prime}$ values for $[{\rm Fe(bpy)_3}]^{3+/2+}$ in the ILs composed of $[{\rm N(Tf)_2}]^-$ are close to that in [C$_2$mim][BF$_4$] while $[{\rm BF_4}]^-$ is much smaller than $[{\rm N(Tf)_2}]^-$.

16.7 Concluding Remarks

ILs are expected as alternative electrolytes for various electrochemical applications including batteries, fuel cells, electroplating, electrolysis, and so on. A number of electrochemical reactions have been reported to be studied in ILs. However, electrochemistry in ILs is not always similar to that in conventional electrolyte solutions. The strong electrostatic interaction among the charged species is one of the major factors, which distinguish ILs from other electrolyte solutions. In conventional electrolyte solutions, however, electrostatic interaction has been known to affect the thermodynamic and transport properties of electroactive species. Theories describing the effects of electrostatic interaction on these properties are available only in diluted electrolyte solutions. ILs are regarded as the concentrated electrolytes, for which no theory has been established yet. The characteristics observed for the electrode reactions of tris(bpy) complexes in some ILs are able to be explained partly by the theories in conventional electrolyte solutions. However, the effects of electrostatic interaction must be taken into account in order to understand the inapplicability of those theories to the ILs. The electrochemical data concerning with thermodynamics and kinetics in ILs have not been accumulated yet. Investigation of fundamental electrochemical reactions in ILs is still important for understanding electrochemistry in ILs and considering potential applications of ILs.

References

1. Fawcett WR (2004) Liquids, solutions, and interfaces from classical macroscopic descriptions to modern microscopic details. Oxford University Press, New York
2. Izutsu K (2009) Electrochemistry in nonaqueous solutions. Wiley, Weinheim
3. Ohno H (2011) Electrochemical aspects of ionic liquids, 2nd edn. Wiley, New Jersey
4. Tachikawa N, Katayama Y, Miura T (2007) Electrode kinetics of some iron complexes in an imide-type room-temperature ionic liquid. J Electrochem Soc 154:F211–F216
5. Toshimitsu Y, Katayama Y, Miura T (2012) Electrode reactions of ruthenium–bipyridine complex in amide-type ionic liquids. Electrochim Acta 82:43–47
6. Katayama Y, Toshimitsu Y, Miura T (2014) Electrode kinetics of the redox reaction of tris (2,2′-bipyridine)nickel complexes in an ionic liquid. Electrochim Acta 131:36–41
7. Katayama Y, Kawahara Y, Toshimitsu Y, Yamato Y, Miura T (2011) Redox reactions of transition metal 2,2′-bipyridine complexes in amide-type ionic liquids. Paper presented at the 62nd Annual Meeting of International Society of Electrochemistry, Niigata, 11–16 September 2011
8. Bartlett PN, Eastwick-Field V (1993) A reinvestigation of the electrochemistry of [Ni(II)(bpy)$_3$(ClO)$_4$)$_2$] in acetonitrile using rotating disc and rotating ring-disc electrodes. Electrochim Acta 38:2515–2523
9. Henne BJ, Bartak DE (1984) Metal-vapor synthesis and electrochemistry of bis(bipyridyl) nickel(0). Inorg Chem 23:369–373
10. Yamagata M, Katayama Y, Miura T (2006) Electrochemical behavior of samarium, europium, and ytterbium in hydrophobic room-temperature molten salt systems. J Electrochem Soc 153: E5–E9

11. Sahami S, Weaver MJ (1981) Entropic and enthalpic contributions to the solvent dependence of the thermodynamics of transition-metal redox couples: Part I. Couples containing aromatic ligands. J Electroanal Chem 122:155–170
12. Barbante GJ, Hogan CF, Hughes AB (2009) Solid state spectroelectrochemistry of microparticles of ruthenium diimine complexes immobilized on optically transparent electrodes. J Solid State Electrochem 13:599–608
13. Prasad R, Scaife DB (1977) Electro-oxidation and electro-reduction of some iron (II), cobalt(II) and nickel(II) polypyridyl complexes in acetonitrile. J Electroanal Chem 84:373–386
14. Richert SA, Tsang PKS, Sawyer DT (1989) Ligand-centered redox processes for manganese, iron and cobalt, MnL_3, FeL_3, and CoL_3, complexes (L = acetylacetonate, 8-quinolinate, picolinate, 2,2′-bipyridyl, 1,10-phenanthroline) and for their tetrakis(2,6-dichlorophenyl) porphinato complexes[M(Por)]. Inorg Chem 28:2471–2475
15. Sato Y, Tanaka N (1969) Polarographic behavior of tris(2,2′-bipyridine)chromium(III) and tris (ethylenediamine)chromium(III) in acetonitrile solutions. Bull Chem Soc Jpn 42:1021–1024
16. Katayama Y, Toshimitsu Y, Miura T (2013) Electrode kinetics of tris(2,2′-bipyridine)ruthenium complexes in 1-ethyl-3-methylimidazolium tetrafluoroborate ionic liquid. J Electrochem Soc 160:H219–H228
17. Katayama Y, Yoshihara M, Miura T (2014) Electrochemical reaction of tris(1,10-phenanthroline)iron complexes in some amide-type ionic liquids. ECS Trans 64(4):109–118
18. Sahami S, Osteryoung RA (1984) Electrochemical and spectroscopic studies of polypyridine complexes of iron(II/III) and ruthenium(II/III) in the aluminum chloride-N-1-butylpyridinium chloride molten salt. Inorg Chem 23:2511–2518
19. McDevitt MR, Addison AW (1993) Medium effects on the redox properties of tris(2,2′-bipyridyl)ruthenium complexes. Inorg Chim Acta 204:141–146
20. Saji T, Yamada T, Aoyagui S (1975) Electron-transfer rate constants for redox systems of Fe(III)/Fe(II) complexes with 2,2′-bipyridine and/or cyanide ion as measured by the galvanostatic double pulse method. J Electroanal Chem Interfacial Electrochem 61:147–153
21. Bard AJ, Faulkner LR (2001) Electrochemical methods—fundamentals and applications, 2nd edn. Wiley, New York
22. Scott AM, Pyati R (2001) Solvent viscosity and interrelated effects on electrochemiluminescence intensity of tris(2,2′-bipyridyl)ruthenium(II). J Phys Chem B 105:9011–9015
23. Macartney DH, Sutin N (1983) Electron-exchange rates of polypyridine complexes: electron-transfer reactions involving the tris(polypyridine)nickel(II/III) couple in acidic aqueous media. Inorg Chem 22:3530–3534
24. Dolidze TD, Khoshtariya DE, Illner P, Kulisiewicz L, Delgado A, van Eldik R (2008) High-pressure testing of heterogeneous charge transfer in a room-temperature ionic liquid: evidence for solvent dynamic control. J Phys Chem B 112:3085–3100
25. Hupp JT, Weaver MJ (1984) Solvent, ligand, and ionic charge effects on reaction entropies for simple transition-metal redox couples. Inorg Chem 23:3639–3644
26. Migita T, Tachikawa N, Katayama Y, Miura T (2009) Thermoelectromotive force of some redox couples in an amide-type room-temperature ionic liquid. Electrochemistry 77:639–641
27. Yamato Y, Katayama Y, Miura T (2013) Effects of the interaction between ionic liquids and redox couples on their reaction entropies. J Electrochem Soc 160:H309–H314

Chapter 17
Electrocatalysis in Room Temperature Ionic Liquids

Andinet Ejigu and Darren A. Walsh

17.1 Introduction

Due to their high thermal and electrochemical stabilities, low vapour pressures, and the ability to readily alter their properties by changing the identity of the constituent ions, RTILs are being used as electrolytes in an increasing number of electrochemical investigations [1–4]. For example, many fundamental studies have explored the effects of changing the compositions of RTILs on parameters such as their viscosities and electrochemical windows, which are often up to 5 V wide [5, 6]. The effects of the RTIL environment on molecular diffusion and electron transfer dynamics have also been studied in detail using simple redox species such as ferrocene [7–9]. Such studies have revealed that molecular diffusion coefficients are often up to three orders of magnitude lower in RTILs than in conventional aqueous and nonaqueous electrolytes due to the relatively high viscosities of most RTILs.

In parallel with fundamental investigations into the electrochemical behaviour of RTILs, many groups have begun to use RTILs as electrolytes in electrochemical energy-conversion and storage devices. The high thermal stabilities of many RTILs have prompted the development of safer Li batteries by substituting RTILs for volatile and flammable organic electrolytes [10, 11]. A number of groups have also begun to use RTILs as electrolytes in intermediate temperature ($>100\ ^\circ$C) fuel cells because, unlike the Nafion electrolytes usually used in proton exchange membrane fuel cells (PEMFCs), many RTILs can conduct protons in the absence of bulk water [12–15]. Due to the wide electrochemical windows of some RTILs, large amounts of energy can be stored in RTIL-based supercapacitors [16]. As these examples show, RTILs undoubtedly exhibit many properties that make them attractive for use in electrochemical energy conversion and storage. However, sluggish mass

A. Ejigu • D.A. Walsh (✉)
School of Chemistry, University of Nottingham, Nottingham NG7 2RD, UK
e-mail: darren.walsh@nottingham.ac.uk

A.A.J. Torriero (ed.), *Electrochemistry in Ionic Liquids*,
DOI 10.1007/978-3-319-15132-8_17

transport in RTIL electrolytes can affect the performances of such devices and the development of less viscous RTILs should aid in solving such issues.

Electrocatalytic reactions are those in which the reaction rate is sensitive to the nature of electrode material and the operating principles of electrochemical energy-conversion devices (with the exception of supercapacitors) often involve electrocatalysis. In such reactions, there is usually a strong interaction between the electrode surface and the reactants, intermediates, or products. In PEMFCs, electrocatalysis of the HOR and ORR are driven at the anode and cathode, respectively, using Pt-based catalysts. In direct methanol fuel cells (DMFCs), methanol is oxidised at Pt-based anode electrocatalysts and the ORR occurs at Pt-based cathode electrocatalysts. The ORR also plays a key role in rechargeable Li-air batteries, occurring at the cathode during discharge, while the oxygen evolution reaction (OER) occurs during charging. Research into electrocatalysis is usually driven by a desire to develop more-effective or less-expensive electrocatalysts, or to understand the fundamental processes occurring at the electrocatalyst/electrolyte interface (or both). For example, a relatively large amount of Pt is required to drive the ORR in PEMFCs and DMFCs due to its sluggishness [17]. It is perhaps not surprising that the ORR in PEMFCs is a sluggish reaction considering that it involves breakage of the O–O double bond and the formation of four O–H bonds:

$$O_2 + 4e^- + 4H^+ \rightarrow 2H_2O; \quad E^0 = 1.23\,\mathrm{V}, \quad \text{where } E^0 \text{ is the standard potential} \tag{17.1}$$

Even at Pt electrocatalysts, the sluggishness of the ORR generates an overpotential of more than 0.4 V (even at moderate current densities), lowering the PEMFC's power output [18]. In an effort to reduce the cost of the Pt used to catalyse the ORR in PEMFCs, a lot of research is aimed at developing new, more-efficient, less-expensive ORR catalysts and Pt alloys are especially promising in this regard [19, 20]. Optimisation of ORR catalysts for use in Li-air batteries also focuses on improving the rate of the ORR (and the OER) through modification of the electrode structure and composition [21, 22]. In contrast, the HOR (Eq. 17.2), which is one of the most studied of all electrochemical reactions, both from fundamental and applied points of view, occurs rapidly at Pt (which is the basis of its exploitation in the standard hydrogen electrode). Consequently, relatively small amounts of Pt are necessary to catalyse the HOR in PEMFCs, so most PEMFC electrocatalyst research focuses on the problematic ORR rather than the HOR.

$$H_2 \rightarrow 2H^+ + 2e^-; \quad E^0 = 0.0\,\mathrm{V} \tag{17.2}$$

It is now generally accepted that optimisation of the composition and/or morphology of electrocatalysts require a deep understanding of the fundamental processes occurring during electrocatalysis. The burgeoning field of surface electrocatalysis, which often combines electrochemical investigations at well-defined electrode surfaces, surface analysis and theoretical chemistry, is yielding new insights into electrocatalysis (primarily in aqueous media), as well as the development of new and more effective electrocatalysts [23].

The unique environment within RTILs offers a number of opportunities for those researchers interested in both fundamental and applied electrocatalysis. The high density of ions and absence of solvent molecules in RTILs makes the environment within RTILs very different from that in which electrocatalysis has traditionally been studied. One could argue that our understanding of electrocatalysis and the role that spectator ions play during electrocatalysis is inapplicable in RTILs and that a new body of knowledge must be developed before we can understand electrocatalysis in RTILs and develop electrocatalysts specifically for use in RTIL-based electrochemical energy-conversion devices. In recent years, a few groups have begun to study reactions such as the HOR, ORR, methanol oxidation reaction (MOR) and CO oxidation reaction (COOR) in RTILs [24–30]. Some of these studies have focussed on the development of electrochemical energy-conversion devices and others have focussed primarily on the fundamental electrochemical and kinetic insights that such measurements can yield. In this chapter, we provide an overview of the work that has been done in this area thus far. As much of this work has involved the use of protic ionic liquids (PILs), and is particularly dependent on their proton conductivity, we begin the chapter with a brief introduction to the chemistry of PILs. As we will see, as well as providing a conductive liquid electrolyte in which electrochemistry and electrocatalysis can take place, the chemistry of PILs (and in particular the proton transfer equilibrium) appears to play a crucial role during some of the technological applications of PILs. The chapter also contains detailed descriptions of the effects of surface adsorbates during a number of electrocatalytic reactions. As we will see, such effects will almost certainly dictate the direction that some of this research takes. Finally, we end the chapter with some concluding remarks and we offer some recommendations for the future of this field of research. Hopefully, by reading this chapter it will become clear that the insights obtained from the work described here may well extend beyond the borders of 'ionic liquid electrochemistry' and aid in our understanding of electrochemistry and electrocatalysis in a broader sense. It should also become apparent that the developments currently under way in this area could potentially contribute to the development of next-generation electrochemical energy-conversion devices.

17.2 Protic Ionic Liquids

Over the past few years, a number of PILs have emerged offering properties that are often different from those offered by the better-known aprotic RTILs. PILs are simply formed by proton transfer from a Brønsted acid to a Brønsted base and can be synthesised relatively easily on a large scale, at low cost, and without the need for further purification. PILs are also often more conductive than most aprotic RTILs [31]. The proton transfer reaction that yields a PIL can be represented

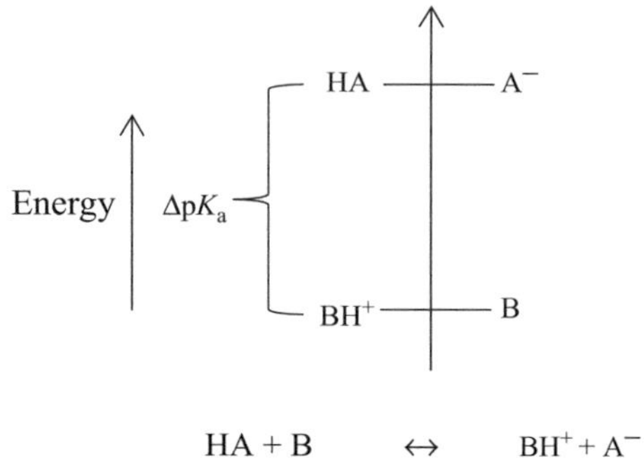

Fig. 17.1 Scheme showing the formation of a protic ionic liquid (PIL) and a representation of the energy gap across which the proton transfer occurs. ΔpK_a is the difference between the pK_a of the acid HA and that of the base B. Adapted with permission from Bautista-Martinez, J A; Tang, L; Belieres, J-P; Zeller, R; Angell, C A; Friesen, C (2009). J. Phys. Chem. C 113:12586-12593. Copyright 2009 American Chemical Society

schematically by Fig. 17.1, which shows how a proton 'falls' from an energy level on the donor acid into the free acceptor level on the base to form a protonated cationic species [32, 33]. In this figure, ΔpK_a is the difference between the pK_a of the acid HA and that of the base B and represents the driving force for proton transfer. Correlations between ΔpK_a and the physicochemical properties of PILs exist and, if ΔpK_a is greater than 15, the resulting PILs are thermally stable and exhibit high ionicity, resulting in high conductivity and low vapour pressures [34]. It has also been suggested that the extent of proton transfer should be such that less than 1 % neutral species are present in order to form a pure or 'good' PIL [35, 36]. While the concentration of protonated species in PILs is usually of the order of 5–10 M, PILs can behave as acids or bases depending on their composition. For example, litmus paper turns dark blue when in contact with triethylammonium acetate, $[HN_{222}][Ac]$, and bright pink when in contact with triethylammonium trifluoromethanesulfonate $[HN_{222}][CF_3SO_3]$ [36].

17.3 Electrocatalysis of the Oxygen Reduction Reaction in RTILs

Due to its importance in PEMFCs and lithium air batteries, the ORR is one of the most widely studied of all electrochemical reactions. The ORR is also of fundamental interest to electrochemists, being a relatively complex electrocatalytic process involving a bond breakage and multiple bond-forming steps (in the case of the $4e^-$ reduction of O_2 to H_2O shown in Eq. 17.1). The ORR has been studied using a variety of electrode materials and RTIL electrolytes, including aprotic RTILs [37, 38], PILs [26, 39], and mixtures of RTILs and weak and strong acids [40]. In aprotic RTILs, the ORR proceeds *via* a $1e^-$ reduction to superoxide [41, 42]:

$$O_2 + e^- \rightarrow O_2^{\bullet -} \tag{17.3}$$

It has been shown that the lifetime of $O_2^{\bullet -}$ depends on the nature of the cations present in the RTIL [43, 44]. Abraham and co-workers studied the influence of Li^+ on the mechanism of the ORR at glassy carbon (GC) and Au in 1-ethyl-3-methylimidazolium bis(trifluoromethanesulfonyl)imide, $[C_2mim][N(Tf_2)]$, using cyclic voltammetry [45]. In neat $[C_2mim][N(Tf_2)]$, the ORR proceeds through a quasireversible $1e^-$ reduction to $O_2^{\bullet -}$ at GC and Au indicating that $[C_2mim]^+$ stabilises $O_2^{\bullet -}$ (the solid line in Fig. 17.2a shows the cyclic voltammogram (CV) obtained at an Au electrode). However, the peak due to $O_2^{\bullet -}$ oxidation (A1) decreased in size significantly when Li^+ ions were added and new peaks appeared at positive potentials (Fig. 17.1a dashed line). The authors explained this behaviour as due to a reaction between $O_2^{\bullet -}$ and Li^+ ions, which yielded LiO_2. Some of this LiO_2 is unstable and can decompose to Li_2O_2 that is oxidised at positive potentials. Similar behaviour was also observed in Li^+-containing pyrrolidinium-based RTILs [43].

The influence of various cations (the tetrabutylammonium ion $[TBA]^+$, Li^+, Na^+, and K^+) on the ORR has also been studied in $[C_2mim][N(Tf_2)]$ [46]. The Lewis acidity of the cations decreases in the order $Li^+ > Na^+ > K^+ > [C_2mim]^+ > [TBA]^+$. As shown in Fig. 17.2b, there is no significant change in the ORR electrochemistry following the addition of $[TBA]^+$ to $[C_2mim][N(Tf_2)]$ and the product of the ORR is $O_2^{\bullet -}$ (in the figure the $[C_2mim]^+$ cation is labelled EMI^+). However, the effect of hard acids on the ORR became more evident upon moving from solutions that contained K^+ to Na^+ and then to Li^+. The response became more electrochemically irreversible as the added cation became more acidic, and the product of the ORR changed from stable superoxides to unstable metal superoxides and metal peroxides. The overpotential for oxidation of the oxides increased as a function of the cation hardness.

Walsh and co-workers studied the ORR in the PIL diethylmethylammonium trifluoromethanesulfonate $[dema][CF_3SO_3]$, at Pt electrodes using rotating ring-disc electrode (RRDE) voltammetry [39]. In this case, the RRDE voltammetry indicated that the ORR proceeded predominantly *via* a $4e^-$ reduction to H_2O. This pathway was confirmed by measuring the increase in the PIL's H_2O content as a function of the electrolysis time. Similarly, Nohira and co-workers observed the $4e^-$ reduction ORR in fluorohydrogenate RTILs, in which protons are transported by the fluorohydrogenate ions [47]. In contrast, Zhao and co-workers studied the ORR in a number of PILs and suggested that the ORR proceeded *via* a $2e^-$ reduction to H_2O_2 based on a digital simulation of the ORR cyclic voltammogram [26]. Friesen's group have examined the effect of proton donors on the mechanism of the ORR [40]. In neat aprotic RTILs, the ORR proceeds *via* a $1e^-$ reduction to $O_2^{\bullet -}$ as expected from the discussion above, while the ORR approached a $4e^-$ reduction to H_2O as the concentration of protic species increased.

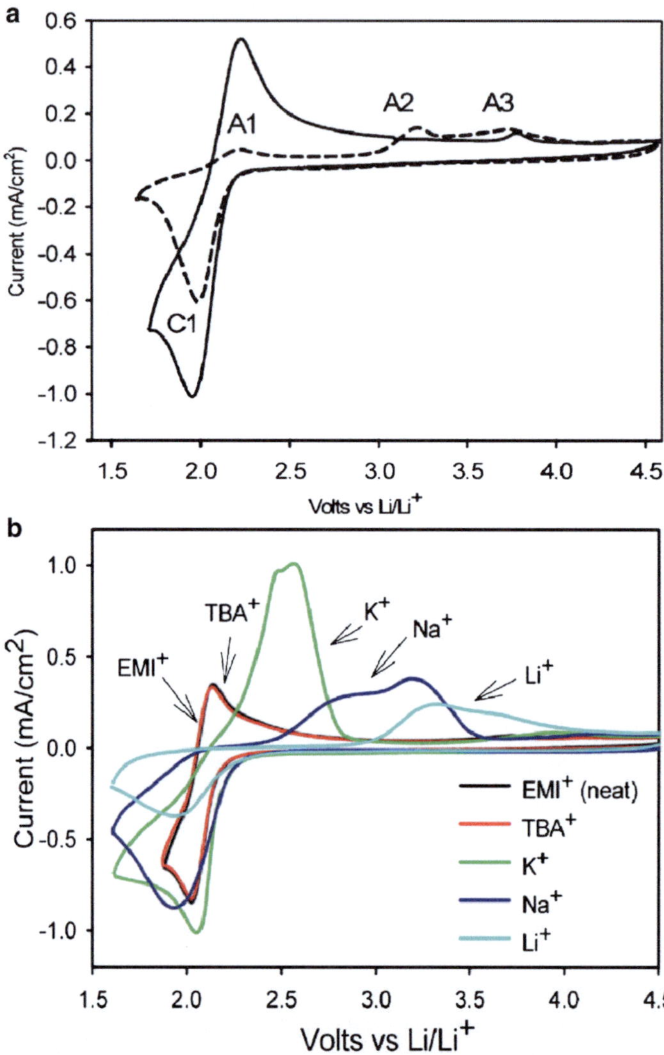

Fig. 17.2 (**a**) CVs of O_2-saturated [C_2mim][N(Tf_2)] with (*dashed line*) and without (*solid line*) 0.025 M Li[N(Tf_2)] recorded at 100 mV/s at an Au electrode. Adapted with permission from Allen, C J; Mukerjee, S; Plichta, E J; Hendrickson, M A; Abraham, K M (2011) J. Phys. Chem. Lett. 2:2420-2424. Copyright 2011 American Chemical Society. (**b**) CVs recorded in [C_2mim][N(Tf_2)] at a GC electrode at 100 mV s^{-1} before and after addition of various ions at a concentration of 0.025 M. EMI^+ is the [C_2mim]$^+$ cation. Reprinted with permission from Allen, C J; Hwang, J; Kautz, R; Mukerjee, S; Plichta, E J; Hendrickson, M A; Abraham, K M (2012) J. Phys. Chem. C 116:20755-20764. Copyright 2012 American Chemical Society

As well as understanding the effects of a particular RTIL's structure on the mechanism of the ORR, it is important to understand the effects of RTIL structure on the reaction rate. For rapid ORR kinetics in PILs, which is technologically important for the development of, e.g. nonhumidified intermediate temperature fuel cells that can operate above the boiling point of H_2O [12, 13, 48], the ideal PIL should be one with a cation that readily releases protons and an anion that is stable above the ORR standard potential and provides hopping sites for protons generated at the anode during H_2 oxidation [49]. Slow ORR kinetics in H_2 fuel cells generate a large overpotential at the cathode that reduces the measured open circuit potential (OCP) and Watanabe et al. recently drew a correlation between the measured OCP of a PIL-based H_2/O_2 fuel cell at 150 °C and ΔpK_a of the PIL [48]. The OCP increased linearly from 0.6 to 1.05 V as ΔpK_a increased from 10 to 18 and then decreased once ΔpK_a passed 18. They attributed the decrease in OCP at high ΔpK_a to the increasing strength of N–H bond, which became dominant and hindered release of the proton.

The correlation between the measured OCP of a PIL-based fuel cells and ΔpK_a does not take into account the effect of the electrode/PIL interfacial structure during electrocatalysis of the ORR. It is well known that strong adsorption of ions affects the rate of the ORR in aqueous electrolytes [50] and one could argue that it is unlikely that ORR kinetics in PILs are controlled solely by proton availability. The role of the Pt/PIL interfacial composition during electrocatalysis of the ORR in PILs was recently explored by Walsh and co-workers [51]. In this study, a significant difference in the onset potential of the ORR was observed upon moving from [dema][CF_3SO_3] (ORR onset potential = 0.7 V) to [dema][$N(Tf_2)$] (ORR onset potential = −0.2 V). CO displacement experiments (for details see Sect. 17.5.1) were used to probe the extent of ion adsorption onto the Pt electrode from the PILs. It was found that in [dema][$N(Tf_2)$] the $[N(Tf_2)]^-$ anion was strongly adsorbed on the Pt surface at potentials above −0.2 V and it was suggested that this ion adsorption hindered the rate of the ORR at the Pt surface (increasing the ORR overpotential) [51]. The adsorption of $[N(Tf_2)]^-$ ions on Pt surfaces had also been observed previously by Munakata and co-workers using in situ FT-IR spectroscopy [52]. Their spectroscopic results (Fig. 17.3) also showed that $[N(Tf_2)]^-$ ions from [dema][$N(Tf_2)$] adsorbed onto Pt (the peaks at 1,180 cm^{-1} and 1,382 cm^{-1} are due to the $-SO_2-N$ group), as did the $[FSI]^-$ anion, but the bulkier $[BETI]^-$ anion did not.

It is well known that adsorbed oxide species play a very important role during electrocatalysis of the ORR in aqueous media and drastically affect the rate of the ORR in PEMFCs [53, 54]. Using spectroscopic and electrochemical analysis, the Walsh group have showed that Pt surfaces become covered with a layer of adsorbed oxides at high potentials in ammonium-based PILs due to the oxidation of trace water in the RTIL [25]. In [dema][CF_3SO_3], adsorption of oxide species at Pt occurred positive of 1.0 V vs. RHE (compared to about 0.8 V in aqueous media). The formation of this oxide layer hindered the rate of the ORR, which then increased as the oxides were electrochemically removed from the surface [25]. Interestingly, analysis of ORR polarisation curves recorded in [dema][CF_3SO_3] also revealed a change in the ORR Tafel slope from 70 mV per decade at low ORR overpotentials to 117 mV per decade at high overpotentials (Fig. 17.4) [39]. The change in the Tafel slope with increasing potential has been attributed to a number of effects, including a change in the adsorption behaviour of reaction

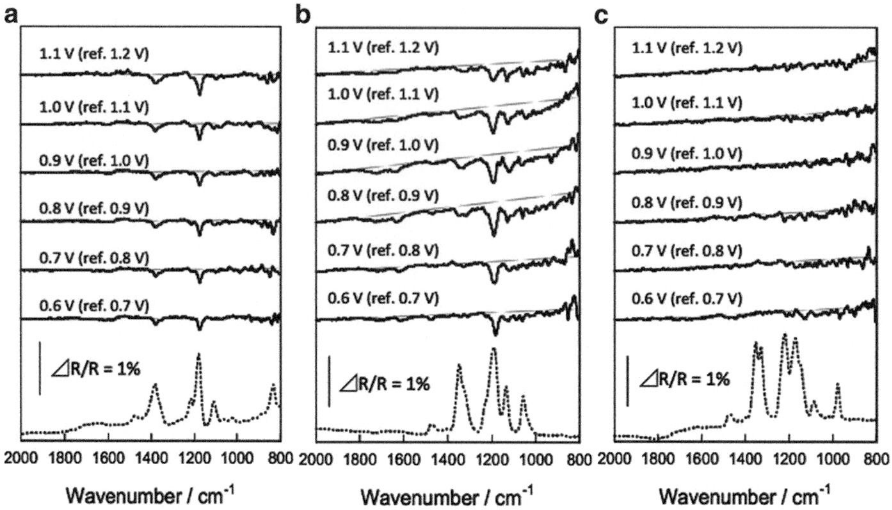

Fig. 17.3 Normalised interfacial FT-IR spectra for (**a**) [dema][FSI], (**b**) [dema][N(Tf$_2$)], and (**c**) [dema][BETI] on a Pt electrode. The spectra were obtained by stepping the electrode potential from 1.2 to 0.6 V in the O_2-saturated PILs. The bottom dashed lines in (**a**), (**b**), and (**c**) are the absorbance spectra of [dema][FSI], [dema][N(Tf$_2$)], and [dema][BETI] obtained using a Pt electrode before applying the potential. Reprinted with permission from Munakata, H; Tashita, T; Haibara, M; Kanamura, K (2010) ECS Trans. 33:463-469. Copyright 2010 Electrochemical Society

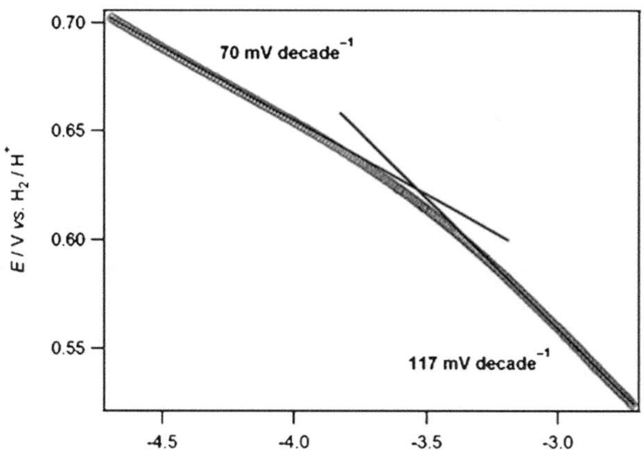

Fig. 17.4 Tafel plot of the applied potential, E (in the range 0.7 V $\geq E \geq$ 0.5 V), vs. log of the mass transport corrected current for the ORR in [dema][CF_3SO_3]. On the x-axis, i is the current at potential E and i_L is the limiting current measured from the ORR polarisation curve. Reprinted from Walsh, D A; Ejigu, A; Smith, J; Licence, P (2013) Phys. Chem. Chem. Phys. 15:7548-7554 with permission of the PCCP Owner Societies

intermediates and a change in the surface coverage of chemisorbed oxygen species formed from water oxidation in the acidic solution [55–58]. Potential-dependent adsorption of oxides causes a change in the Tafel slope due to site-blocking and electronic effects of the adsorbed species [58]. As discussed above, significant oxidation of trace water in [dema][CF_3SO_3] was observed at appreciable rates only at potentials positive of 1.0 V. Given that the potential of the Pt electrode used in that study did not exceed 1.0 V, it is unlikely that adsorption of intermediate species from water oxidation is the cause of the changing Tafel slope [39]. Instead, it was suggested that adsorbed intermediates from the ORR are the origin of the species causing the low Tafel slope at low overpotentials. The main conclusion from this work is that Pt-oxygen binding appears to play a key role during the ORR in PILs, as is also the case in aqueous media. Given that PILs are being actively considered for use in intermediate temperature fuel cells, it may be that the use of Pt alloys for the ORR could increase the rate of the ORR in PILs significantly and increase the cell performance. Indeed, it will be interesting to see whether a correlation between the Pt–O binding strength and the rate of the ORR in PILs generates a 'volcano' curve, as it does in aqueous media [59]. Of course, one of the main reasons for using such RTILs in fuel cells is to increase the operating temperature above 100°C and it will also be interesting to see how Pt-alloy electrocatalysts perform during such reactions at elevated temperatures.

17.4 Electrocatalysis of the H_2 Oxidation and H_2 Evolution Reactions in RTILs

As discussed above, the HOR and the closely related hydrogen evolution reaction (HER; the reverse of Eq. 17.2) are of fundamental and applied interest to electrochemists and each has been studied in RTILs [33, 38, 60–62]. Compton and co-workers reported that the appearance and position of the HER peak during HOR/HER electrochemistry in RTILs depends on the stability of the protonated anion (the RTIL anions accommodate the protons generated during the HOR) and on the equilibrium position of the protonation reaction [60]. For example, the HOR is quasireversible in $[N(Tf_2)]^-$-based RTILs while the reduction process is not observed at all in 1-hexyl-3-methyl imidazolium chloride and 1-butyl-3-methyl imidazolium nitrate, indicating that the electrochemically generated protons react with the anions of the RTIL [60, 61].

Compton and co-workers also studied the kinetics and mechanism of the HER in [C_2mim][$N(Tf_2)$] using cyclic voltammetry at Au, Mo, Ni, Ti, and Pt electrodes [63]. In this study, H[$N(Tf_2)$] was added to the RTIL as the proton source and the CVs recorded are shown in Fig. 17.5. The HER at Pt began at about −0.25 V vs. Ag/Ag^+ and some oxidation of H_2 was visible during the reverse sweep (the reaction was quasireversible). However, use of each of the other metals resulted in an irreversible HER with negligible H_2 oxidation visible during the reverse sweeps.

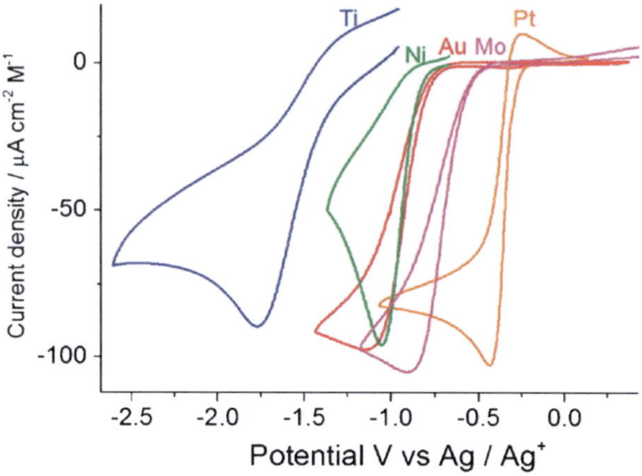

Fig. 17.5 CVs recorded for the hydrogen evolution reaction in solutions of H[N(Tf_2)] dissolved in [C_2mim][N(Tf_2)] (at concentrations between 12 and 95 mM) using a range of metal electrodes. Reprinted from Meng, Y; Aldous, L; Belding, S R; Compton, R G (2012) Phys. Chem. Chem. Phys. 14:5222-5228 with permission of the PCCP Owner Societies

The results shown in Fig. 17.5 clearly illustrate the effect of changing the electrode material on the HER kinetics. As the Pt electrode was changed for Mo, Au, and so on, the entire HER wave shifted in a negative potential direction and the authors were able to determine standard heterogeneous electron transfer rate constants, k^0, for the HER at each electrode. k^0 decreased from 1.3×10^{-3} cm/s at Pt to 2.6×10^{-8} cm/s at Ti. While such k^0 values are lower than those measured for the HER in aqueous solution, the general trend in activities of the metals for the HER is approximately the same in the RTIL as in aqueous media. However, differences in the HER in the RTIL system and in aqueous media were revealed when the mechanism of the HER was considered. In general, the HER is initiated by the Volmer step (Eq. 17.4, in which H_{ads} is an adsorbed H atom) and this is followed by either the electrochemical Heyrovsky reaction (Eq. 17.5) or the Tafel reaction (Eq. 17.6). In aqueous media, the slow step is H_{ads} migration and transfer coefficients, α, close to 1.5 are measured [63]. However, analysis of HER Tafel plots recorded in [C_2mim][N(Tf_2)] yielded α values less than 0.5, indicating that the rate-determining step at each electrode in the RTIL was the Volmer step [63].

$$\mathrm{Pt} + \mathrm{H}^+ + \mathrm{e}^- \rightarrow \mathrm{Pt} - \mathrm{H}_{\mathrm{ads}} \tag{17.4}$$

$$\mathrm{Pt} - \mathrm{H}_{\mathrm{ads}} + \mathrm{H}^+ + \mathrm{e}^- \rightarrow \mathrm{Pt} + \mathrm{H}_2 \tag{17.5}$$

$$\mathrm{Pt} - \mathrm{H}_{\mathrm{ads}} + \mathrm{Pt} - \mathrm{H}_{\mathrm{ads}} \rightarrow 2\mathrm{Pt} + \mathrm{H}_2 \tag{17.6}$$

Recently, Navarro-Suarez and co-workers studied the activities of different single-crystal Pt facets for the HOR in a number of RTILs [62]. This study revealed that the rate of the HOR was not only dependent on the nature of RTIL anions, but also on the crystallographic orientation of Pt surfaces. For example, Pt(111) was the most active surface in 1-ethyl-3-methyl imidazolium ethyl sulphate [C_2mim][$EtSO_4$], while Pt(100) was most active in [C_2mim][N(Tf_2)]. Pt(100) was the least active surface in [C_2mim][$EtSO_4$], [C_4mim][BF_4], and [C_4mim][CF_3SO_3], with the activity decreasing in the order Pt(111) > Pt(110) > Pt(100). The suggested reason for this trend was that anion adsorption is weaker at more compact surfaces so the HOR could occur at lower overpotentials on more compact surfaces. However, this suggestion does not explain the high activity of Pt(100) in $[C_2mim][N(Tf_2)]$. In this case, the authors speculated that some sort of surface reconstruction occurs when the Pt(100) surface comes in contact with the $[N(Tf_2)]^-$ anion.

Using a series of ten different (predominantly ammonium-based) PILs, the groups of Angell and Friesen showed that there is a gap between the HOR and HER onset potentials, which can be correlated to ΔpK_a [33]. The protons produced during the HOR in the PILs can only occupy the conjugate base sites, which are of much higher energy than the conjugate acid sites (see Fig. 17.1). Thus, the authors suggested that adding excess base to the PIL may sustain the HOR at more negative potentials, when the proton carriers are the free bases. However, such PIL/base mixtures may not be practical if one wishes to perform the HOR at elevated temperatures, as most amines tend to have low boiling points. It may be that doping of the PIL with thermally stable bases (e.g. tetramethylguanidine) could be a solution to such an issue. The work by Angell and Friesen has also illustrated the effect of ion adsorption onto the electrocatalyst surface during the HOR [33]. Figure 17.6 shows a rapid decay in the electrocatalytic HOR current as the potential is scanned positive of about 1.2 V in triethylammonium trifluoromethanesulfonate, $[tea][CF_3SO_3]$. This decrease in current was attributed to the formation of an adlayer of PIL anions on the Pt surface. Later work by Walsh and co-workers, involving the use of electrochemical and X-ray photoelectron spectroscopy techniques, showed that the decrease in the HOR current at positive potentials in PILs can be attributed to the formation of surface oxides (Pt–O-type species), which can form due to the presence of trace water that persists in PILs, even after extensive drying [25].

While it is well known that underpotential deposition of hydrogen atoms occurs during potential cycling of Pt electrodes in aqueous solutions, and can play a

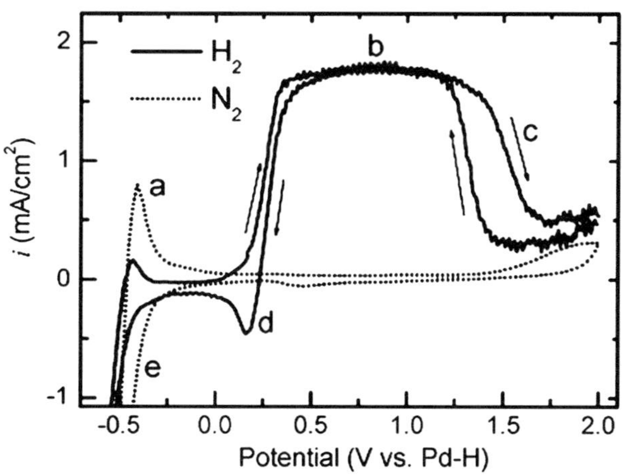

Fig. 17.6 Voltammograms recorded at activated Pt immersed in (*solid line*) H_2-saturated $[tea][CF_3SO_3]$ and (*dashed line*) the N_2-saturated PIL. Reprinted with permission from Bautista-Martinez, J A; Tang, L; Belieres, J P; Zeller, R; Angell, C A; Friesen, C (2009) J. Phys. Chem. C 113:12586-12593. Copyright 2009 American Chemical Society

significant role during electrocatalytic reactions [64], only recently has such behaviour been described in RTILs. Compton and co-workers first demonstrated that hydrogen atoms adsorb onto Pt surfaces during the voltammetry of benzoic acid in imidazolium-based aprotic RTILs [65]. The adsorption and desorption of hydrogen atoms during potential cycling of Pd electrodes in the PIL [dema][CF_3SO_3] has also been reported [66]. Ejigu and Walsh investigated the electrochemical adsorption of hydrogen atoms from PILs onto Pt electrodes using a rotating Pt disc electrode and demonstrated that hydrogen atom adsorption is sensitive to the nature of the PIL anions [51]. During potential cycling, adsorption of a monolayer of hydrogen atoms onto Pt was observed in [CF_3SO_3]-based PILs. This was confirmed by measuring the charge under the adsorbed hydrogen oxidation peak, which corresponds to that expected for oxidation of complete monolayer of H atoms. In contrast, no adsorption was detected in an $[N(Tf_2)]^-$-based PIL and this was attributed to the strong adsorption of the $[N(Tf_2)]^-$ anion on Pt surface as discussed in the previous section (and in the following section).

17.5 Overview of the Roles of Adsorbates During Electrocatalysis in RTILs

Surface adsorbates play a number of key roles during electrocatalysis in conventional aqueous electrolytes. For example (and as discussed previously), the large overpotential for the ORR in PEMFCs (and in acidic aqueous solutions) is attributed in large part to coverage of Pt electrocatalysts by adsorbed oxides and a significant amount of research has been devoted to tuning the Pt-oxygen binding strength of Pt alloys in order to speed up the rate of the ORR [19, 20, 59, 67, 68]. The OER at Pt occurs on oxide-covered Pt and the rate of the OER depends on the oxide layer thickness [69]. During the MOR in DMFCs, an adsorbed oxide layer is necessary for the complete oxidation of methanol to CO_2 [70]. It has been suggested that the MOR takes place *via* several steps producing adsorbed intermediates, the most recalcitrant of which is adsorbed CO (CO_{ads}). The complete oxidation of methanol to CO_2 is believed to involve a Langmuir-Hinshelwood mechanism, in which CO_{ads} reacts with surface-bound adsorbed oxides (often identified as OH_{ads}) to yield CO_2 [71–73]. The general processes can be represented as follows:

$$CH_3OH + Pt \rightarrow Pt - CO_{ads} + 4H^+ + 4e^- \quad (17.7)$$

$$Pt + H_2O \rightarrow Pt - OH_{ads} + H^+ + e^- \quad (17.8)$$

$$Pt - CO_{ads} + Pt - OH_{ads} \rightarrow 2Pt + CO_2 + H^+ + e^- \quad (17.9)$$

Net reaction

$$CH_3OH + H_2O \rightarrow CO_2 + 6H^+ + 6e^- \quad (17.10)$$

One should note that poisoning of PEMFC anode catalysts by CO is also a severe problem as CO is found to some extent in most H_2 gas supplies, as H_2 is usually produced by steam reforming of CH_4 (and CO is a by-product). It has been reported that a CO content as low as 10 ppm in H_2 fuel will result in the poisoning of Pt electrocatalysts [74]. As shown in Eqs. 17.8 and 17.9, the formation of OH_{ads} by water oxidation at the Pt surface is necessary for the oxidative removal of adsorbed CO. However, the formation of Pt–OH only occurs appreciably above 0.8 V vs. RHE [75]. This factor is considered to be the origin of the high overpotentials for the MOR and COOR and, often, a second metal that can provide oxide species at low potentials is added to Pt electrocatalysts to reduce such overpotentials. For example, Pt-based alloys containing elements such as Ru, Mo, W, and Sn have been used in attempts to speed up the electrocatalysis of methanol [70, 76, 77]. The Pt-Ru alloy (1:1 atomic ratio) is the most active binary catalyst and is most frequently used as the anode catalyst in DMFCs [78]. Ru is more easily oxidised than Pt and is able to form Ru-oxide adsorbates at ~0.2 V vs. RHE, thereby promoting the oxidation of CO to CO_2, as summarised in Eqs. 17.11–17.13:

$$CH_3OH + Pt \rightarrow Pt - CO_{ads} + 4H^+ + 4e^- \quad (17.11)$$

$$Ru + H_2O \rightarrow Ru - OH_{ads} + H^+ + e^- \quad (17.12)$$

$$Pt - CO_{ads} + Ru - OH_{ads} \rightarrow Pt + Ru + CO_2 + H^+ + e^- \quad (17.13)$$

17.5.1 The CO Oxidation and Methanol Oxidation Reactions in RTILs

Walsh et al. studied the COOR and MOR at Pt electrodes in the PIL [dema][CF_3SO_3] as a function of temperature [24]. The results showed that oxidation of trace H_2O provides the adsorbed oxygen species necessary for the COOR and MOR as illustrated by Eqs. 17.7–17.10. Due to the relative scarcity of water in the PIL, the overpotential for CH_3OH oxidation was drastically higher than that observed in aqueous electrolytes but decreased when the water content of the RTIL was increased. This study also suggested that Pt anodes in any future PIL-based intermediate temperature fuel cells will likely be poisoned by CO ([dema][CF_3SO_3] is currently the most promising PIL for fuel cells [12]), considering that a high overpotential is required to oxidise CO_{ads}. These results also indicate that developments in electrocatalysis of COOR and MOR may also benefit from the inclusion into the electrocatalyst of readily oxidised metals such as Ru to the Pt as illustrated by Eqs. 17.11–17.13.

Using various Pt single-crystal electrodes, Sanchez-Sanchez and co-workers recently showed that the rate of the COOR in RTILs depends on the nature of the RTIL anions [30]. They compared the COOR in [C_2mim][BF_4] and [C_2mim][N(Tf_2)] and suggested that the higher COOR onset potential in [C_2mim][N(Tf_2)] was due to adsorption of the [N(Tf_2)]$^-$ anions onto the Pt surface. They also speculated that the hydrophobicity of [C_2mim][N(Tf_2)] plays a role in slowing the rate of CO

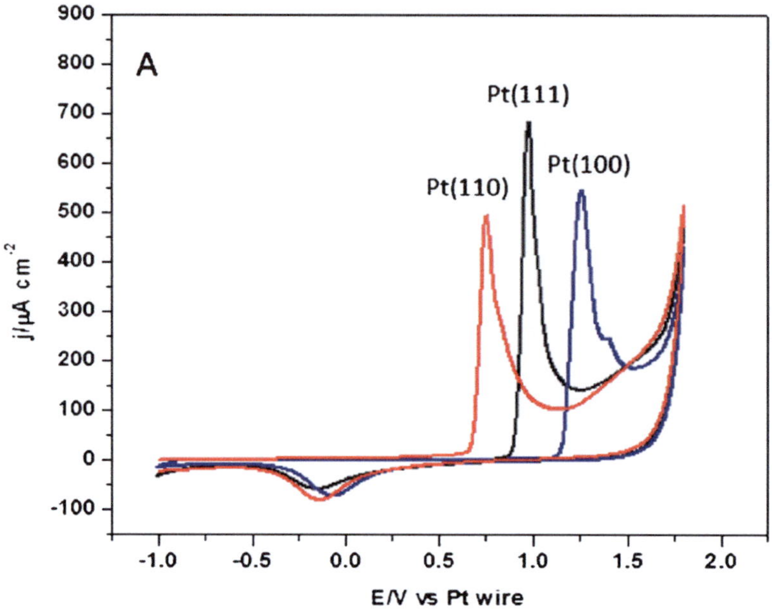

Fig. 17.7 Stripping voltammetry of CO adlayers on the three basal plane Pt electrodes in deaerated [C_2mim][BF_4]. The COOR overpotential is lowest at the Pt(110) surface and highest at the (100) surface. Reprinted with permission from Hanc-Scherer, F A; Sanchez-Sanchez, C M; Ilea, P; Herrero, E (2013) ACS Catal. 3:2935-2938. Copyright 2013 American Chemical Society

oxidation as less water is available for the nucleation of hydroxyl ions. Their work also indicated that the rate of the COOR at Pt in each RTIL depends on the crystallographic orientation of the electrode surface. The COOR was fastest at the Pt(110) facets and slowest at Pt(100) facets (see Fig. 17.7). The reactivity difference of CO was attributed to the differing strength of CO or oxide adsorption on the different surfaces.

Recently Ejigu and Walsh studied the COOR in PILs containing [CF_3SO_3]$^-$ and [N(Tf_2)]$^-$ anions [51]. In each PIL, CO was pre-adsorbed onto the Pt surface in a potential region where H_{ads} was expected to present on the Pt surface (as discussed in Sect. 17.4). An oxidation pre-peak was observed prior to the main CO oxidation peak in the [CF_3SO_3]$^-$-based PIL, and no pre-peak was observed in the PIL that contained the [N(Tf_2)]$^-$ anion (see Fig. 17.8 for a comparison of the data recorded in [dmea][CF_3SO_3], where [dmea] is the dimethylethylammonium cation, and [dema][N(Tf_2)]). The appearance of the COOR pre-peak in the [CF_3SO_3]$^-$-based liquids was explained as being due to the presence of H_{ads} on the Pt surface during adsorption of CO. Oxidation of these H_{ads} atoms at low potentials results in the generation of free Pt sites, which then facilitate the nucleation of hydroxyl species to promote early CO oxidation [64]. Due to the strong adsorption of the [N(Tf_2)]$^-$ anion on Pt, hydrogen atom adsorption is suppressed and hence no pre-peak was observed during CO oxidation.

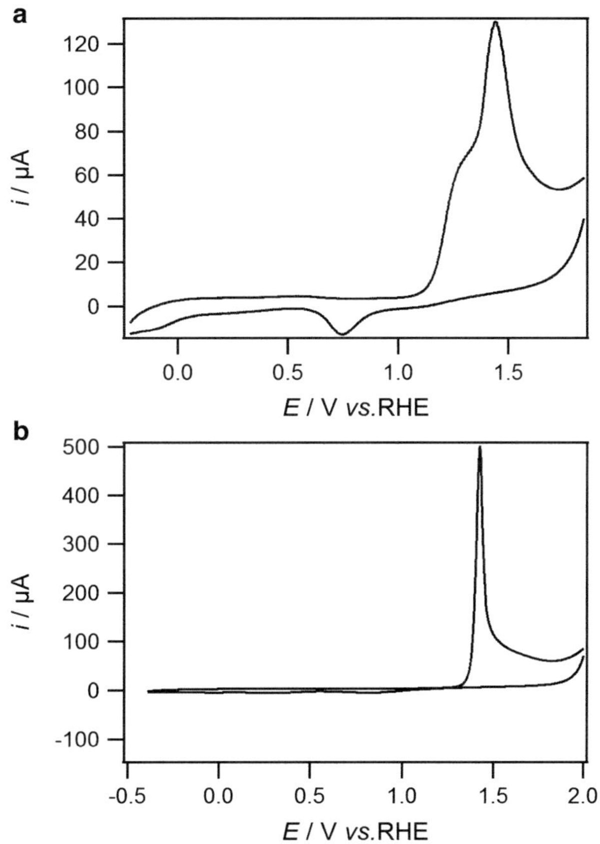

Fig. 17.8 Cyclic voltammograms recorded at a Pt electrode in (**a**) CO-saturated [dmea][CF_3SO_3] and (**b**) CO-saturated [dema][N(Tf_2)]. Prior to performing the scan from the starting potentials (at the negative potential limit), the Pt surface was held at 0.1 V for 40 min. CO oxidation occurred with a peak potential of about 1.5 V and a pre-peak was observed in [dmea][CF_3SO_3]. Adapted with permission from Ejigu, A; Walsh, D A (2014) J. Phys. Chem. C 118:7414-7422. Copyright 2014 American Chemical Society

Ejigu and Walsh also used the 'CO displacement' technique to determine whether the cations or anions of the PILs adsorbed onto Pt surface [51]. In this method, CO is introduced to an electrochemical cell while the working electrode potential is held at a given value. As no charge is considered to pass during CO adsorption onto Pt, any current that flows as CO reaches the Pt surface is due to the displacement of adsorbed ions. If a positive charge is measured during CO adsorption, then the species displaced by CO are adsorbed cations, as shown in Eq. 17.14. Similarly, the displaced species are considered to be anionic if a negative charge is measured during CO adsorption (Eq. 17.15).

$$\text{Pt} - \text{cation} + \text{CO} \rightarrow \text{Pt} - \text{CO} + (\text{cation})^+ + e^- \tag{17.14}$$

$$\text{Pt} - \text{anion} + \text{CO} + e^- \rightarrow \text{Pt} - \text{CO} + (\text{anion})^- \tag{17.15}$$

Use of this method allows access to the potential of zero total charge (pztc) by simply determining the potential at which the displacement charge density vs. potential curve crosses zero. The pztc is the potential at which the charge density on the electrode plus any charge transferred during ion adsorption is zero and differs from the potential of zero charge (pzc), which is the potential when the truly free, excess charge density on the electrode is zero [79]. Measurement of the pztc gives one a reference point at which surface of the electrodes changes from having a majority of adsorbed cations to a majority of anions (which is crucial to reactions such as the ORR, where adsorbed ions govern the reaction rate [80]). We should point out here that measurements of the potential of zero charge in RTILs are usually carried out by measuring the differential capacitance as a function of

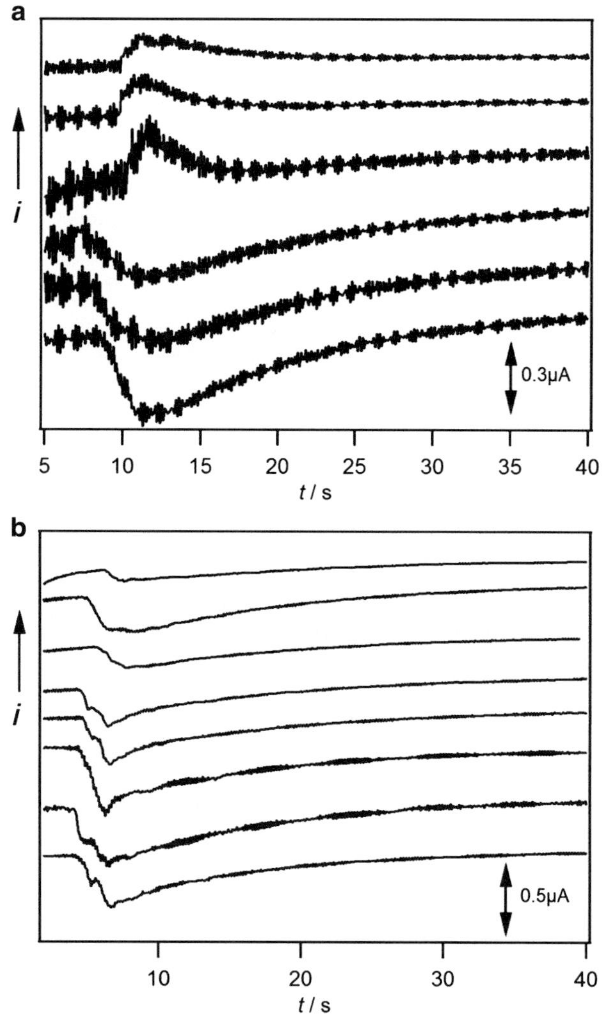

Fig. 17.9 Current-time transients recorded during CO displacement experiments at a Pt electrode in (**a**) [dema][CF_3SO_3] and (**b**) [dema][$N(Tf_2)$]. CO was introduced into the cell after the background current reached a negligible, steady-state value and the resulting transient current increase or decrease is shown. In (**a**), the applied potentials were (from *top* to *bottom*) 0.211, 0.241, 0.251, 0.281, 0.321 and 0.346 V. In (**b**), the applied potentials were (from *top* to *bottom*) −0.2, −0.16, −0.14, −0.1, −0.05, 0.0, 0.1, and 0.2 V. Reprinted with permission from Ejigu, A; Walsh, D A (2014) J. Phys. Chem. C 118:7414-7422. Copyright 2014 American Chemical Society

potential but complications arise when the cation and anion sizes are different and ions adsorb onto the electrode surface [81–84]. Figure 17.9 shows the current-time (*i–t*) curves recorded as CO was dosed into an electrochemical cell containing a Pt working electrode held at various potentials when immersed in the PILs [dema][CF_3SO_3] and [dema][N(Tf_2)] [51]. When the CO dosing potentials were $\leq$0.251 V in [dema][CF_3SO_3], the transient current increase was positive, indicating that the displaced ions were cations (Fig. 17.9a) and it was suggested that the species was H_{ads}. The transient current increased in the negative direction when CO was dosed at potentials $\geq$0.281 indicating that adsorbed [CF_3SO_3]$^-$ anions desorbed as CO reached the Pt. Using this method, the pztc of Pt in [dema][CF_3SO_3] and [dmea][CF_3SO_3] were determined to be 0.271 V and 0.289 V, respectively. In contrast, the CO displacement behaviour in [dema][N(Tf_2)] was quite different from that observed in [dema][CF_3SO_3]. Anion displacement was observed at all potentials studied (Fig. 17.9b), indicating that there was a strong interaction between Pt surfaces and the [N(Tf_2)]$^-$ anion. It was suggested that binding of the [N(Tf_2)] ions to the Pt surface was responsible for hindering both hydrogen adsorption and early onset of CO oxidation, and for the high overpotential of the ORR in [dema][N(Tf_2)] (as discussed in Sect. 17.3).

17.5.2 Hydrazine Oxidation in RTILs

Another notable electrocatalytic reaction where surface oxides take part is the electrooxidation of hydrazine (N_2H_4). The use of hydrazine as a fuel in the 'direct hydrazine fuel cell' has attracted research interest because the hydrazine/O_2 fuel cell exhibits a high open circuit potential of 1.61 V (as opposed to 1.23 V in H_2/O_2 fuel cell) and its theoretical energy-conversion efficiency ($\Delta G/\Delta H$) is 100 % (as opposed to 83 % in H_2/O_2 fuel cell) [85]. The electrooxidation of hydrazine occurs *via* a 4-electron oxidation process yielding molecular N_2 [86]:

$$N_2H_4 \rightarrow N_2 + 4H^+ + 4e^- \tag{17.16}$$

Pt is known to be the best electrocatalyst for hydrazine oxidation but, even at Pt, hydrazine oxidation exhibits a significant activation overpotential, which can vary by as much as 0.5 V depending on the electrode pretreatment [87]. Compton and co-workers described the role that surface oxides play during oxidation of hydrazine at Pt ultramicroelectrodes in aqueous media [86]. It was shown that hydrazine oxidation is activated significantly when oxides are present on the Pt surface. During cyclic voltammetry, hydrazine oxidation at Pt oxides (oxides that remained on the Pt surface from preceding voltammetric scans) occurred at potentials 400 mV more negative than at oxide-free Pt. The reactivity of Pt surface oxides formed in RTILs towards hydrazine oxidation was reported for the first time recently and Fig. 17.10 shows the data obtained in the PIL [dmea][CF_3SO_3] [88]. During the first potential cycle, hydrazine oxidation coincides with the onset of Pt

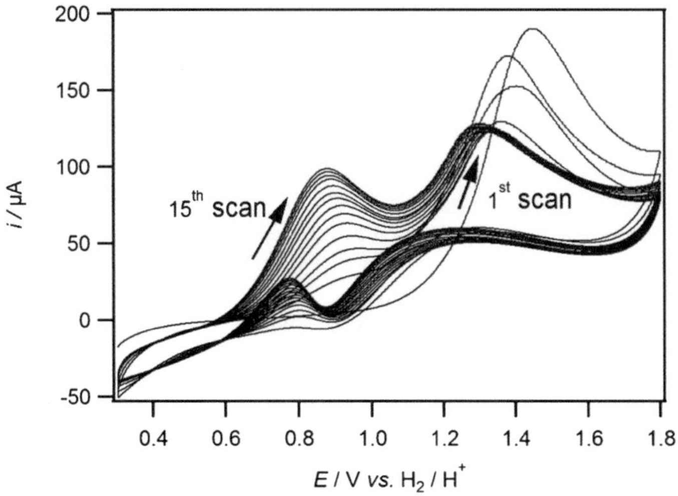

Fig. 17.10 Cyclic voltammograms recorded at a 5 mm diameter Pt disc electrode in 50 mM N_2H_4 in dimethylethylammonium trifluoromethanesulfonate at 45 °C. The potential of the working electrode was cycled between 0.3 V (initial potential) and 1.8 V. Reprinted with permission from Walsh, D A; Ejigu, A; Muhammad, S; Licence, P (2014) ChemElectroChem 1:281-288. Copyright 2014 Wiley VCH

oxide formation (1.0 V). However, during subsequent scans, oxides that persisted on the Pt surface activated the oxidation of hydrazine and reduced the hydrazine oxidation overpotential by ~0.5 V. This was demonstrated by the appearance of a new oxidation peak between 0.6 and 1.0 V that increased in height with successive scans. The increase in the hydrazine oxidation current during consecutive cycles was attributed to the increase in the amount of oxides on the surface due to the quasireversibility of oxide formation.

17.6 Concluding Remarks

The literature is replete with claims that RTILs will be used in more efficient and safer electrochemical energy-conversion devices. However, many of the studies in which such claims are made have only considered physicochemical properties of the RTILs and their effects on, e.g. mass transfer or electrochemical windows of RTILs. Others have built prototype devices and tested their performance but have not optimised the electrocatalyst composition and/or structure. It is our opinion that many of these proposed devices may only be realised if the kinetics of the electrochemical reactions occurring in such devices are studied with a view to optimising the electrodes' compositions; indeed, it is widely claimed that sluggish kinetics of the ORR in aqueous media are largely to blame for the slow uptake of conventional PEMFC technology. However, despite a relatively late start (relative to the broader field of electrochemistry in ionic liquids), it is highly encouraging that the field of electrocatalysis in ionic liquids is now developing at pace.

Much of the work done thus far has involved the use of Pt electrocatalysts and it should be clear from our discussion that surface adsorbates play crucial roles in a number of electrocatalytic reactions. Adsorbate (such as surface-oxide) effects in the 'traditional' field of electrocatalysis in aqueous media have usually been tackled by developing Pt-alloy electrocatalysts (for the HOR, ORR, MOR, and COOR) and it will be interesting to see how Pt-alloy (as well as non-Pt) electrocatalysts perform during electrocatalysis in RTILs. Compton has already made a step in this direction, comparing the activity of Pt electrocatalysts for the HER in RTILs to that of other metals. In the case of the ORR in RTILs, it will be interesting to explore whether Pt alloys are more active than Pt and whether a 'volcano' relationship between the electrocatalyst composition and activity can be identified, as it has for the ORR in aqueous media. In addition, given that the COOR and MOR coincide with oxidation of Pt surfaces, it may be natural to assume that inclusion of a readily oxidisable metal into the Pt electrocatalyst can aid in lowering the reaction overpotential but such work is yet to be done.

Finally, we wish to make a comment about the ions that constitute RTILs and how their adsorption onto electrocatalysts may affect electrocatalyst performance. In some respects, it is remarkable that the area of electrochemistry in RTILs has traditionally been divided into work addressing the structure and composition of the electrode/RTIL interface [89–91] and that addressing the effects of the physicochemical properties of RTILs on their electrochemistry. We believe that only a marriage of these areas of research into a new area called 'surface electrocatalysis in RTILs' will lead to a thorough understanding of electrocatalysis in RTILs. Some of the work described in this chapter is the first to probe this area of RTIL electrochemistry and has indicated that ion adsorption effects can be quite significant during electrocatalysis in RTILs. This result is perhaps not surprising considering that RTILs are composed entirely of ions and contain no solvent molecules to hinder access of ions to the electrocatalyst surface. We expect that, as the field of surface electrocatalysis in RTILs develops, it will lead not only to us increasing our understanding of RTIL electrochemistry and electrocatalysis in general, but may also lead to the realisation of next-generation electrochemical energy conversion devices that exploit the beneficial properties of RTILs.

References

1. Macfarlane DR, Forsyth M, Howlett PC, Pringle JM, Sun J, Annat G, Neil W, Izgorodina EI (2007) Ionic liquids in electrochemical devices and processes: managing interfacial electrochemistry. Acc Chem Res 40:1165–1173
2. Armand M, Endres F, MacFarlane DR, Ohno H, Scrosati B (2009) Ionic-liquid materials for the electrochemical challenges of the future. Nat Mater 8:621–629

3. MacFarlane DR, Tachikawa N, Forsyth M, Pringle JM, Howlett PC, Elliott GD, Davis JH Jr, Watanabe M, Simon P, Angell CA (2014) Energy applications of ionic liquids. Energy Environ Sci 7:232–250
4. Fedorov MV, Kornyshev AA (2014) Ionic liquids at electrified interfaces. Chem Rev 114:2978–3036
5. O'Mahony AM, Silvester DS, Aldous L, Hardacre C, Compton RG (2008) Effect of water on the electrochemical window and potential limits of room-temperature ionic liquids. J Chem Eng Data 53:2884–2891
6. Buzzeo MC, Evans RG, Compton RG (2004) Non-haloaluminate room-temperature ionic liquids in electrochemistry—a review. Chemphyschem 5:1106–1120
7. Lovelock KRJ, Cowling FN, Taylor AW, Licence P, Walsh DA (2010) Effect of viscosity on steady-state voltammetry and scanning electrochemical microscopy in room temperature ionic liquids. J Phys Chem B 114:4442–4450
8. Walsh DA, Lovelock KRJ, Licence P (2010) Ultramicroelectrode voltammetry and scanning electrochemical microscopy in room-temperature ionic liquid electrolytes. Chem Soc Rev 39:4185–4194
9. Zhao C, Burrell G, Torriero AAJ, Separovic F, Dunlop NF, MacFarlane DR, Bond AM (2008) Electrochemistry of room temperature protic ionic liquids. J Phys Chem B 112:6923–6936
10. Tsunashima K, Yonekawa F, Sugiya M (2009) Lithium secondary batteries using a lithium nickelate-based cathode and phosphonium ionic liquid electrolytes. Electrochem Solid-State Lett 12:A54–A57
11. Lewandowski A, Świderska-Mocek A (2009) Ionic liquids as electrolytes for li-ion batteries—an overview of electrochemical studies. J Power Sources 194:601–609
12. Lee SY, Ogawa A, Kanno M, Nakamoto H, Yasuda T, Watanabe M (2010) Nonhumidified intermediate temperature fuel cells using protic ionic liquids. J Am Chem Soc 132:9764–9773
13. Nakamoto H, Watanabe M (2007) Brønsted acid-base ionic liquids for fuel cell electrolytes. Chem Commun 19(24):2539–2541
14. Liu S, Zhou L, Wang P, Shao Z, Yi B (2013) Nonhumidified high temperature H_2/Cl_2 fuel cells using protic ionic liquids. J Mater Chem A 1:4423–4426
15. Liu S, Zhou L, Wang P, Zhang F, Yu S, Shao Z, Yi B (2014) Ionic-liquid-based proton conducting membranes for anhydrous H_2/Cl_2 fuel-cell applications. ACS Appl Mater Interfaces 6:3195–3200
16. Balducci A, Dugas R, Taberna PL, Simon P, Plee D, Mastragostino M, Passerini S (2007) High temperature carbon-carbon supercapacitor using ionic liquid as electrolyte. J Power Sources 165:922–927
17. Stephens IEL, Bondarenko AS, Gronbjerg U, Rossmeisl J, Chorkendorff I (2012) Understanding the electrocatalysis of oxygen reduction on platinum and its alloys. Energy Environ Sci 5:6744–6762
18. Nørskov JK, Rossmeisl J, Logadottir A, Lindqvist L, Kitchin JR, Bligaard T, Jonsson H (2004) Origin of the overpotential for oxygen reduction at a fuel-cell cathode. J Phys Chem B 108:17886–17892
19. Stamenkovic V, Mun BS, Mayrhofer KJJ, Ross PN, Marković NM, Rossmeisl J, Greeley J, Nørskov JK (2006) Changing the activity of electrocatalysts for oxygen reduction by tuning the surface electronic structure. Angew Chem Int Ed Engl 45:2897–2901
20. Stamenkovic VR, Fowler B, Mun BS, Wang GF, Ross PN, Lucas CA, Marković NM (2007) Improved oxygen reduction activity on $Pt_3Ni(111)$ via increased surface site availability. Science 315:493–497
21. Ren X, Zhang SS, Tran DT, Read J (2011) Oxygen reduction reaction catalyst on lithium/air battery discharge performance. J Mater Chem 21:10118–10125
22. Lu Y-C, Gasteiger HA, Shao-Horn Y (2011) Catalytic activity trends of oxygen reduction reaction for nonaqueous Li-Air batteries. J Am Chem Soc 133:19048–19051
23. Greeley J, Marković NM (2012) The road from animal electricity to green energy: combining experiment and theory in electrocatalysis. Energy Environ Sci 5:9246–9256

24. Ejigu A, Johnson L, Licence P, Walsh DA (2012) Electrocatalytic oxidation of methanol and carbon monoxide at platinum in protic ionic liquids. Electrochem Commun 23:122–124
25. Johnson L, Ejigu A, Licence P, Walsh DA (2012) Hydrogen oxidation and oxygen reduction in protic ionic liquids. J Phys Chem C 116:18048–18056
26. Khan A, Lu X, Aldous L, Zhao C (2013) Oxygen reduction reaction in room temperature protic ionic liquids. J Phys Chem C 117:18334–18342
27. Barnes AS, Rogers EI, Streeter I, Aldous L, Hardacre C, Wildgoose GG, Compton RG (2008) Unusual voltammetry of the reduction of O_2 in $[C_4dmim][N(Tf)_2]$ reveals a strong interaction of $O_2^{\bullet -}$ with the $[C_4dmim]^+$ cation. J Phys Chem C 112:13709–13715
28. Buzzeo MC, Klymenko OV, Wadhawan JD, Hardacre C, Seddon KR, Compton RG (2003) Voltammetry of oxygen in the room-temperature ionic liquids 1-ethyl-3-methylimidazolium bis((trifluoromethyl)sulfonyl)imide and hexyltriethylammonium bis((trifluoromethyl)sulfonyl)imide: one-electron reduction to form superoxide. Steady-state and transient behavior in the same cyclic voltammogram resulting from widely different diffusion coefficients of oxygen and superoxide. J Phys Chem A 107:8872–8878
29. Snyder J, Fujita T, Chen MW, Erlebacher J (2010) Oxygen reduction in nanoporous metal-ionic liquid composite electrocatalysts. Nat Mater 9:904–907
30. Hanc-Scherer FA, Sanchez-Sanchez CM, Ilea P, Herrero E (2013) Surface-sensitive electrooxidation of carbon monoxide in room temperature ionic liquids. ACS Catal 3:2935–2938
31. Lu X, Burrell G, Separovic F, Zhao C (2012) Electrochemistry of room temperature protic ionic liquids: a critical assessment for use as electrolytes in electrochemical applications. J Phys Chem B 116:9160–9170
32. Yoshizawa M, Xu W, Angell CA (2003) Ionic liquids by proton transfer: vapor pressure, conductivity, and the relevance of ΔpK_a from aqueous solutions. J Am Chem Soc 125:15411–15419
33. Bautista-Martinez JA, Tang L, Belieres JP, Zeller R, Angell CA, Friesen C (2009) Hydrogen redox in protic ionic liquids and a direct measurement of proton thermodynamics. J Phys Chem C 113:12586–12593
34. Miran MS, Kinoshita H, Yasuda T, Susan MABH, Watanabe M (2012) Physicochemical properties determined by ΔpK_a for protic ionic liquids based on an organic super-strong base with various Brønsted acids. Phys Chem Chem Phys 14:5178–5186
35. MacFarlane DR, Seddon KR (2007) Ionic liquids—progress on the fundamental issues. Aust J Chem 60:3–5
36. Angell CA, Byrne N, Belieres J-P (2007) Parallel developments in aprotic and protic ionic liquids: physical chemistry and applications. Acc Chem Res 40:1228–1236
37. Evans RG, Klymenko OV, Saddoughi SA, Hardacre C, Compton RG (2004) Electroreduction of oxygen in a series of room temperature ionic liquids composed of group 15-centered cations and anions. J Phys Chem B 108:7878–7886
38. Rollins JB, Conboy JC (2009) Kinetics and thermodynamics of hydrogen oxidation and oxygen reduction in hydrophobic room-temperature ionic liquids. J Electrochem Soc 156:B943–B954
39. Walsh DA, Ejigu A, Smith J, Licence P (2013) Kinetics and mechanism of oxygen reduction in a protic ionic liquid. Phys Chem Chem Phys 15:7548–7554
40. Switzer EE, Zeller R, Chen Q, Sieradzki K, Buttry DA, Friesen C (2013) Oxygen reduction reaction in ionic liquids: the addition of protic species. J Phys Chem C 117:8683–8690
41. Katayama Y, Onodera H, Yamagata M, Miura T (2004) Electrochemical reduction of oxygen in some hydrophobic room-temperature molten salt systems. J Electrochem Soc 151:A59–A63
42. Katayama Y, Sekiguchi K, Yamagata M, Miura T (2005) Electrochemical behavior of oxygen/superoxide ion couple in 1-butyl-1-methylpyrrolidinium bis(trifluoromethylsulfonyl) imide room-temperature molten salt. J Electrochem Soc 152:E247–E250
43. De Giorgio F, Soavi F, Mastragostino M (2011) Effect of lithium ions on oxygen reduction in ionic liquid-based electrolytes. Electrochem Commun 13:1090–1093

44. Monaco S, Arangio AM, Soavi F, Mastragostino M, Paillard E, Passerini S (2012) An electrochemical study of oxygen reduction in pyrrolidinium-based ionic liquids for lithium/oxygen batteries. Electrochim Acta 83:94–104
45. Allen CJ, Mukerjee S, Plichta EJ, Hendrickson MA, Abraham KM (2011) Oxygen electrode rechargeability in an ionic liquid for the Li-air battery. J Phys Chem Lett 2:2420–2424
46. Allen CJ, Hwang J, Kautz R, Mukerjee S, Plichta EJ, Hendrickson MA, Abraham KM (2012) Oxygen reduction reactions in ionic liquids and the formulation of a general ORR mechanism for Li–air batteries. J Phys Chem C 116:20755–20764
47. Kiatkittikul P, Yamaguchi J, Taniki R, Matsumoto K, Nohira T, Hagiwara R (2014) Influence of cationic structures on oxygen reduction reaction at Pt electrode in fluorohydrogenate ionic liquids. J Power Sources 266:193–197
48. Miran MS, Yasuda T, Susan MA, Dokko K, Watanabe M (2013) Electrochemical properties of protic ionic liquids: correlation between open circuit potential for H_2/O_2 cells under non-humidified conditions and ΔpK_a. RSC Adv 3:4141–4144
49. Rana UA, Forsyth M, MacFarlane DR, Pringle JM (2012) Toward protic ionic liquid and organic ionic plastic crystal electrolytes for fuel cells. Electrochim Acta 84:213–222
50. Marković NM, Ross PN (2002) Surface science studies of model fuel cell electrocatalysts. Surf Sci Rep 45:121–229
51. Ejigu A, Walsh DA (2014) The role of adsorbed ions during electrocatalysis in ionic liquids. J Phys Chem C 118:7414–7422
52. Munakata H, Tashita T, Haibara M, Kanamura K (2010) Synthesis of new protic ionic liquids for fuel cells on the basis of in-situ FT-IR measurements. ECS Trans 33:463–469
53. Subramanian NP, Greszler TA, Zhang J, Gu W, Makharia R (2012) Pt-oxide coverage-dependent oxygen reduction reaction (ORR) kinetics. J Electrochem Soc 159:B531–B540
54. Stamenkovic V, Schmidt TJ, Ross PN, Marković NM (2002) Surface composition effects in electrocatalysis: kinetics of oxygen reduction on well-defined Pt_3Ni and Pt_3Co alloy surfaces. J Phys Chem B 106:11970–11979
55. Marković N, Gasteiger H, Ross PN (1997) Kinetics of oxygen reduction on Pt(hkl) electrodes: implications for the crystallite size effect with supported Pt electrocatalysts. J Electrochem Soc 144:1591–1597
56. Murthi VS, Urian RC, Mukerjee S (2004) Oxygen reduction kinetics in low and medium temperature acid environment: correlation of water activation and surface properties in supported Pt and Pt alloy electrocatalysts. J Phys Chem B 108:11011–11023
57. Paik CH, Jarvi TD, O'Grady WE (2004) Extent of PEMFC cathode surface oxidation by oxygen and water measured by CV. Electrochem Solid-State Lett 7:A82–A84
58. Wang JX, Marković NM, Adzic RR (2004) Kinetic analysis of oxygen reduction on Pt(111) in acid solutions: intrinsic kinetic parameters and anion adsorption effects. J Phys Chem B 108:4127–4133
59. Stamenkovic VR, Mun BS, Arenz M, Mayrhofer KJJ, Lucas CA, Wang GF, Ross PN, Marković NM (2007) Trends in electrocatalysis on extended and nanoscale Pt-bimetallic alloy surfaces. Nat Mater 6:241–247
60. Silvester DS, Aldous L, Hardacre C, Compton RG (2007) An electrochemical study of the oxidation of hydrogen at platinum electrodes in several room temperature ionic liquids. J Phys Chem B 111:5000–5007
61. Silvester DS, Ward KR, Aldous L, Hardacre C, Compton RG (2008) The electrochemical oxidation of hydrogen at activated platinum electrodes in room temperature ionic liquids as solvents. J Electroanal Chem 618:53–60
62. Navarro-Suárez AM, Hidalgo-Acosta JC, Fadini L, Feliu JM, Suárez-Herrera MF (2011) Electrochemical oxidation of hydrogen on basal plane platinum electrodes in imidazolium ionic liquids. J Phys Chem C 115:11147–11155
63. Meng Y, Aldous L, Belding SR, Compton RG (2012) The hydrogen evolution reaction in a room temperature ionic liquid: mechanism and electrocatalyst trends. Phys Chem Chem Phys 14:5222–5228

64. Kucernak AR, Offer GJ (2008) The role of adsorbed hydroxyl species in the electrocatalytic carbon monoxide oxidation reaction on platinum. Phys Chem Chem Phys 10:3699–3711
65. Meng Y, Norman S, Hardacre C, Compton RG (2013) The electroreduction of benzoic acid: voltammetric observation of adsorbed hydrogen at a platinum microelectrode in room temperature ionic liquids. Phys Chem Chem Phys 15:2031–2036
66. Tremblay J, Nguyen NL, Rochefort D (2013) Hydrogen absorption by a palladium electrode from a protic ionic liquid at temperatures exceeding 100°C. Electrochem Commun 34:102–104
67. Tripkovic DV, Strmcnik D, van der Vliet D, Stamenkovic V, Marković NM (2008) The role of anions in surface electrochemistry. Faraday Discuss 140:25–40
68. Adzic RR, Zhang J, Sasaki K, Vukmirovic MB, Shao M, Wang JX, Nilekar AU, Mavrikakis M, Valerio JA, Uribe F (2007) Platinum monolayer fuel cell electrocatalysts. Top Catal 46:249–262
69. Reier T, Oezaslan M, Strasser P (2012) Electrocatalytic oxygen evolution reaction (OER) on Ru, Ir, and Pt catalysts: a comparative study of nanoparticles and bulk materials. ACS Catal 2:1765–1772
70. Gasteiger HA, Markovic N, Ross PN, Cairns EJ (1993) Methanol electrooxidation on well-characterized Pt-Ru alloys. J Phys Chem 97:12020–12029
71. Iwasita T (2002) Electrocatalysis of methanol oxidation. Electrochim Acta 47:3663–3674
72. Batista EA, Malpass GRP, Motheo AJ, Iwasita T (2004) New mechanistic aspects of methanol oxidation. J Electroanal Chem 571:273–282
73. Antolini E, Salgado JRC, Gonzalez ER (2006) The methanol oxidation reaction on platinum alloys with the first row transition metals: the case of Pt–Co and –Ni alloy electrocatalysts for DMFCs: a short review. Appl Catal Environ 63:137–149
74. Bose S, Kuila T, Nguyen TXH, Kim NH, Lau K-t, Lee JH (2011) Polymer membranes for high temperature proton exchange membrane fuel cell: recent advances and challenges. Prog Polym Sci 36:813–843
75. Koper MTM (ed) (2009) Fuel cell catalysis: a surface science approach. Wiley, Hoboken
76. Aricò AS, Srinivasan S, Antonucci V (2001) DMFCs: from fundamental aspects to technology development. Fuel Cells 1:133–161
77. Götz M, Wendt H (1998) Binary and ternary anode catalyst formulations including the elements W, Sn and Mo for PEMFCs operated on methanol or reformate gas. Electrochim Acta 43:3637–3644
78. Liu H, Song C, Zhang L, Zhang J, Wang H, Wilkinson DP (2006) A review of anode catalysis in the direct methanol fuel cell. J Power Sources 155:95–110
79. Frumkin AN, Petrii OA (1975) Potentials of zero total and zero free charge of platinum group metals. Electrochim Acta 20:347–359
80. Mayrhofer KJJ, Blizanac BB, Arenz M, Stamenkovic VR, Ross PN, Markovic NM (2005) The impact of geometric and surface electronic properties of Pt-catalysts on the particle size effect in electrocatalysis. J Phys Chem B 109:14433–14440
81. Lockett V, Sedev R, Ralston J, Horne M, Rodopoulos T (2008) Differential capacitance of the electrical double layer in imidazolium-based ionic liquids: influence of potential, cation size, and temperature. J Phys Chem C 112:7486–7495
82. Islam MM, Alam MT, Okajima T, Ohsaka T (2009) Electrical double layer structure in ionic liquids: an understanding of the unusual capacitance–potential curve at a nonmetallic electrode. J Phys Chem C 113:3386–3389
83. Fedorov MV, Kornyshev AA (2008) Towards understanding the structure and capacitance of electrical double layer in ionic liquids. Electrochim Acta 53:6835–6840
84. Fedorov MV, Kornyshev AA (2008) Ionic liquid near a charged wall: structure and capacitance of electrical double layer. J Phys Chem B 112:11868–11872
85. Asazawa K, Yamada K, Tanaka H, Oka A, Taniguchi M, Kobayashi T (2007) A platinum-free zero-carbon-emission easy fuelling direct hydrazine fuel cell for vehicles. Angew Chem Int Ed Engl 46:8024–8027

86. Aldous L, Compton RG (2011) The mechanism of hydrazine electro-oxidation revealed by platinum microelectrodes: role of residual oxides. Phys Chem Chem Phys 13:5279–5287
87. Bard AJ (1963) Chronopotentiometric oxidation of hydrazine at a Pt electrode. Anal Chem 35:1602–1607
88. Walsh DA, Ejigu A, Muhammad S, Licence P (2014) The formation and role of oxide layers on Pt during hydrazine oxidation in protic ionic liquids. ChemElectroChem 1:281–288
89. Islam MM, Alam MT, Ohsaka T (2008) Electrical double-layer structure in ionic liquids: a corroboration of the theoretical model by experimental results. J Phys Chem C 112:16568–16574
90. Kornyshev AA (2007) Double-layer in ionic liquids: paradigm change? J Phys Chem B 111:5545–5557
91. Aliaga C, Santos CS, Baldelli S (2007) Surface chemistry of room-temperature ionic liquids. Phys Chem Chem Phys 9:3683–3700

Chapter 18
Oxygen Reduction Reaction in Ionic Liquids: An Overview

Cristina Pozo-Gonzalo

18.1 Introduction

Due to the importance of electrochemical reduction of oxygen in life processes such as biological respiration and industrial applications including fuel cells or batteries, that reduction process and mechanism have been investigated over the last decades. The mechanism of the oxygen reduction reaction is highly complicated and dependent on a large number of factors, such as the electrode material, catalyst, solvent electrolyte and proton activity [1].

The reduction of oxygen was first studied in aqueous media; and its electrocatalysis has been widely investigated, and reviewed previously [1, 2]. In a general and simplified manner, the oxygen reduction reaction (ORR) in acidic and alkaline aqueous solutions occurs by three main pathways, leading to hydrogen peroxide or water as the final product. This implies that the electron-transfer steps are accompanied by homogeneous or heterogeneous reactions: (1) the direct 4-electron reduction pathway from O_2 to H_2O (Eqs. 18.1 and 18.4), (2) the peroxo route from O_2 to H_2O, generating hydrogen peroxide (H_2O_2) as an intermediate (Eqs. 18.2, 18.3 and 18.5, 18.6) and (3) the 2-electron reduction pathway from O_2 to hydrogen peroxide (H_2O_2) (Eqs. 18.2 and 18.5).

In basic media:

$$O_2 + 2H_2O + 4e^- \longrightarrow 4OH^- \quad (18.1)$$

$$O_2 + H_2O + 2e^- \longrightarrow HO_2^- + OH^- \quad (18.2)$$

$$HO_2^- + H_2O + 2e^- \longrightarrow 3OH^- \quad (18.3)$$

C. Pozo-Gonzalo (✉)
ARC Centre of Excellence for Electromaterials Science, IFM-Institute for Frontier Materials, Deakin University, 221 Burwood Hwy, Burwood, VIC 3125, Australia
e-mail: cpg@deakin.edu.au

A.A.J. Torriero (ed.), *Electrochemistry in Ionic Liquids*,
DOI 10.1007/978-3-319-15132-8_18

In acidic media:

$$O_2 + 4H^+ + 4e^- \longrightarrow 2H_2O \tag{18.4}$$

$$O_2 + 2H^+ + 2e^- \longrightarrow H_2O_2 \tag{18.5}$$

$$H_2O_2 + 2H^+ + 2e^- \longrightarrow 2H_2O \tag{18.6}$$

Song and Zhang collected detailed kinetic information and investigated the mechanisms for the ORR in aqueous media depending on the nature of the working electrode and catalysts [2].

The chemical reactions involved in the mechanisms in aqueous media are very fast, making it difficult to detect the intermediates and therefore, to evaluate the reaction mechanism in detail. The use of aprotic non-aqueous solvents will facilitate the understanding of the mechanism by stabilising intermediates such as the superoxide anions ($O_2^{\bullet -}$) [3].

Sawyer et al. have devoted their research to the understanding of oxygen and its electrogenerated species in non-aqueous aprotic solvents [1, 4–6]. They have also published a comprehensive review comparing thermodynamics and kinetics aspects of oxygen and its electrogenerated species in aqueous, and in non-aqueous, aprotic solvents [1].

Oxygen is reduced through a 1-electron reduction pathway from O_2 to superoxide anion ($O_2^{\bullet -}$) in a reversible manner in non-aqueous aprotic solvents as exemplified in Eq. 18.7.

$$O_2 + e^- \rightleftharpoons O_2^{\bullet -} \tag{18.7}$$

The superoxide anion is a highly reactive species because of its nucleophilic nature. For this reason non-aqueous aprotic solvents such as acetonitrile (ACN), dimethylformamide (DMF), dimethylsulfoxide (DMSO) and pyridine have been selected for studies involving the $O_2/O_2^{\bullet -}$ redox process due to the stabilisation of superoxide in such media [4].

Furthermore, the superoxide anion can react as a strong Brønsted base in the presence of a proton source (Brønsted acid). Therefore, the addition of a Brønsted acid (HA) to a non-aqueous aprotic solvent will bring a change in the electrochemical mechanism from an one-electron reversible reduction process (Eq. 18.7) to an irreversible two-electron reduction (Eqs. 18.7 and 18.8) process yielding to hydrogen peroxide as the final product.

$$O_2^{\bullet -} + 2H^+ + e^- \rightarrow H_2O_2 \tag{18.8}$$

However, Eq. 18.8 is not a straightforward reaction but a consequence of several chemical and electrochemical reactions, the first step being the protonation of the superoxide, as shown in Eq. 18.9.

$$O_2^{\bullet -} + HA \rightleftharpoons HO_2^{\bullet} + A^- \qquad (18.9)$$

The resulting $HO_2^{\bullet}$ from Eq. 18.9 may undergo various reactions in conventional organic solvents, depending on the solvent nature. Andrieux et al. [3], reported the kinetics of the mechanism of the reduction of oxygen in aprotic solvents and in the presence of protic additives. In summary, the hydroperoxyl, $HO_2^{\bullet}$, can undergo different mechanisms:

1. Reduction at the electrode surface
2. Reduction by $O_2^{\bullet -}$ in the solution
3. H-atom abstraction from the solvent
4. Disproportionation reaction

The disproportionation reaction (Eq. 18.10) has been for some time the most commonly accepted mechanism in aprotic non-aqueous media. However, in media such as DMSO, $HO_2^{\bullet}$ will follow Eq. 18.11 and a subsequent protonation step [3].

$$HO_2^{\bullet} + HO_2^{\bullet} \rightarrow H_2O_2 + O_2 \qquad (18.10)$$

$$O_2^{\bullet -} + HO_2^{\bullet} \rightarrow HO_2^{-} + O_2 \qquad (18.11)$$

Che et al. [7] performed a comprehensive analysis of the kinetic parameters of the superoxide anion in different aprotic solvents (e.g. DMSO, ACN and DMF) and in the presence of water. Evidence for a reaction between the superoxide anion, $O_2^{\bullet -}$, and the hydroperoxyl, $HO_2^{\bullet}$ (Eq. 18.11), was found rather than the disproportionation reaction of $HO_2^{\bullet}$ radicals (Eq. 18.10).

It has also been reported that the rate of disproportionation for $HO_2^{\bullet}$ depends on the electrolytic media (i.e. chlorobenzene > ACN > H_2O > DMF > DMSO) being slowest in the case of DMSO [1].

Ionic liquids (ILs) are defined as salt-like materials that are liquids below 100 °C. They are a promising type of electrolyte and depending on the cation–anion combination can offer many advantages such as chemical and electrochemical stability, low flammability, ionic conductivity or negligible vapour pressure [8].

The reduction of oxygen in ILs has been studied during the last 20 years as an alternative to the aqueous and non-aqueous media described above due to their outstanding properties and their potential application in energy storage devices.

Thus, the goal of this chapter is to compile the main findings in the reduction of oxygen from a fundamental point of view, and its reaction mechanisms within ILs. The chapter will be divided according to families of ionic liquids depicted in Scheme 18.1 (e.g. imidazolium, pyrrolidinium, quaternary ammonium and phosphonium) since differences in the physical properties, structure and stability of ionic liquids will affect the stability of the electrogenerated species of oxygen.

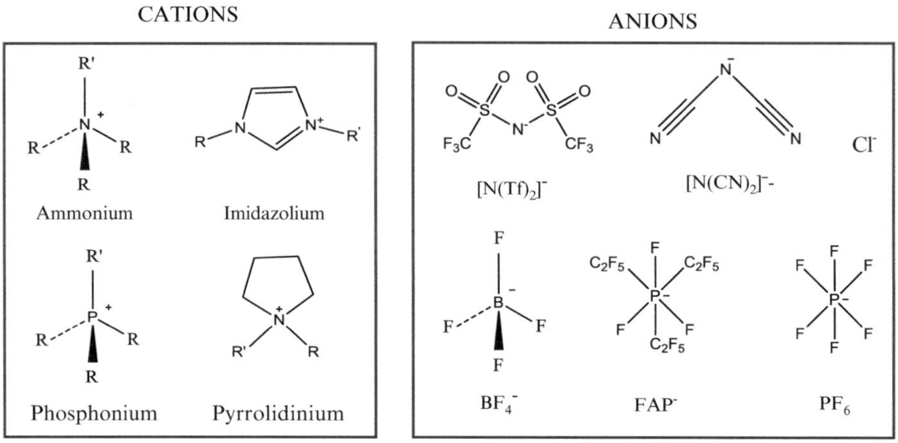

Scheme 18.1 Ionic liquid cations and anions that will be reviewed in this chapter

18.2 Mechanism in Imidazolium-Based Ionic Liquids

The oxygen reduction reaction (ORR) in ionic liquids (ILs) has been studied for more than two decades. In 1991, Carter et al. [9] reported, for the first time, the one-electron reduction of oxygen in 1-ethyl-3-methylimidazolium chloride [C_2mim]Cl (Scheme 18.2), mixed with aluminium chloride at a glassy-carbon disc electrode. The electrogenerated superoxide species, however, was unstable and no reverse oxidation process was observed in cyclic voltammetry at scan rates of 5–200 mV s^{-1} following the reduction of oxygen. This irreversibility was attributed to introduction of a proton source when the molten salt was saturated with O_2. Despite the efforts made to remove protic impurities, the superoxide could not be stabilised under these working conditions.

Interestingly, AlNashef et al. [10] reported the first evidence of the electrochemical generation of stable superoxide ions in the ionic liquid, 1-*n*-butyl-3-methylimidazolium hexafluorophosphate ([C_4mim][PF_6]) (Scheme 18.2). However, the attachment of a methyl group to the carbon in the 2-position ([C_4dmim][PF_6]) led to an irreversible reduction process (Fig. 18.1) which in this work was attributed to the presence of impurities in the IL. The activity of the superoxide anions towards imidazolium-based ILs will be addressed below.

Carter [9] and AlNashef [10] research works were focused on imidazolium-based ILs with similar structures (e.g. 1-*n*-alkyl-3-methylimidazolium), differing only in the length of the alkyl chain at the 1-position and the nature of the anion (Cl^- vs. PF_6^-), however, only the latest study showed reversibility of the $O_2/O_2^{\bullet -}$ redox couple. Therefore, these contradictory results indicated that a deeper knowledge of the influences of the cations and anions of the imidazolium-based ILs on the oxygen reduction reaction was still needed.

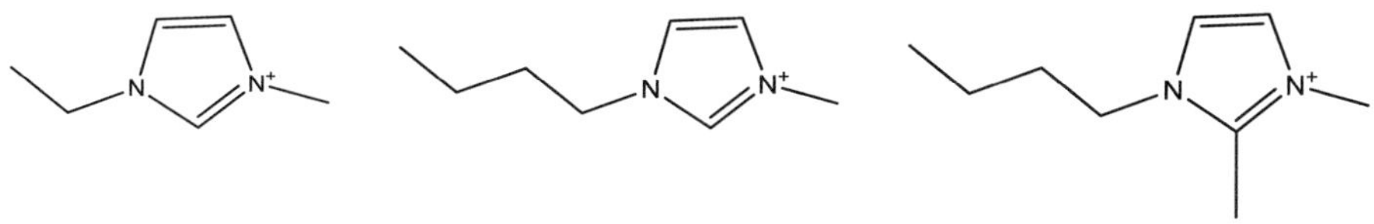

$[C_2mim]^+$ $[C_4mim]^+$ $[C_4dmim]^+$

Scheme 18.2 Series of the first imidazolium-based ionic liquids studied for oxygen reduction reaction in the literature

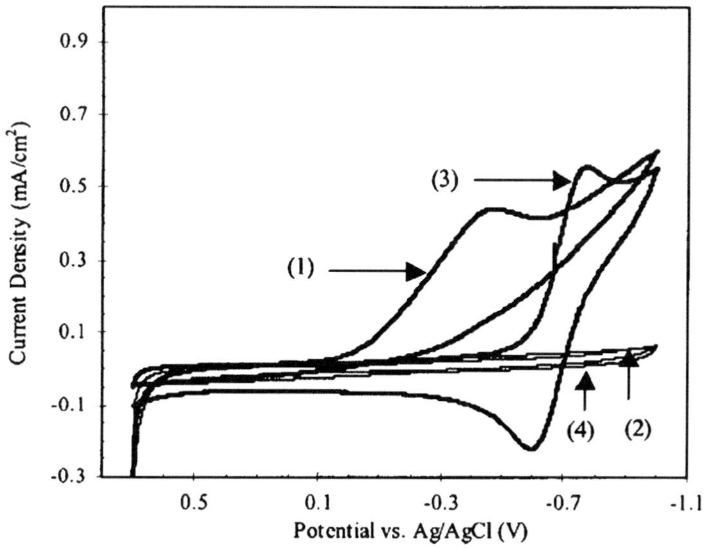

Fig. 18.1 Cyclic voltammogram of ($[C_4dmim][PF_6]$) (1) saturated with oxygen and (2) nitrogen and $[C_4mim][PF_6]$ (3) saturated with oxygen and (4) nitrogen. Scan rate: 100 mV s^{-1} (Reprinted from ref [10] with permission from the American Chemical Society)

The thermodynamic and kinetic parameters of the $O_2/O_2^{\bullet-}$ redox couple in a series of 1-*n*-alkyl-3-methylimidazolium tetrafluoroborate ILs was then studied by Zhang et al. [11] at glassy carbon (GC), gold (Au) and platinum (Pt) working electrodes. A quasireversible redox reaction corresponding to the generation of superoxide from oxygen via one-electron transfer was observed for the 1-*n*-alkyl-3-methylimidazolium tetrafluoroborates, regardless of the nature of the working electrode (Fig. 18.2). For the $O_2/O_2^{\bullet-}$ redox couple, the formal potential ($E^{0'}$) values were similar for all media and working electrodes examined, while the heterogeneous standard rate constant (k^0) depended on the nature of the working electrode, with $k^0(Pt) < k^0(Au) < k^0(GC)$ possible because of stronger interactions between the superoxide anion and the metal surface resulting in smaller k^0 values [11].

Katayama investigated the reversibility of the $O_2/O_2^{\bullet-}$ redox couple in terms of the ratio of the anodic and cathodic current densities, (j_p^a/j_p^c) in a series of aliphatic, aromatic and alicyclic bis(trifluoromethanesulfonyl)amide-based ILs

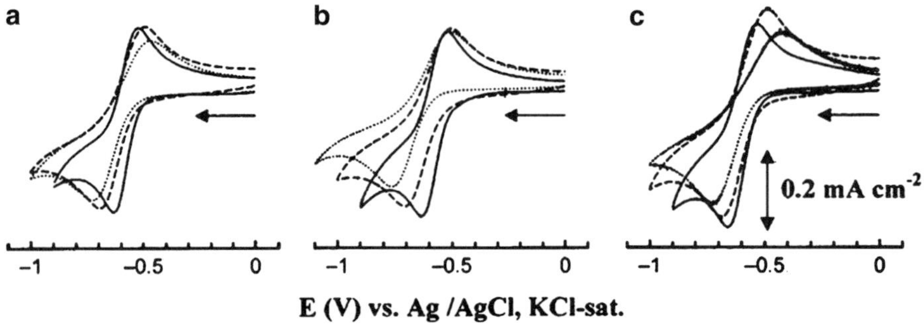

Fig. 18.2 Cyclic voltammogram obtained at GC (*solid line*), Au (*dash line*) and Pt (*dotted line*) electrodes in a O_2-saturated (**a**) $[C_2mim][BF_4]$ (**b**) $[C_3mim][BF_4]$ (**c**) $[C_4mim][BF_4]$. Scan rate 100 mV s^{-1} (Reprinted from ref [11] with permission from The Electrochemical Society)

(e.g. ammonium, imidazolium and pyrrolidinium-based ILs) [12]. A ratio of 0.48 was obtained in the 1-ethyl-3-methylimidazolium bis(trifluoromethanesulfonyl)amide $[C_2mim][N(Tf)_2]$, while 0.37 was obtained for 1,2-dimethyl-3-propylimidazolium bis(trifluoromethanesulfonyl)amide $[C_3dmim][N(Tf)_2]$, suggesting that there is an increased activity of superoxide anion towards imidazolium cations substituted in the carbon in the 2-position. Given the prior purification and distillation of the IL $[C_3dmim][N(Tf)_2]$, it is then plausible that the irreversibility of $O_2/O_2^{\bullet-}$ redox couple for the similar ionic liquid ($[C_4dmim][PF_6]$) previously mentioned, it is in fact due to the impurities present in the IL as assumed by AlNashef et al. [10].

These differences in electrochemical behaviour for O_2-saturated imidazolium-based IL exemplify the importance of highly pure ionic liquids and the possible side reactions between impurities and oxygen electrogenerated species.

Further information about the nature of the reactivity of the superoxide anion towards the imidazolium cation was attained by determining the electron distribution in organic cations such as imidazolium, pyrrolidinium and quaternary ammonium (Fig. 18.3) calculated by Gaussian98 with 6–31G/UHF. The atoms with highest positive charge will be susceptible to attack by the reactive superoxide anion.

Figure 18.3 shows that positive charges are located on the 2, 4 and 5 positions of the imidazolium rings in the case of $[C_2mim]^+$ and $[C_3dmim]^+$ and therefore, these positions are sites for attack by nucleophilic reagents. It is also notable that there are no positively charged carbon atoms in pyrrolidinium, C_4mpyr, and quaternary ammonium $[N_{2226}]^+$, cations, consistent with the stability of the superoxide in these ionic liquids, as observed by cyclic voltammetry (discussed further in Sects. 18.3 and 18.4). Therefore, it seems that $O_2^{\bullet-}$ is stable in the presence of both aliphatic $[N_{2226}]^+$, and alicyclic $[C_4mpyr]^+$ cations and unstable to some degree in the presence of aromatic imidazolium cations $[C_2mim]^+$ and $[C_3dmim]^+$ [12].

It is important to remark at this stage that the asymmetry in the heights of the forward and reverse peaks in the cyclic voltammograms measured under O_2-saturated conditions could be due partly to the reaction of the superoxide anion with imidazolium cation. However, other physical properties (e.g. viscosity,

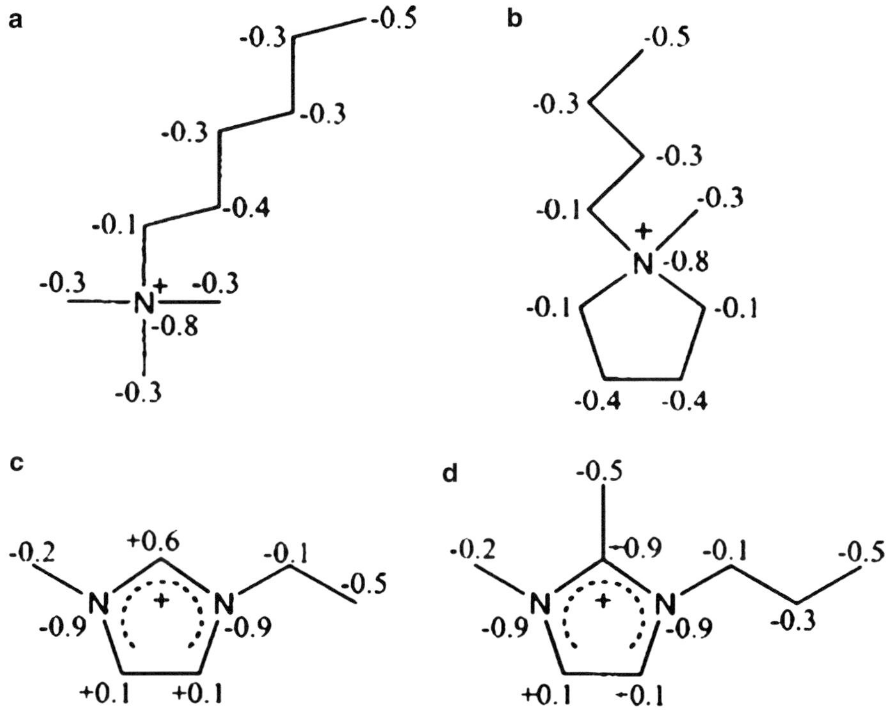

Fig. 18.3 Mulliken charges of carbon and nitrogen atoms in (**a**) $[N_{2226}]^+$, (**b**) $[C_4mpyr]^+$, (**c**) $[C_2mim]^+$ and (**d**) $[C_3dmim]^+$ calculated by Gaussian98 with 6–31G/UHF (Reprinted from ref [12] with permission from The Electrochemical Society)

conductivity) and transport properties (diffusion coefficient of oxygen and superoxide) could also have an important impact on the electrochemical behaviour. The influences of these parameters will be explained in more detail in the Sect. 18.3 when describing quaternary ammonium ILs.

Katayama et al. [12] were the first authors to study the reduction of superoxide anion, $O_2^{\bullet-}$, in a series of ILs (imidazolium, pyrrolidinium, and quaternary ammonium cations) with the hydrophobic anion, $[N(Tf)_2]^-$ which until now, it had only been reported in the presence of certain aprotic solvents, such as DMSO [13].

The bis(trifluoromethanesulfonyl)imide anion was selected due to the limited solubility of water which minimises the reaction between superoxide and water allowing superoxide anion to reduce further.

When analysing the cyclic voltammograms in Fig. 18.4 corresponding to $[C_2mim][N(Tf)_2]$, two reduction processes (C_1 and C_2) and one well-defined oxidation process (A_1) are observed. The chemically reversible process (C_1/A_1) is attributed the $O_2/O_2^{\bullet-}$ redox couple in good agreement with previous findings in the literature.

Interestingly, the current density of the oxidation wave (A_1) is dependent on the potential where the cathodic scan is reversed. For instance, when the potential is reversed at more negative values (e.g. ~−2.3 V) a significant decrease in the current

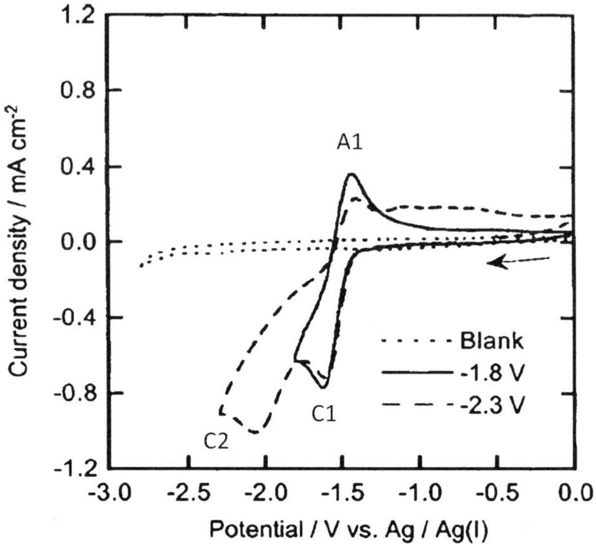

Fig. 18.4 Cyclic voltammograms of a gold electrode in $[C_2mim][N(Tf)_2]$ before and after saturation with oxygen at 25 °C. Scan rate 50 mV s^{-1} (Reprinted from ref [12] with permission from The Electrochemical Society)

density for the anodic process (A_1) is observed compared to the voltammogram obtained when reversing the potential at −1.8 V. Another interesting feature in the cyclic voltammograms in Fig. 18.4 is the lack of anodic distinct peaks observed at potentials more positive than −1.3 V.

The nature of the second reduction process at ~−2.1 V (Fig. 18.4) in the ORR in $[C_2mim][N(Tf)_2]$ which one may expect to be attributable to the $O_2^{\bullet-}/O_2^{2-}$ redox couple is in fact ascribed to the reduction of the ion-paired $[C_2mim]^+ \cdots O_2^{\bullet-}$ [12] as supported for the evidences abovementioned.

Islam et al. [13] also claimed the reduction of the ion-paired species (i.e. $[C_2mim]^+ \cdots O_2^{\bullet-}$), rather than the reduction of superoxide anions in the ORR of $[C_2mim][BF_4]$. This statement was corroborated by observations of the ORR in DMSO solutions upon addition of $[C_2mim][BF_4]$ which shifted the cathodic process to more positive potentials (Fig. 18.5).

It is worth to mention that a shift in the $E^{0'}$ value for a redox couple, in an aprotic solvents or an anhydrous ionic liquids, can be caused by changes in the polarity of the solvent, nature of the electrolyte or the presence of acidic additives; these changes are associated with nonspecific solvation energies, ion pairing and protonation equilibria, respectively [13, 14].

The reported positive shift in the ORR potential in a O_2-saturated DMSO solution upon increasing additions of 1-ethyl-3-methylimidazolium tetrafluoroborate, $[C_2mim][BF_4]$ reported Islam et al. was ascribed to an ion-pairing phenomenon which will corroborate the formation of the ion pairing $[C_2mim]^+ \cdots O_2^{\bullet-}$. Such phenomenon was more dominant than either the solvation or protonation, because the potential shift occurred for only one of the reduction processes, rather than in both processes.

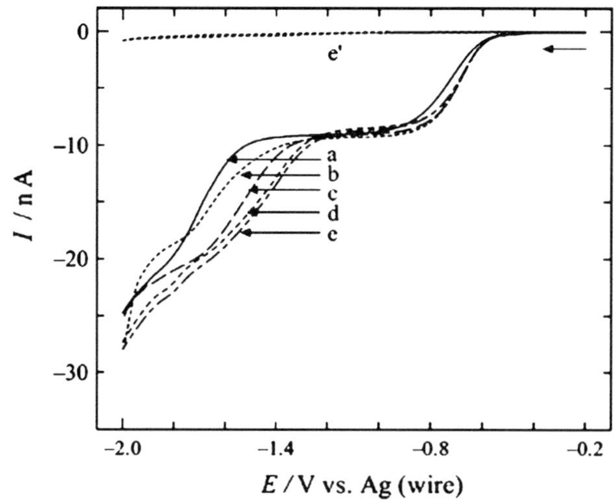

Fig. 18.5 Steady-state voltammograms obtained for the reduction of O_2 at the Au microelectrode (diameter = 10 μm) in DMSO solutions containing 0.1 M tetraethylammonium perchlorate in the absence (**a**) and presence of (**b**) 0.007, (**c**) 0.032, (**d**) 0.096, and (**e** and **e**′) 0.13 M [C_2mim][BF_4]. The voltammograms (**a–e**) and (**e**′) were measured in O_2 and N_2-saturated solutions, respectively. Scan rate 10 mV s^{-1} (Reprinted from ref [13] with permission from The American Chemical Society)

Finally, Barnes et al. [15] quantitatively simulated the reduction mechanism of oxygen in [C_4dmim][$N(Tf)_2$] IL in the presence and in the absence of a follow-up chemical reaction involving formation of $[C_4dmim]^+ \cdots O_2^{\bullet-}$ ion pairs (Eq. 18.12).

$$O_2^{\bullet-} + [C_4dmim]^+ \rightleftharpoons [C_4dmim]^+ \cdots O_2^{\bullet-} \quad (18.12)$$

An excellent fitting between the experimental and simulated voltammograms for the ORR was achieved when a follow-up homogeneous step (associated with the formation of the ion pairs between the imidazolium cation and the superoxide anion) was included.

18.2.1 Influence of Protic Additives on the ORR in Imidazolium-Based Ionic Liquids

The role of additives with different proton activities in the mechanisms of the ORR have been reported in the literature for several imidazolium-based ionic liquids, and they are in concordance with those of previously reported on aprotic solvents (e.g. DMF, DMSO, ACN) upon addition of protic additives [13]. As an example,

$$O_2 + e^- \rightleftharpoons O_2^{\bullet -} \quad (18.7)$$

$$O_2^{\bullet -} + HCl \rightleftharpoons HO_2^{\bullet} + Cl^- \quad (18.13)$$

$$2HO_2^{\bullet} \longrightarrow H_2O_2 + O_2 \quad (18.14)$$

Scheme 18.3 Proposed mechanism for the ORR in an imidazolium-based ionic liquid in the presence of a strong acid (e.g. HCl)

the addition of protic additives has a strong effect on the electrochemistry of oxygen reduction in terms of the onset potential and the mechanism of the ORR.

In the report by Carter et al. [9], the electrogenerated superoxide species was found to be unstable in the IL 1-ethyl-3-methylimidazolium chloride mixed with aluminium chloride. The irreversibility was attributed to the introduction of a proton source when the molten salt was saturated with O_2. Removal of the protons from the melt under vacuum was unsuccessful. Scheme 18.3 was proposed in that work to explain the detrimental effect of strong protic additives on the stability of the superoxide anion.

The mechanism depicted in Scheme 18.3 which involves the disproportionation of $HO_2^{\bullet}$ radicals was the more accepted one at that time [3]. Posteriorly, the role of a proton source in the oxygen reduction reaction was evaluated in a similar ionic liquid, $[C_2mim][BF_4]$ (Scheme 18.2), in the presence of 2.1 mM and 2.64 M of water [11]. The increase in water concentration modified the electrochemistry of the oxygen reduction reaction from a reversible reduction process corresponding to the $O_2/O_2^{\bullet -}$ redox couple to an irreversible cathodic process. In summary, the main features observed upon addition of water were: (1) an increase of the current density due to more favourable mass transport condition (increased fluidity and conductivity in the medium), (2) shift in potential for the reduction process to more positive values caused by changes to the protonation equilibria and the solvation of the electrogenerated species [13], and (3) loss of reversibility for the reduction process.

Therefore, it was concluded that superoxide anion is not stable in $[C_2mim][BF_4]$ containing 2.64 M of water, and the electron transfer involved in this mechanism is increased from a quasireversible 1-electron reduction process in the presence of 2.1 mM of water to an irreversible 2-electron reduction process in the presence of 2.64 M of water as depicted in Fig. 18.6.

These findings indicated that when water is present in aprotic solvents, it participates in the reduction of oxygen and it will be detrimental to the reversibility of the $O_2/O_2^{\bullet -}$ redox couple. The first step of the proposed mechanism starts with the reversible reduction of oxygen to superoxide (Eq. 18.7), followed by the protonation step (Eq. 18.15) and the subsequent disproportionation step (Eq. 18.11) (Scheme 18.4) [3, 7].

As mentioned before, the first ORR mechanism for ionic liquids reported in the literature was extrapolated from the literature on aprotic solvents. However, the role of protic additives in the ORR in the case of ILs was not fully understood leading to some discrepancies in the mechanisms depicted in Schemes 18.3 and 18.4 for similar imidazolium-based ILs.

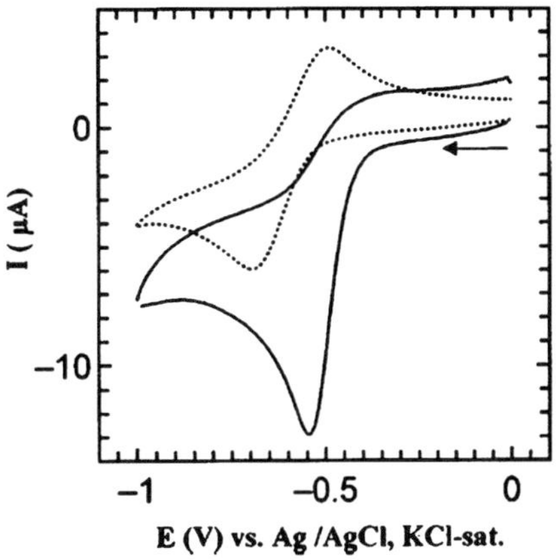

Fig. 18.6 CV of gold electrode in O_2-saturated $[C_2mim][BF_4]$ containing less than ca. 2.1 mM H_2O (*dotted line*) and 2.64 M H_2O (*solid line*). Scan rate 100 mV s^{-1} (Reprinted from ref [11] with permission from The Electrochemical Society)

$$O_2^{\bullet-} + H_2O \rightleftharpoons HO_2^{\bullet} + OH^- \quad (18.15)$$

$$O_2^{\bullet-} + HO_2^{\bullet} \longrightarrow O_2 + HO_2^- \quad (18.11)$$

Scheme 18.4 Proposed mechanism for the ORR in an imidazolium-based ionic liquid in the presence of water

In aprotic conventional organic solvents, it has been reported that the electrogenerated $HO_2^{\bullet}$ may undergo various reactions depending on the conditions. Andrieux et al. [3] reported the kinetics of the mechanism as a function of oxygen and acid concentration and nature. In summary, the $HO_2^{\bullet}$ can undergo:

1. Reduction at the electrode surface
2. Reduction by $O_2^{\bullet-}$ in the solution
3. H-atom abstraction from the solvent
4. Disproportionation reaction

For instance, the radical $HO_2^{\bullet}$ could either: (1) react with another $HO_2^{\bullet}$ as described by Chin et al. [16] in acetonitrile or dimethylformamide-based media in the presence of strong acids, or (2) react with another superoxide in dimethylsulfoxide in the presence of weak acids as reported by Andreiux et al. [3]. Therefore, the mechanism for the ORR in aprotic solvents in the presence of protic additives depends on the activity of the proton source [17].

René et al. [18] studied the role of different weak acids in $[C_4mim][N(Tf)_2]$ and compared those ORR mechanisms with those in DMF solutions. Under these conditions, the superoxide behaves as a base in the imidazolium-based ionic liquid and can be protonated by a weak acid, like phenol. The protonated species ($HO_2^{\bullet}$) is immediately reduced in solution by another superoxide as described by the Eq. 18.13.

More recently, Switzer et al. [17] reported the role of different protic additives (with pK_a values ranging across 30 units, from water to triflic acid) in $[C_4dmim][N(Tf)_2]$. It was concluded that the oxygen reduction performance can be tuned, depending on the proton activity of the additives. The effect of the addition of protic additives is somehow similar to those described by Zhang et al. [11] in $[C_2mim][BF_4]$ in the presence of large quantities of water. Enhanced performance of the oxygen reduction mechanism was observed in both the onset of the oxygen reduction process and the number of electrons exchanged from 1 electron in the neat aprotic $[C_4dmim][N(Tf)_2]$ to 4 electrons in the mixtures with protic additives.

18.3 Mechanism in Quaternary Ammonium-Based Ionic Liquids

The first reported work on the electrochemistry of the ORR in quaternary ammonium-based ionic liquids was in 2003 by Buzzeo et al. [19] who aimed to clarify the asymmetry observed in the cyclic voltammograms corresponding to the $O_2/O_2^{\bullet-}$ redox couple for some ionic liquids. The diffusion coefficients of oxygen and the electrogenerated superoxide anion ($O_2^{\bullet-}$) in $[N_{2226}][N(Tf)_2]$ (Scheme 18.5) were determined by chronoamperometric measurements at 20 °C and, interestingly, were found to differ by a factor of over 30 (e.g. $D_{O_2} = 1.48 \times 10^{-10} m^2 s^{-1}$, $D_{O_2^-} = 4.66 \times 10^{-12} m^2 s^{-1}$). The differences in diffusion coefficients for oxygen and superoxide anions could play in important role on the pronounced asymmetry in the heights of the forward and reverse peaks in the cyclic voltammograms measured under O_2-saturated conditions similar to those observed in Fig. 18.4.

To learn more about the stability of superoxide in the presence of quaternary-alkyl-ammonium cations, Evans et al. [20] performed a comprehensive study of the oxygen reduction reaction in several ILs containing highly fluorinated anions, including the quaternary ammonium ionic liquids $[N_{2226}][N(Tf)_2]$.

In their work, the concentration and diffusion coefficient of oxygen in $[N_{2226}][N(Tf)_2]$ was determined at 35 °C. Under these conditions, the diffusion coefficients of oxygen and superoxide anion differ by only one order of magnitude (e.g. $D_{O_2} = 3.2 \times 10^{-10} m^2 s^{-1}$, $D_{O_2^-} = 0.145 \times 10^{-10} m^2 s^{-1}$) [20]. These differences in diffusion coefficients were claimed to be related to the strength of the coulombic forces between the ions of the ionic liquids and either the neutral oxygen or the superoxide anion. Although it is well known that diffusion rates increase with temperature, this research work did not explain why the diffusion coefficients, D_{O_2} and $D_{O_2^-}$, became similar when the temperature was increased.

Scheme 18.5 Most common quaternary ammonium-based ionic liquid studied in oxygen reduction reactions

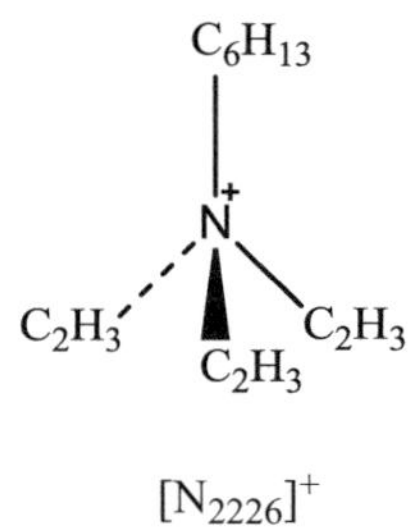

$[N_{2226}]^+$

Further research work on $[N_{2226}][N(Tf)_2]$ by Katayama et al. [12] using a Au macroelectrode showed an almost reversible $O_2/O_2^{\bullet -}$ reaction pair with a ratio anodic to cathodic current densities, (j_p^a/j_p^c) of 0.97 suggesting that $O_2^{\bullet -}$ is largely stable in this ionic liquid. It is important to mention that a stable salt comprising superoxide with ammonium cations, $[(CH_3)_4N][O_2]$ has previously been reported [5], and therefore the stability of superoxide anion in quaternary ammonium-based systems will be expected.

The Mulliken charges of carbon and nitrogen atoms depicted in Fig. 18.3 corroborated the far greater stability of the quaternary ammonium $[N_{2226}][N(Tf)_2]$ against the superoxide anion, compared with that of the imidazolium cations. This work concluded that the superoxide anion is more stable in the presence of aliphatic and alicyclic cations such as quaternary ammonium or pyrrolidinium than in the presence of aromatic rings, such as those found in imidazolium cations [12]. Furthermore, a second reduction process corresponding to the reduction of superoxide to peroxide was reported when the potential was swept to more negative values (Eq. 18.16). Until then, the reduction of superoxide to peroxide anions had only been reported in the presence of certain aprotic solvents, such as DMSO [13].

$$O_2^{\bullet -} + e^- \rightarrow O_2^{2-} \quad (18.16)$$

18.4 Mechanism in Pyrrolidinium-Based Ionic Liquids

Despite the electrochemical stability of pyrrolidinium-based ionic liquids and its stability towards superoxide [21], only a few research works on ORR mechanisms have been reported from a fundamental point of view. Different studies on the oxygen reduction in $[C_4mpyr][N(Tf)_2]$ demonstrated an electrochemically reversible one-electron process following Eq. 18.7 [20, 22].

Katayama et al. [12] reported the ORR of a neat $[C_4mpyr][N(Tf)_2]$ showing a reversible $O_2/O_2^{\bullet -}$ redox couple (C_1/A_1) with a ratio of the anodic to cathodic current densities, (j_p^a/j_p^c) of 0.93 similar to that described for the ammonium $[N_{2226}]^+$ analogue (Fig. 18.7). Furthermore, a second reduction process C_2 corresponding to the reduction of superoxide to peroxide anion was also observed when the potential is swept to more negative values (~−2.7 V) in a similar manner as in the imidazolium- or quaternary-ammonium ILs described in the previous sections.

Interestingly, in this case the reverse oxidation wave A_1 is independent of the end potential of the cathodic scan as opposed to the electrochemical behaviour in O_2-saturated imidazolium-based analogue IL. After scanning the cathodic sweep limit to more negative values (~−2.7 V) an additional oxidation process (A_2) is observed, which was attributed to the oxidation of peroxide anion to oxygen (Eq. 18.17).

$$O_2^{2-} \rightarrow O_2 + 2e^- \quad (18.17)$$

The oxygen and superoxide solubilities and diffusion coefficients in $[C_4mpyr]$ $[N(Tf)_2]$ were reported by Katayama et al. [23] and compared with a series of

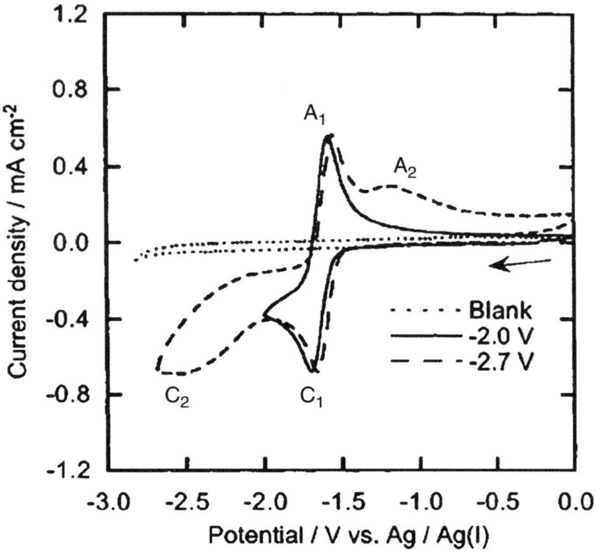

Fig. 18.7 Cyclic voltammogram of oxygen in $[C_4mpyr][N(Tf)_2]$ obtained with a gold electrode. Scan rate: 50 mV s^{-1} (Reprinted from ref [12] with permission from The Electrochemical Society)

ionic liquids. In a nutshell, the oxygen solubility in $[C_4mpyr][N(Tf)_2]$ (13.6 mM) is close to that in $[N_{2226}][N(Tf)_2]$ (8.9 mM) but generally higher than in some imidazolium-based ILs as exemplified in Table 18.1 [19].

18.5 Mechanism in Quaternary Phosphonium-Based Ionic Liquids

Evans et al. [20] studied the influence of highly fluorinated anions, $[N(Tf)_2]^-$ and $[FAP]^-$, on the oxygen reduction reaction for ILs containing trihexyl(tetradecyl) phosphonium cation $[P_{66614}]^+$ (Scheme 18.6). The oxygen reduction reaction studies were performed with a 10 μm diameter ultramicroelectrodes.

Figure 18.8 depicted the cyclic voltammograms in a O_2-saturated $[P_{66614}][FAP]$ using Pt and GC ultramicroelectrodes. The steady-state wave corresponding to the reduction process of oxygen is similar to those observed in quaternary ammonium and pyrrolidinium-containing ILs. However, the reverse wave corresponding to the oxidation process is broader and similar to those in aprotic solvents following the addition of a proton source attributed to an ECE mechanism.

Since, the ionic liquids under study were completely dried, the most likely proton source was claimed to originate from the phosphonium cation after α-proton abstraction by superoxide anion in analogy with the Witting alkene synthesis to yield a phosphorous ylide with the general structure $R_3P{=}CHR'$.

Scheme 18.6 Structure of trihexyl(tetradecyl) phosphonium cation

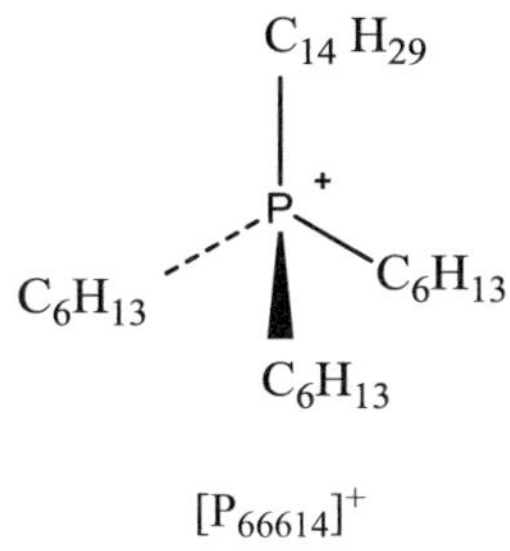

Fig. 18.8 Cyclic voltammograms showing the reduction of oxygen in $[P_{66614}][FAP]$ at Au and Pt electrodes (diameter = 10 μm). Scan rate: 500 mV s^{-1} (Reprinted from ref [20] with permission from The American Chemical Society)

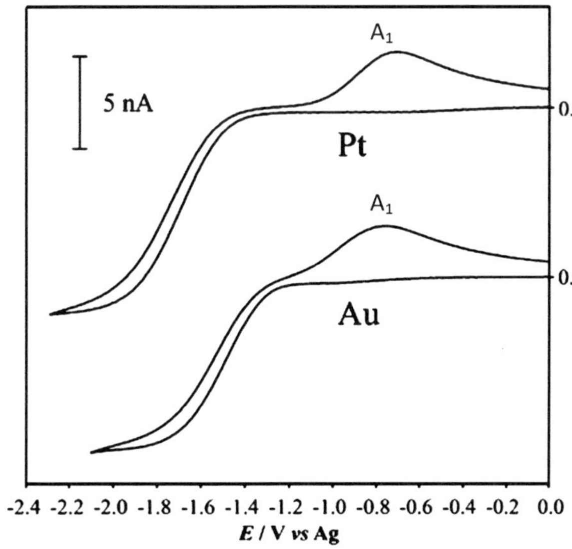

Therefore, a possible mechanism for quaternary phosphonium ILs was proposed, involving several electrochemical and chemical steps, in a similar manner to that described by Carter [9] for the ORR in 1-ethyl-3-methylimidazolium chloride mixed with aluminium chloride in the presence of hydrochloric acid. The first starting step involves the reduction of oxygen to yield superoxide as described in Eq. 18.7, followed by α-proton abstraction (Eq. 18.18) and subsequent disproportionation reaction of $HO_2^{\bullet}$ (Eq. 18.10).

$$O_2^{\bullet-} + \left[(C_6H_{13})_3PCH_2C_{13}H_{27}\right]^+ \rightarrow HO_2^{\bullet} + (C_6H_{13})_3P = CHC_{13}H_{27} \quad (18.18)$$

The nature of the oxidation peak (A_1) observed in Fig. 18.8 was hypothesised and several options were discussed (e.g. oxidation of phosphorous ylide, perhydroxyl radical or oxygen-derived species produced in follow-up reactions) [20].

In order to confirm the abovementioned theories a weak acid, 25 mM solution of catechol was added to the ammonium-based IL, $[N_{2226}][N(Tf)_2]$ expecting to reproduce the shape of the cyclic voltammogram depicted in Fig. 18.8 due to the similarities in composition between the ammonium and phosphonium-based ILs. Unfortunately, those attempts were unsuccessful.

Thus, it seems at that time that the electrochemistry of oxygen in quaternary phosphonium ILs is not straightforward, given the number of species involved and the unknown nature of those species. On the contrary, a few years later the generation of a stable superoxide was reported by Hayyan et al. [24] in $[P_{66614}][N(Tf)_2]$.

More recently, Pozo-Gonzalo et al. [25] reported, for the first time, a stable electrogenerated superoxide ion in the phosphonium-based ionic liquid, $[P_{66614}]Cl$, in the presence of large quantities of water (e.g. 1.5 and 4.5 wt%), leading to a chemically reversible $O_2/O_2^{\bullet-}$ redox couple (Fig. 18.9a), instead of the irreversible disproportionation reaction (Eq. 18.19) which is usually observed even in the presence of small quantities of water [26]. It is important to highlight that Eq. 18.19 is a consequence of several chemical and electrochemical equations.

$$2O_2^{\bullet-} + H_2O \rightarrow O_2 + HOO^- + OH^- \tag{18.19}$$

The stability of the superoxide anion under these conditions was attributed to the ion-pairing interactions between the superoxide anion and the phosphonium cation. This interaction has been further analysed by computational quantum chemistry, which indicated a shorter intermolecular distance and higher bond angle (i.e. $r = 3.128$ Å, $\theta = 106.8°$) between the superoxide anion and the phosphonium cation, in comparison with those of the quaternary ammonium cation (i.e. $r = 3.434$ Å, $\theta = 109.8°$) known for its superior stabilisation towards superoxide, as described in the Sect. 18.3. Furthermore, NMR experiments have been performed to assess the degree of long-term degradation of $[P_{66614}]^+$, in the presence of superoxide and peroxide species showing no chemically distinct degradation products [27].

Figure 18.9b depicted the electrochemical behaviour of the O_2-saturated $[P_{66614}]$ [Cl] in the presence of 1.5 wt% of water. When the potential was scanned up to -1.9 V, two cathodic (C_1, C_2) and two anodic (A_1, A_2) processes were observed. A tentative mechanism was established for the ORR in phosphonium-based ILs according to previous research works on mechanisms of the reduction of oxygen in the presence of weak proton additives and water [3, 18] as already exemplified in Scheme 18.4.

$$O_2 + 1e^- \rightleftharpoons O_2^{\bullet-} \tag{18.7}$$

$$O_2^{\bullet-} + H_2O \rightleftharpoons HO_2^{\bullet} + OH^- \tag{18.15}$$

$$HO_2^{\bullet} + O_2^{\bullet-} \rightleftharpoons HO_2^- + O_2 \tag{18.13}$$

$$HO_2^- + H_2O \rightarrow H_2O_2 + OH^- \tag{18.20}$$

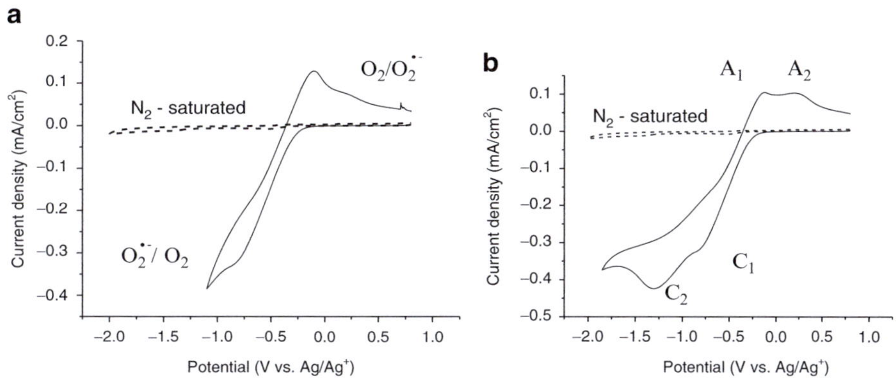

Fig. 18.9 Cyclic voltammograms corresponding to the oxygen reduction reaction using a glassy carbon in a N_2 (*dotted line*) and O_2-saturated (*solid line*) $[P_{66614}][Cl]$ in the presence of 1.5 wt% of water (**a**) scanned up to −1 V and (**b**) scanned up to −1.9 V. Scan rate: 10 mV s^{-1}

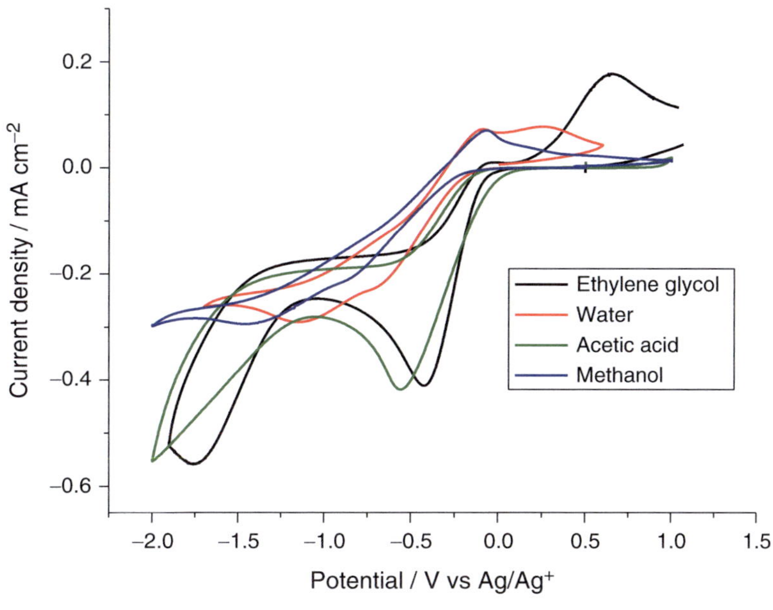

Fig. 18.10 Cyclic voltammograms for ORR on GC in O_2-saturated $[P_{66614}][Cl]$ containing 2.5 mmol additive/g IL. Scan rate: 10 mV s^{-1}

The roles of proton donors of similar proton activities (e.g. water, methanol and ethylene glycol), as well as weak acid in a phosphonium-based IL $[P_{66614}][Cl]$, was also studied from the ORR perspective [28].

$[P_{66614}][Cl]$/ethylene glycol mixtures showed better ORR performance, in terms of onset potential and current density, than mixtures containing water or methanol (Fig. 18.10). This enhancement in performance, similar to that produced by the

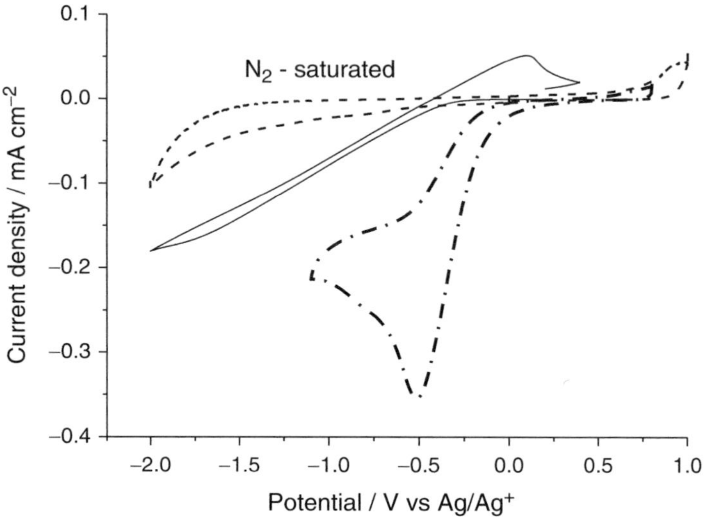

Fig. 18.11 Cyclic voltammograms of the O_2-saturated $[P_{66614}]Cl$ in the absence (*solid line*) and in the presence of 1.8 wt% HCl (*dash dotted line*). N_2-saturated $[P_{6,6,6,14}]Cl$ in the presence of 1.8 wt % HCl (*dashed line*)

weak acid, was attributed to the self-stabilisation of ethylene glycol after proton removal. It seems that the proton activity of water in $[P_{66614}][Cl]$ is inferior than that of ethylene glycol in the same IL which was corroborated by experimental calculation of the hydrogen-bond donating ability using the Kamlet-Taft parameters [29, 30].

Interestingly, the oxygen reduction reaction in $[P_{66614}][Cl]$/ethylene glycol mixture using a glassy carbon working electrode follows an overall 4-electron pathway (i.e. two 2-electron processes) which is of importance in the performance of air cathodes in electrochemical devices.

Thus, the role of strong acidic moieties in the stability of the $O_2/O_2^{\bullet-}$ redox couple in $[P_{66614}]Cl$ has been studied via addition of hydrochloric acid, (a common side product in phosphonium-based ILs synthesis) [31] by cyclic voltammetry. Figure 18.11 depicts that in the presence of HCl the reduction process for oxygen is irreversible [28] as opposed to the pure $[P_{66614}]Cl$ where chemical reversibility of the electrogenerated superoxide anion is observed which exemplify of the detrimental effect of HCl on the reversibility of the O_2 reduction process.

Therefore, it seems plausible that the instability of the superoxide anion $O_2^{\bullet-}$ in $[P_{66614}]^+$, that has previously been observed by other researchers [20], could be related to the presence of residual acidic impurities (Fig. 18.11).

18.6 Calculation of Diffusion and Concentration of Oxygen in Ionic Liquids

The diffusion and concentration of oxygen are important parameters from the oxygen reduction reaction point of view. Thus, the calculation of these parameters in ionic liquids can be obtained using different methodologies reported in the literature.

1. Calculation by Chronoamperometry using an ultramicroelectrode based on the time-dependent current response, *i*, following a potential step [19]. Theoretical transients are calculated according to the Aoki model which is given by the expression:

$$i = -4nFDcrf(\tau)$$

where n is the number of electrons transferred, F is the Faraday constant (96,485 C mol^{-1}), r is the radius of the microdisc, D is the diffusion coefficient of the electroactive species, c is the concentration of the electroactive species, and the dimensionless parameter, τ, is defined as $\tau = 4Dt/{r_e}^2$.

2. The diffusion coefficient and concentration of oxygen in neat ionic liquids can be determined by cyclic voltammetry using an ultramicroelectrode and chronoamperometry using a macroelectrode [23]. The steady-state current, i_{ss}, with an ultramicroelectrode having radius r_0 is given by the following equation:

$$i_{ss} = 4nFDCr_0 \tag{18.21}$$

The current density is proportional to the square root of time as represented by Cottrell's equation:

$$j(t) = \frac{nFCD^{1/2}}{\pi^{1/2}t^{1/2}} \tag{18.22}$$

Therefore, the diffusion coefficient (D) and solubility of oxygen (C) can be calculated from the DC and $D^{1/2}\,C$ obtained from Eqs. 18.21 and 18.22.

3. Hydrodynamic chronocoulometry (HCC) can be used to simultaneously determine the diffusion coefficient (D) and solubility of oxygen (C) [11]. The relationship between the charge (Q) and time (t) is given in the following equations;

$$Q = Q_\delta + I_L t \tag{18.23}$$

where

$I_L = nFADC/\delta$
$Q_\delta = 0.3764\ nFADC\delta$
$\delta = 1.610D^{1/3}\,\upsilon^{1/6}\,\omega^{-1/2}\,(1 + 0.2980\ \mathrm{Sc}^{-1/3} + 0.14514\ \mathrm{Sc}^{-2/3})$

where I_L is the Levich limiting current, Q_δ is the charge passed in the electrolysis of O_2 contained initially in the hydrodynamic boundary layer, which has the thickness δ, n is the number of electrons involved in the electrode reaction, F is the Faraday constant, A is the electrode surface area, and υ is the kinematic viscosity of the solution. Sc is the Schmidt number ($Sc = \upsilon/D$), and ω is the electrode angular velocity ($\omega = 2\pi f$, where f is the frequency of the electrode rotation).

A summary of the solubility and diffusion coefficient of oxygen, and diffusion coefficient of superoxide reported in the literature is included in Tables 18.1 and 18.2. It is important to highlight that the solubility of oxygen in water is lower than in organic solvents and that is due to the nonpolar character of the Dioxygen [1].

Table 18.1 Diffusion coefficient and concentration of oxygen in ionic liquid

System	$10^6\, D_{O_2}$ ($cm^2\ s^{-1}$)	C ($mol\ cm^{-3}$)	η (mPa s)	Temperature (°C)	References
Imidazolium					
$[C_2mim][N(Tf)_2]$	7.3	3.9	32.1	20	[32]
$[C_2mim]BF_4$	17	1.1	37	25	[11]
$[C_3mim]BF_4$	13	0.97	103	25	[11]
$[C_4mim]BF_4$	12	1.1	180	25	[11]
$[C_4mim]PF_6$	2.2	3.6	312	25	[10]
Quaternary ammonium					
$[N_{2226}][N(Tf)_2]$	1.48	11.6	220	20	[19, 32]
$[N_{2226}][N(Tf)_2]$	3.2	8.9	220[a]	35	[20]
Pyrrolidinium					
$[C_4mpyr][N(Tf)_2]$	1.8	13.6	89	25	[23]
$[C_4mpyr][N(Tf)_2]$	5.2	6.1	95[a]	35	[20]
Quaternary phosphonium					
$[P_{66614}][N(Tf)_2]$	7.5	6.0	–	35	[20]
$[P_{66614}]FAP$	6.1	7.8	–	35	[20]
$[P_{66614}]TPTP$	9.5	8.0	–	25	[33]
$[P_{66614}][N(Tf)_2]$	1.7	9.8	369	20	[27]
$[P_{66614}][N(CN)_2]^-$	2.1	9.1	474	20	[27]

[a]Measured at 20 °C

Table 18.2 Diffusion coefficient of superoxide anion in ionic liquid

System	$10^8\, D_{O_2^{\bullet-}}$ ($cm^2\ s^{-1}$)	Temperature (°C)	References
$[C_2mim][N(Tf)_2]$	2.7	20	[32]
$[C_2mim][N(Tf)_2]$	2.7	20	[32]
$[N_{2226}][N(Tf)_2]$	4.66	20	[19]
$[N_{2226}][N(Tf)_2]$	14.5	35	[20]
$[C_4mpyr][N(Tf)_2]$	0.86	25	[23]
$[C_4mpyr][N(Tf)_2]$	0.34	35	[20]

18.7 Conclusions

Electrocatalytic oxygen reduction reactions (ORR) in a wide range of room temperature ionic liquid systems which encompass imidazolium, pyrrolidinium, ammonium, and phosphonium-based ILs have been summarised in this chapter with the goal to compile the state of the art. Early researches on the ORR in ILs have mainly focused on the one reduction process of O_2 leading to the superoxide anion ($O_2^{\bullet -}$) as the final product, and one of the major conclusions extracted from those is that most of the ionic liquids studied are able to support oxygen reduction reaction to some degree under dry conditions and in the absence of protic moieties. Detailed studies on the reversibility of the $O_2/O_2^{\bullet -}$ redox couple in terms of the ratio of the anodic and cathodic current densities (j_p^a/j_p^c) revealed that superoxide is more stable in aliphatic and alicyclic quaternary ammonium-based ILs than in imidazolium-based ILs. Theoretical calculations pointed out the positive charge located at the 2-position of the imidazolium rings as the reason for inferior stability towards superoxide anion, as that carbon is prone to nucleophilic attack by the superoxide anions.

Thus, the electrochemical stability of aliphatic and alicyclic quaternary ammonium-based ILs, together with the enhanced stability towards superoxide anion in comparison with imidazolium-based ionic liquids, make them good candidates for practical applications.

On the other hand, phosphonium-based ILs have not been explored in much detail despite the reversibility observed in ionic liquids such as [P_{66614}][Cl] even in the presence of large quantities of water.

The influence of protic additives in the ORR has been extensively studied in numerous ionic liquids and it is normally detrimental to the stability of the superoxide anion under such conditions. In fact, early research work in the literature reported the instability of superoxide in imidazolium-based ionic liquid which was in fact due to the acidic moieties present in such ILs.

However, recently the addition of protic additives has been explored to enhance the electrocatalytic performance of the ORR in relation with the onset reduction potential, current density and number of electrons exchanged during the process, which is important from an applied point of view. This outcome has an important impact on the design of electrolytes for energy storage devices and it opens up new research venues to explore.

Acknowledgement The author gratefully acknowledges financial support from the Australian Research Council (ARC) through the ARC Centre of Excellence for Electromaterials Science (ACES).

References

1. Sawyer DT, Sobkowiak A, Roberts JL (1995) Electrochemistry for chemists. Wiley, New York
2. Song C, Zhang J (2008) Electrocatalytic oxygen reduction reaction. In: Zhang J (ed) PEM fuel cell electrocatalysts and catalyst layers. Springer, London, pp 89–134
3. Andrieux CP, Hapiot P, Saveant JM (1987) Mechanism of superoxide ion disproportionation in aprotic-solvents. J Am Chem Soc 109(12):3768–3775
4. Sawyer DT, Chiericato G, Angelis CT, Nanni EJ, Tsuchiya T (1982) Effects of media and electrode materials on the electrochemical reduction of dioxygen. Anal Chem 54(11):1720–1724
5. Sawyer DT, Valentine JS (1981) How super is superoxide? Acc Chem Res 14(12):393–400
6. Merritt MV, Sawyer DT (1970) Electrochemical studies of the reactivity of superoxide ion with several alkyl halides in dimethyl sulfoxide. J Org Chem 35(7):2157–2159
7. Che Y, Tsushima M, Matsumoto F, Okajima T, Tokuda K, Ohsaka T (1996) Water-induced disproportionation of superoxide ion in aprotic solvents. J Phys Chem 100(51):20134–20137
8. MacFarlane DR, Tachikawa N, Forsyth M, Pringle JM, Howlett PC, Elliott GD, Davis JH, Watanabe M, Simon P, Angell CA (2014) Energy applications of ionic liquids. Energy Environ Sci 7(1):232–250
9. Carter MT, Hussey CL (1991) Electrochemical reduction of dioxygen in room-temperature imidazolium chloride-aluminium chloride molten salts. Inorg Chem 30:1149–1151
10. AlNashef IM, Leonard ML, Kittle MC, Matthews MA, Weidner JW (2001) Electrochemical generation of superoxide in room-temperature ionic liquids. Electrochem Solid-State Lett 4(11):D16–D18
11. Zhang D, Okajima T, Matsumoto F, Ohsaka T (2004) Electroreduction of dioxygen in 1-n-alkyl-3-methylimidazolium tetrafluoroborate room-temperature ionic liquids. J Electrochem Soc 151(4):D31–D37
12. Katayama Y, Onodera H, Yamagata M, Miura T (2004) Electrochemical reduction of oxygen in some hydrophobic room-temperature molten salt systems. J Electrochem Soc 151(1):A59–A63
13. Islam MM, Ohsaka T (2008) Roles of ion pairing on electroreduction of dioxygen in imidazolium-cation-based room-temperature ionic liquid. J Phys Chem C 112(4):1269–1275
14. Gupta N, Linschitz H (1997) Hydrogen-bonding and protonation effects in electrochemistry of quinones in aprotic solvents. J Am Chem Soc 119(27):6384–6391
15. Barnes AS, Rogers EI, Streeter I, Aldous L, Hardacre C, Wildgoose GG, Compton RG (2008) Unusual voltammetry of the reduction of O-2 in C(4)dmim N(Tf)(2) reveals a strong interaction of O-2(center dot-) with the C(4)dmim (+) cation. J Phys Chem C 112(35):13709–13715
16. Chin DH, Chiericato G, Nanni EJ, Sawyer DT (1982) Proton-induced disproportionation of superoxide ion in aprotic media. J Am Chem Soc 104(5):1296–1299
17. Switzer EE, Zeller R, Chen Q, Sieradzki K, Buttry DA, Friesen C (2013) Oxygen reduction reaction in ionic liquids: the addition of protic species. J Phys Chem C 117(17):8683–8690
18. Rene A, Hauchard D, Lagrost C, Hapiot P (2009) Superoxide protonation by weak acids in imidazolium based ionic liquids. J Phys Chem B 113(9):2826–2831
19. Buzzeo MC, Klymenko OV, Wadhawan JD, Hardacre C, Seddon KR, Compton RG (2003) Voltammetry of oxygen in the room-temperature ionic liquids 1-ethyl-3-methylimidazolium bis((trifluoromethyl)sulfonyl)imide and hexyltriethylammonium bis((trifluoromethyl)sulfonyl)imide: one-electron reduction to form superoxide. Steady-state and transient behavior in the same cyclic voltammogram resulting from widely different diffusion coefficients of oxygen and superoxide. J Phys Chem A 107(42):8872–8878
20. Evans RG, Klymenko OV, Saddoughi SA, Hardacre C, Compton RG (2004) Electroreduction of oxygen in a series of room temperature ionic liquids composed of group 15-centered cations and anions. J Phys Chem B 108(23):7878–7886

21. MacFarlane DR, Meakin P, Sun J, Amini N, Forsyth M (1999) Pyrrolidinium imides: a new family of molten salts and conductive plastic crystal phases. J Phys Chem B 103 (20):4164–4170
22. Allen CJ, Hwang J, Kautz R, Mukerjee S, Plichta EJ, Hendrickson MA, Abraham KM (2012) Oxygen reduction reactions in ionic liquids and the formulation of a general ORR mechanism for Li–air batteries. J Phys Chem C 116(39):20755–20764
23. Katayama Y, Sekiguchi K, Yamagata M, Miura T (2005) Electrochemical behavior of oxygen/superoxide ion couple in 1-butyl-1-methylpyrrolidinium bis(trifluoromethylsulfonyl) imide room-temperature molten salt. J Electrochem Soc 152(8):E247–E250
24. Hayyan M, Hashim MA, AlNashef IM, Tan XM, Chooi KL (2010) Generation of superoxide ion in trihexyl (tetradecyl) phosphonium bis (trifluoromethylsulfonyl) imide room temperature ionic liquid. J Appl Sci 10(12):1176–1180
25. Pozo-Gonzalo C, Torriero AAJ, Forsyth M, MacFarlane DR, Howlett PC (2013) Redox chemistry of the superoxide ion in a phosphonium-based ionic liquid in the presence of water. J Phys Chem Lett 4(11):1834–1837
26. Costentin C, Evans DH, Robert M, Savéant J-M, Singh PS (2005) Electrochemical approach to concerted proton and electron transfers. Reduction of the water−superoxide ion complex. J Am Chem Soc 127(36):12490–12491
27. Pozo-Gonzalo C, Howlett PC, Hodgson JL, Madsen LA, MacFarlane DR, Forsyth M (2014) Insights into the reversible oxygen reduction reaction in a series of phosphonium-based ionic liquids. Phys Chem Chem Phys 16:25062–25070
28. Pozo-Gonzalo C, Virgilio C, Yan Y, Howlett PC, Byrne N, MacFarlane DR, Forsyth M (2014) Enhanced performance of phosphonium based ionic liquids towards 4 electrons oxygen reduction reaction upon addition of a weak proton source. Electrochem Commun 38:24–27
29. Shukla SK, Khupse ND, Kumar A (2012) Do anions influence the polarity of protic ionic liquids? Phys Chem Chem Phys 14(8):2754–2761
30. Crowhurst L, Mawdsley PR, Perez-Arlandis JM, Salter PA, Welton T (2003) Solvent-solute interactions in ionic liquids. Phys Chem Chem Phys 5(13):2790–2794
31. Khoo T, Somers A, Torriero AAJ, MacFarlane DR, Howlett PC, Forsyth M (2013) Discharge behaviour and interfacial properties of a magnesium battery incorporating trihexyl(tetradecyl) phosphonium based ionic liquid electrolytes. Electrochim Acta 87:701–708
32. Buzzeo MC, Klymenko OV, Wadhawan JD, Hardacre C, Seddon KR, Compton RG (2004) Kinetic analysis of the reaction between electrogenerated superoxide and carbon dioxide in the room temperature ionic liquids 1-ethyl-3-methylimidazolium bis(trifluoromethylsulfonyl) imide and hexyltriethylammonium bis(trifluoromethylsulfonyl)imide. J Phys Chem B 108 (12):3947–3954
33. Hayyan M, Mjalli FS, AlNashef IM, Hashim MA (2012) Generation and stability of superoxide ion in tris(pentafluoroethyl)trifluorophosphate anion-based ionic liquids. J Fluor Chem 142:83–89

Part V
Applications

Chapter 19
Ionic Liquids in Surface Protection

Joaquín Arias-Pardilla, Tulia Espinosa, and María Dolores Bermúdez

19.1 Ionic Liquids in Surface Protection

Room temperature ionic liquids (ILs) are salts that are liquid at or below 373 K. Although there exist millions of theoretical combinations of anions and cations, some of the most commonly studied species contain the following cation moieties (ammonium, phosphonium, imidazolium and pyridinium) (Fig. 19.1 and Table 19.1):

A variety of anions have been described. Many of them contain boron, sulphur or phosphorus and fluorine in their formulations: $[BF_4]^-$, $[CH_3BF_3]^-$, $[CF_3BF_3]^-$, $[n\text{-}C_3F_7BF_3]^-$, $[n\text{-}C_4F_9BF_3]^-$, $[PF_6]^-$, $[(C_2F_5)_3PF_3]^-$, $[CF_3SO_3]^-$ and $[N(SO_2CF_3)_2]^-$ (Table 19.2).

Unfortunately, these fluorine-containing anions are susceptible to hydrolysis in the presence of water or moisture from the atmosphere in contact with ILs. The reactions are of the general type described for $[BF_4]^-$ and $[PF_6]^-$ [1, 2]:

$$BF_4^- + 4H_2O \rightarrow B(OH)_4^- + 4HF \tag{19.1}$$

$$PF_6^- + 4H_2O \rightarrow H_2PO_4^- + 6HF \tag{19.2}$$

The evolution of the toxic and highly corrosive HF must be avoided in ILs for corrosion protection or tribological uses.

The present chapter on the application of ionic liquids (ILs) in surface protection will be divided into two sections:

- Ionic liquids in corrosion protection and coatings
- Ionic liquids in lubrication and tribology

J. Arias-Pardilla (✉) • T. Espinosa • M.D. Bermúdez
Grupo de Ciencia de Materiales e Ingeniería Metalúrgica, Departamento de Ingeniería de Materiales y Fabricación, Universidad Politécnica de Cartagena, Campus de la Muralla del Mar., Cartagena 30202, Spain
e-mail: joaquin.arias@upct.es

A.A.J. Torriero (ed.), *Electrochemistry in Ionic Liquids*,
DOI 10.1007/978-3-319-15132-8_19

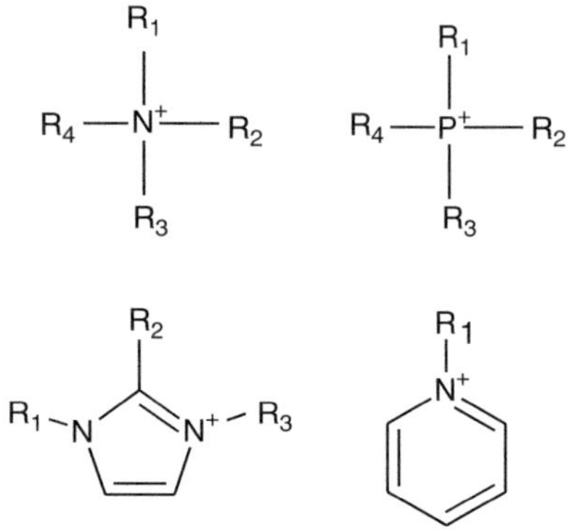

Fig. 19.1 Some of the most widely studied IL cationic moieties

Table 19.1 Cations of aprotic ILs used in corrosion protection

Name	Abbreviations	Structures
N-methyl-*N*-propylpyrrolidinium	$[C_3mPyr]^+$	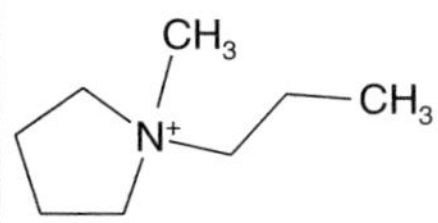
N-methyl-*N*-butylpyrrolidinium	$[C_4mPyr]^+$	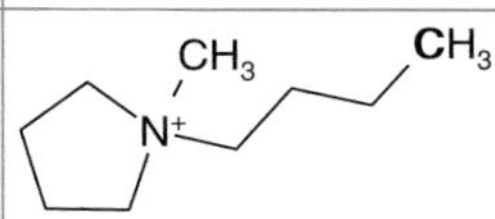
Tri(hexyl)tetradecylphosphonium	$[P_{6\,6\,6\,14}]^+$	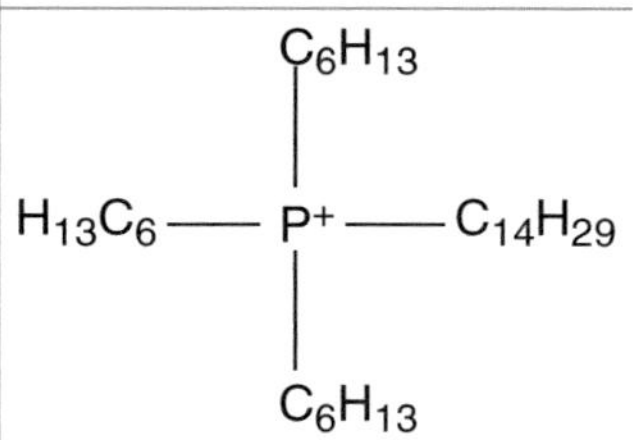
1-Ethyl-3-methylimidazolium	$[C_2mim]^+$	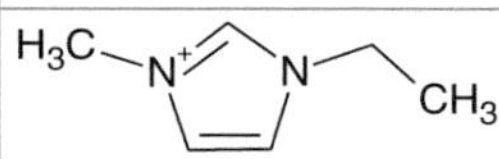
1-Butyl-3-methylimidazolium	$[C_4mim]^+$	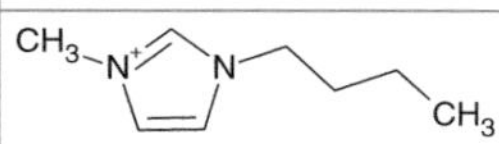
1-Hexyl-3-methylimidazolium	$[C_6mim]^+$	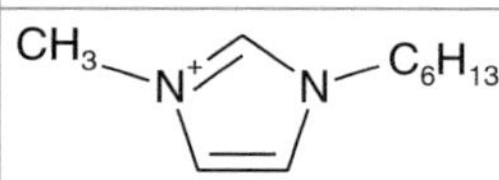
1-Octyl-3-methylimidazolium	$[C_8mim]^+$	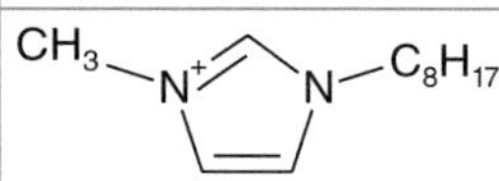
1-Butyl-2-methyl-3-methylimidazolium	$[C_4dmim]^+$	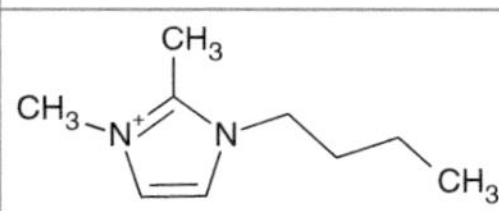

(continued)

Table 19.1 (continued)

Name	Abbreviations	Structures
1,3-Dioctadecylimidazolium	$[dC_{18}im]^+$	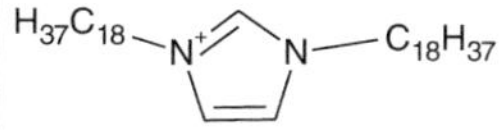
1-Allyl-3-methylimidazolium	$[Amim]^+$	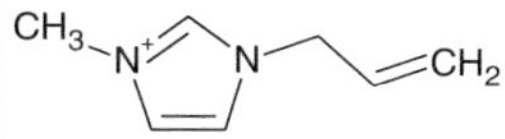
1-Vinyl-3-butylimidazolium	$[C_4vim]^+$	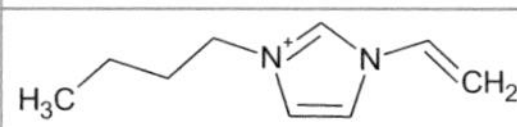
1-Vinyl-3-octylimidazolium	$[C_8vim]^+$	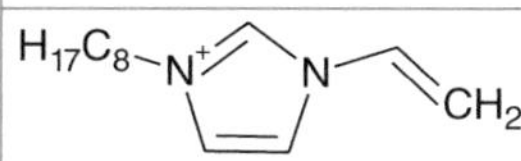
1-Vinyl-3-dodecylimidazolium	$[C_{12}vim]^+$	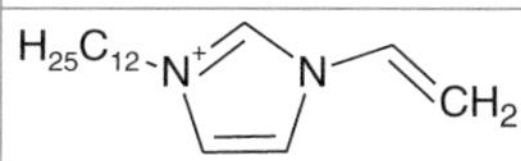
1-Vinyl-3-octadecylimidazolium	$[C_{18}vim]^+$	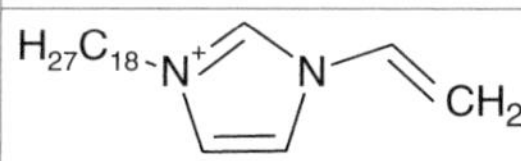
1-Vinyl-3-docosylimidazolium	$[C_{22}vim]^+$	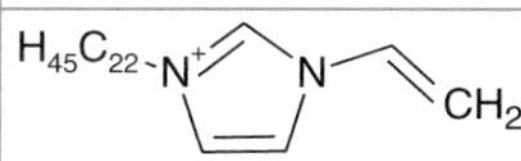
3-(4-Ethoxy-4-oxobutyl)-1-ethylimidazol-3-ium	$[C_2C_4OOC_2im]^+$	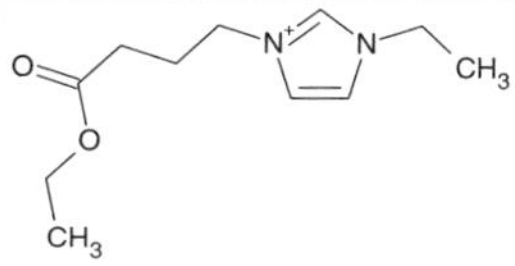
1-Ethyl-3-phenylethylimidazol-3-ium	$[C_2PhC_2im]^+$	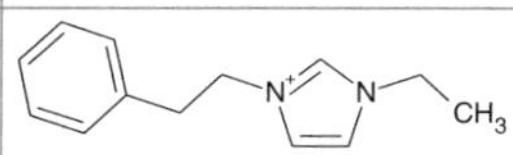
3-(3-Phenylpropyl)-1-propyl-1*H*-imidazol-3-ium	$[C_3PhC_3im]^+$	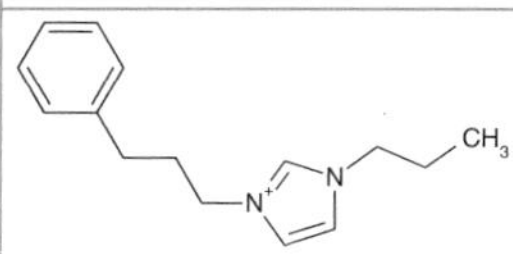
3-(4-Phenoxybutyl)-1-propyl-1*H*-imidazol-3-ium	$[C_3PhOC_3im]^+$	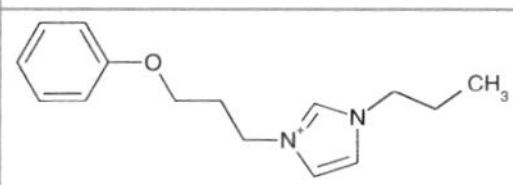
N-Octadecylpyridinium	$[C_{18}py]^+$	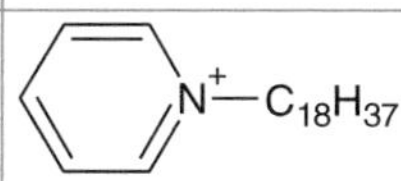
1,1′-(Propane-1,3-diyl) bis (4-methylpyridinium)	$[C_3bPyr]^{2+}$	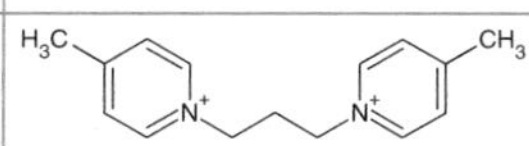
1,1′-(Butane-1,4-diyl) bis (4-methylpyridinium)	$[C_4bPyr]^{2+}$	
Trimethylphenylammonium	$[N_{3\ 3\ 3\ p}]^+$	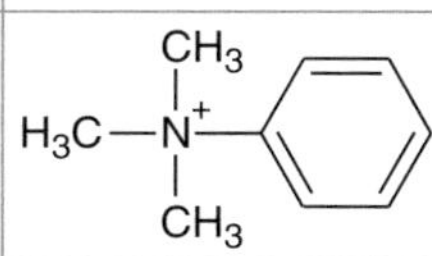

Table 19.2 Anions of aprotic ILs used in corrosion protection

Name	Abbreviations	Structures
Bis(trifluoromethylsulfonyl)imide	$[N(Tf)_2]^-$	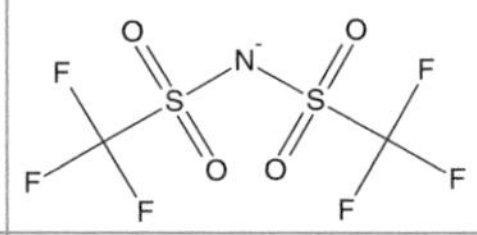
Bis 2,4,4-trimethylpentylphosphinate	$[(i\text{-}C_8)_2PO_2]^-$	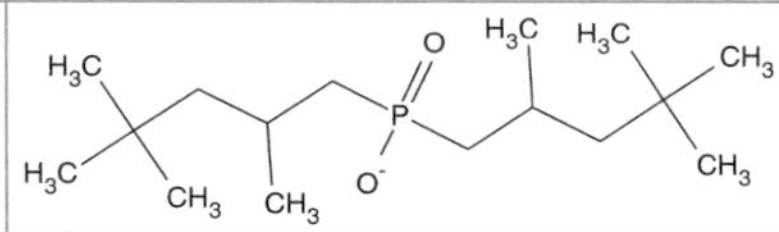
Trifluoromethanesulfonate	$[CF_3SO_3]^-$	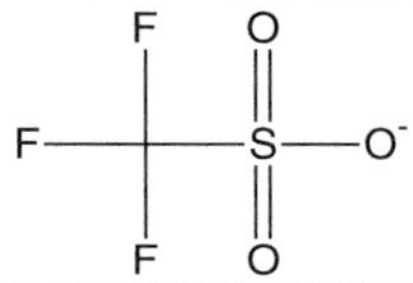
Dimethylphosphate	$[Me_2PO_4]^-$	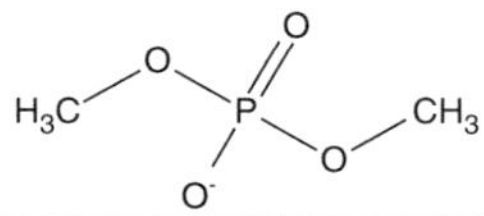
Dibutylphosphate	$[DBP]^-$	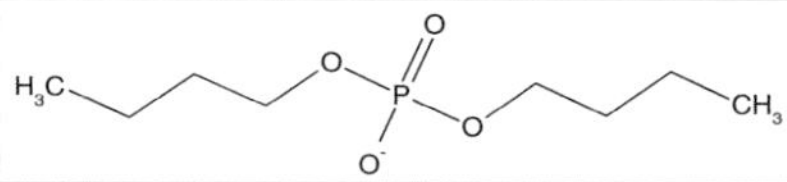
Diphenylphosphate	$[DPP]^-$	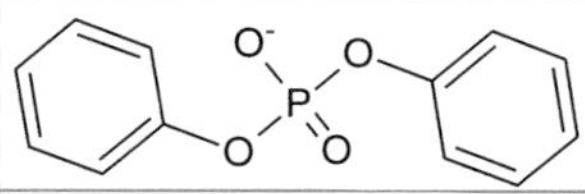
Bis(2-Ethylhexyl)phosphate	$[(EH)_2PO_4]^-$	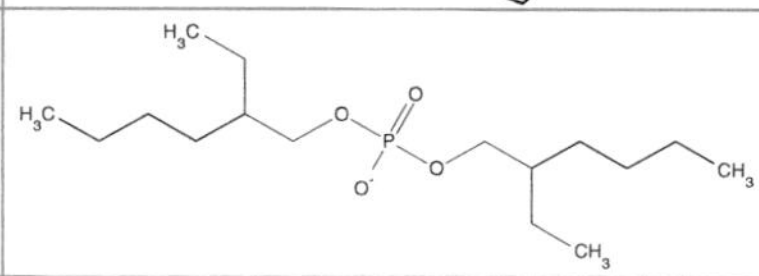
Diisobutyldithiophosphinate	$[(i\text{-}C_4)_2PS_2]^-$	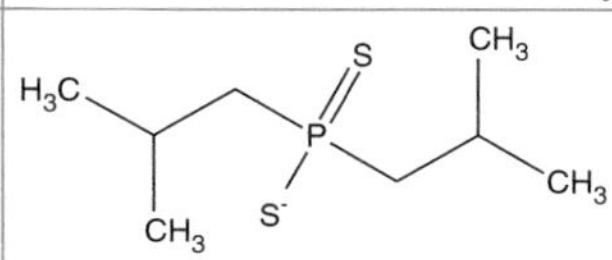
O,O′-Diethyldithiophosphate	$[(OC_2)_2PS_2]^-$	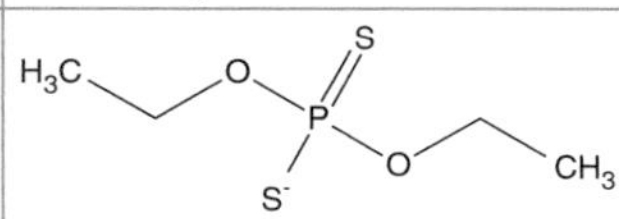
Ethylphosphonate	$[EtPO_3H]^-$	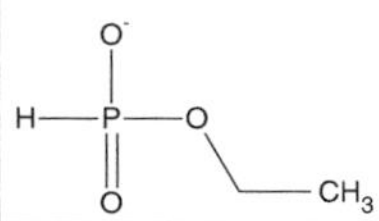
Chloride	$[Cl]^-$	Cl^-
Bromide	$[Br]^-$	Br^-
Nitrate	$[NO_3]^-$	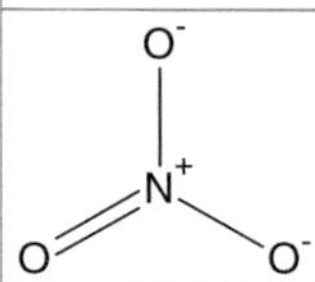

(continued)

Table 19.2 (continued)

Name	Abbreviations	Structures
Tetrafluoroborate	$[BF_4]^-$	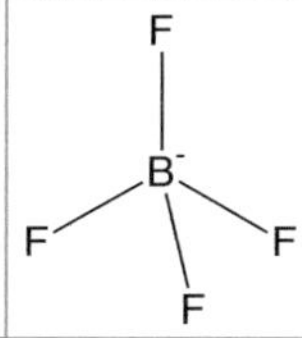
Hexafluorophosphate	$[PF_6]^-$	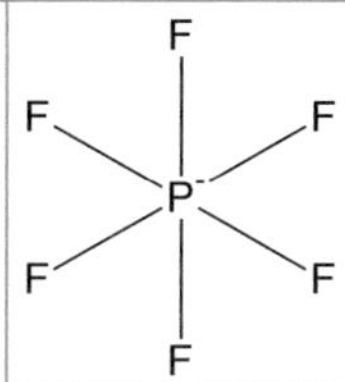
Dicyanamide	$[N(CN)_2]^-$	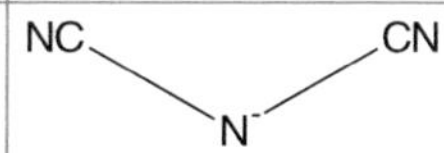
Hydrogensulphate	$[HSO_4]^-$	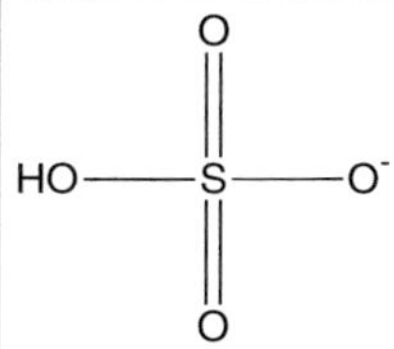
Methylsulphate	$[MeSO_4]^-$	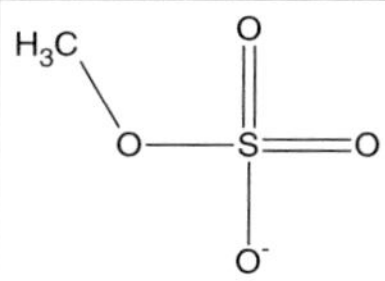
Ethylsulphate	$[EtSO_4]^-$	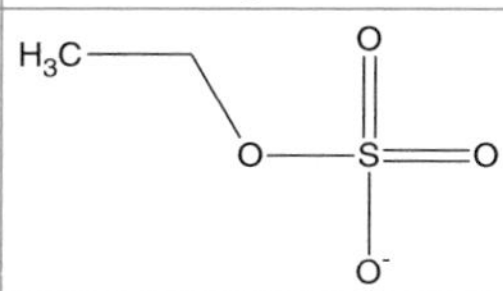
Octylsulphate	$[C_8H_{17}SO_4]^-$	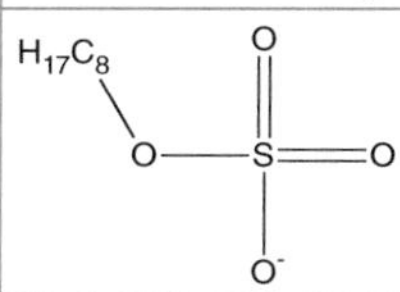
Tosylate	$[TOS]^-$	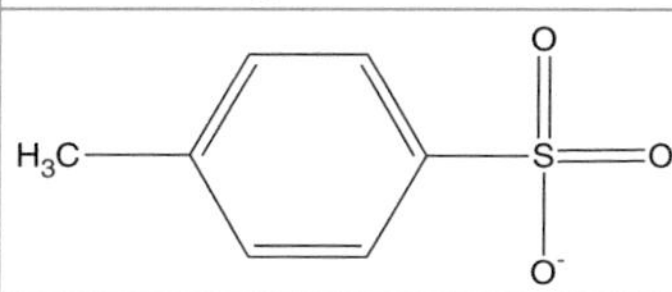
Saccharinate	$[SAC]^-$	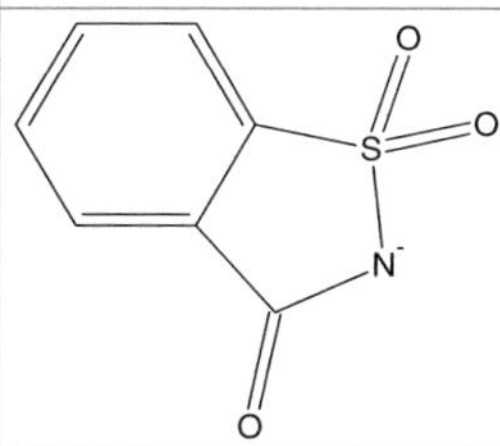

Both sections are intimately related as they depend on physicochemical interactions on ionic liquid molecules with materials surfaces.

19.1.1 Corrosion Protection

One of the most effective methods to prevent corrosion is to isolate the substrate from corrosive environments by means of a corrosion-resistant coating. In this way, ILs have been extensively studied and used as electrolyte during the electrochemical deposition of metals to obtain metallic protective films against corrosion [3–5]. But some ionic liquids can also form a corrosion-resistant thin film over a metallic substrate in absence of metal salt, as corrosion inhibitors.

This use of ILs as corrosion inhibitors starts after the observation that an Al-alloy showed a very low corrosion rate (<0.1 mm·year^{-1}) in the presence of different ILs, especially in the absence of water [6]. The work of Howlett et al. [7] can be considered a pioneer in the field. In this work, surface films produced from $[N(Tf)_2]^-$ anion-based ILs on Li were studied. From then on, many different works can be found in the literature using ILs based on different anions and different metallic substrates. These results will be reviewed below, classifying them as a function of the substrate materials.

19.1.1.1 Lithium and Lithium Alloys

Lithium is a metal of great importance because the most common type of cell used in consumer applications uses metallic lithium as anode. When $[C_4mPyr][N(Tf)_2]$ and $[C_3mPyr][N(Tf)_2]$ were investigated as solvents in a lithium metal battery, the formation of a film was observed [8, 9] when current was applied. At deposition rates up to 1.75 mA·cm^{-2} the deposit was uniform and nondendritic, whereas above 1.75 mA·cm^{-2} the deposit became dendritic and efficiency decayed. When the film was characterised [8] using XPS, DRIFTS, Raman spectroscopy and EIS, it was found to be composed of an outer layer of $[N(Tf)_2]^-$ reduction products, such as LiS_2O_4, $Li_2NSO_2CF_3$, $LiSO_3CF_3$, Li_2SO_3 and Li_2S, as well as Li_2O, LiOH, Li_2CO_3 and LiF (Fig. 19.2). Under this outer layer, LiF and Li_2O predominate and finally Li_2O appears to become the principal component. These products were proposed by Aurbach et al. [10] as reduction products of $[N(Tf)_2]^-$ and were also corroborated by ab initio calculations [11]. In any case, the amount of product formed depends greatly on the presence and concentration of H_2O, because a different behaviour was observed when similar studies were performed in the absence of water [12]. In the same way, a solid electrolyte interface with improved properties was recently obtained using $[C_4mPyr][N(Tf)_2]$ in a Li- S battery protecting lithium metal from continuous attack by soluble polysulfides [13], although not detailed study of the composition was performed.

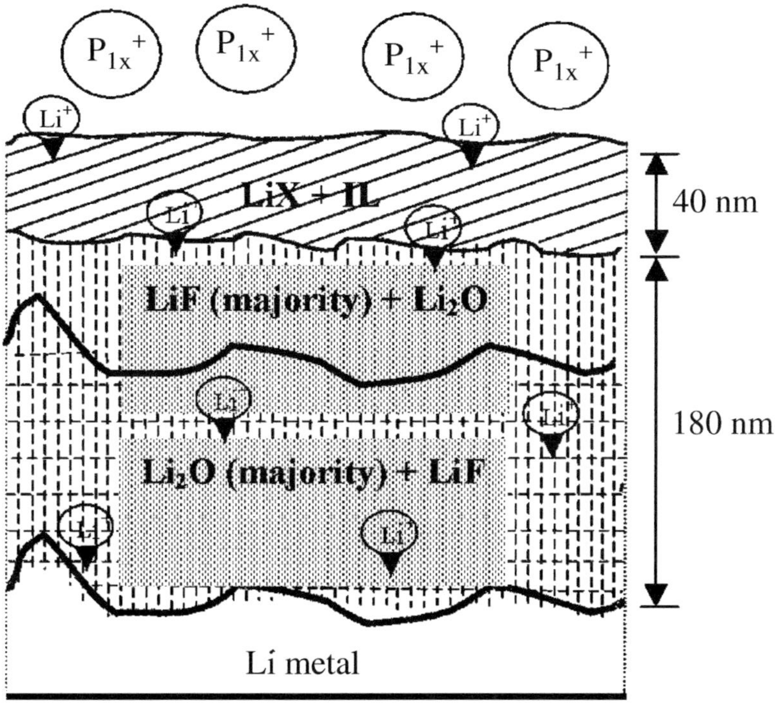

Fig. 19.2 Model of solid electrolyte interface structure on the Li surface. LiX: X can be S_2O_4, F, SO_3CF_3, NSO_2CF_3, SO_3, CO_3, S, O or OH. Reproduced with permission from [8], The Electrochemical Society

19.1.1.2 Magnesium and Magnesium Alloys

In spite of their reactivity, magnesium and magnesium alloys, have nowadays become critical in order to supply the demand for light-metal alloys. The use of Mg- based alloys is not yet widespread because resistance to corrosion is not adequate for many applications. Different ILs have been studied both with magnesium and different magnesium alloys showing promising corrosion protection.

The first protective surface films were obtained on Mg alloy AZ31 (composition: 96 % Mg, 3 % Al, 1 % Zn) in the presence of different ILs based on the $[N(Tf)_2]^-$ anion and different cations such as $[C_2mim]^+$, $[C_3mPyr]^+$ or $[P_{6\,6\,6\,14}]^+$ [14]. In this case no current was applied, the film was obtained by immersing the substrate in the IL at room temperature for specified times and deposition occurred spontaneously. The best inhibition corrosion rate values, by both weight loss and electrochemical measurements, were obtained using the phosphonium cation. The generated film provides improved corrosion resistance against humid environments as well as in the presence of chloride, shifting the open-circuit potential (OCP) to more noble potentials by more than 500 mV and reducing the corrosion current density by 50 times compared to an untreated sample.

The film also shows an apparent self-healing behaviour observed through electrochemical impedance tests, after applying a current density. But self-healing properties were observed only on those samples which have a good continuous coating and smooth surface. Those samples with excessively thick, nonuniform, and highly cracked films showed no protection compared to an untreated sample, despite very high initial corrosion resistance. The critical thickness value where the film becomes less effective is unknown but expected to be in the 100 nm thick range. The film structure was bilayered as in the case of lithium metal for other $[N(Tf)_2]^-$ anion ILs (Fig. 19.2), and was studied by XPS and time -of -flight secondary ion mass spectrometry (ToF- SIMS) [15].

Given these results, the same ionic liquid was used with pure Mg [16], obtaining similar behaviour. In this case open-circuit potential of Mg was measured, with a value near to −1.5 V vs. SCE, a value sufficiently negative to cause the reduction reaction of the anion $[N(Tf)_2]^-$. As in the case of Mg alloy AZ31 a larger exposition time to IL does not provide a film of improved properties as confirmed by electrochemical impedance, in fact most homogeneous films were obtained during only 1 h immersion experiments, whilst at long immersion time experiments (>10 h) low quality films are obtained due to reaction of substrate with IL and also with the residual water in the IL. In 1 h immersion experiment films, EDX studies were unable to conclude if MgF_2 was present on the multilayered structure of the film, due to instrumental difficulties and the thin thickness of the film, whereas in thick films needle structures of hydrated $Mg(OH)_2$ are observed. Thin films show improved corrosion in the presence of chloride by increasing impedance in 3 orders of magnitude.

Coming back to AZ31 alloy, the effect of applied potential in the protective film generated in the presence of IL was also studied [17] using $[P_{6\ 6\ 6\ 14}]$ $[(i\text{-}C_8)_2PO_2]$. Different potential values were used but if a 100 mV anodic potential from OCP was applied, EIS data indicated an initial increase of the resistance until about 4 h of treatment, followed by a significant decrease until a constant value, with an inverse evolution of capacitance. At the initial stage, the increase of the film thickness isolates the Mg alloy surface but with increasing treatment time the film becomes defective.

Interaction between imidazolium-based ionic liquids and the AZ91D magnesium alloy (composition 90 % Mg, 9 % Al, 1 % Zn) was studied, using $[C_4mim]^+$ cation and $[N(Tf)_2]^-$ [18] or $[CF_3SO_3]^-$ [19] as anions. In first case, exposure was conducted both at room temperature and at 200 °C. The reactivity of $[C_4mim][N(Tf)_2]$ was negligible at room temperature but after 9 days of exposure at 200 °C, the surface became dark and profilometric measurements showed increasing average roughness values. After 30 days of immersion although diffuse corrosion was not detectable, some crevice corrosion spots appeared on the surface.

Using XPS and EDX, it was determined that two different processes contribute to the formation of the interface layer. The first one is the corrosion of the metals constituting the alloy with the formation of oxides and fluorides, and the second one, becoming dominant after a few days of exposure, is the adsorption of IL degradation products onto the surface forming a homogeneous layer composed by C, N and F. This layer greatly reduces the extended corrosion process, but alloy corrosion continues at much lower rate via crevice corrosion.

Using dry $[C_4mim][CF_3SO_3]$ the measured corrosion rate is 3–4 orders of magnitude lower than in aqueous 3.5 % NaCl solution implying a surface passivation probably related with $CF_3SO_3^-$ anions but no clear results were obtained, because no sulphur from the anion or nitrogen from the cation was detected by EDX. In any case, by addition of different water quantities, it was determined that the corrosion rate is strongly dependent of water concentration in the IL.

Corrosion protection in Cl^- containing aqueous solutions of Mg alloy ZE41 (Mg-Zn-Rare Earth (RE)-Zr, nominal composition ~4 wt% Zn, 1.7 wt% RE (Ce), ~0.6 % Zr, remaining balance, Mg) [20] pretreated with $[P_{6\,6\,6\,14}]^+$ based ILs with different phosphate and phosphinate anions using different conditions: 1 h or 20 h at room temperature or 50 °C and 10 min, was studied. All of the ILs, with exception of $[DBP]^-$ anion-based IL, offer some degree of protection to the ZE41 alloy decreasing corrosion depth. The best results were obtained with $[P_{6\,6\,6\,14}]^+$ $[DPP]^-$ decreasing surface corrosion and corrosion depth significantly compared to the control sample. Profilometric measurements indicated the presence of a film of approximately 100 nm thick after IL pretreatment. Using electrochemical techniques [21] such as EIS the surface film evolution was studied at open-circuit potential or applying negative/positive potential. It was found that applying a potential bias of −200 mV versus OCP during a 24- h period produces surface films of significantly higher resistance, whereas positive potentials produced a film of reduced resistance compared to open-circuit conditions. ToF- SIMS images indicated the deposition of a homogeneous film under potential conditions, showing better corrosion performance against pitting corrosion using immersion testing in 0.001 NaCl aqueous solution.

The chemical reactions (19.3)–(19.9) were proposed to occur at the surface regardless of any applied potential together with the electrochemical reduction of $[DPP]^-$ anion involving cleavage of one of the phenyl groups.

$$2H_2O + 2e^- \rightleftharpoons H_2 + 2OH^- \tag{19.3}$$

$$O_2 + 4e^- + 2H_2O \rightleftharpoons 4OH^- \tag{19.4}$$

$$Mg \rightleftharpoons Mg^{2+} + 2e^- \tag{19.5}$$

$$Mg + 2OH^- \rightleftharpoons Mg(OH)_2 + 2e^- \tag{19.6}$$

$$M + xH_2O \rightleftharpoons M(OH)_x + x/2H_2(\text{where M is Ce, Zr, Zn or Mg}) \tag{19.7}$$

$$Mg^{2+} + 2DPP^- \rightleftharpoons Mg(DPP)_2 \tag{19.8}$$

$$Mg^{2+} + DPP^- + OH^- \rightleftharpoons Mg(DPP)(OH) \tag{19.9}$$

The different morphology of the superficial films generated with/without potential was observed with ToF- SIMS experiments. In the case of a sample at 24 OCP, $Mg(OH)_2$, $[C_6H_5O]^-$ and $[PO_3]^-$ species coming from IL reaction were observed primarily at the grain boundary, while in a potential applied sample, similar species were more uniformly distributed.

In the case of ZE41 alloy in contact with $[P_{6\ 6\ 6\ 14}][N(Tf)_2]$ in OCP conditions [22] the formation of a superficial film is observed with SO_3CF_2, CF_2 or even F^- species in the form of metal fluorides (MgF_x, ZnF_x or ZrF_x) coming from reduction processes of the $[N(Tf)_2]^-$ anion as in the case of AZ31 alloy.

Corrosion protection using an electrodeposited thin aluminium layer was also investigated [23–25]. This aluminium layer enhances the corrosion resistance of the substrate due to the formation of a stable oxide film on its surface. Aluminium was deposited on different magnesium alloys as AZ91D or AZ31 from different ILs containing aluminium chloride ($AlCl_3$) and $[C_2mim][Cl]$ or $[N_{3\ 3\ 3\ p}][Cl]$. Deposition potential or current should be carefully selected to obtain the optimum film characteristics. Using $AlCl_3/[C_2mim][Cl]$ IL the layer deposited at lower current density was more compact and uniform when compared to that obtained at a higher current density, which presented a loose structure and small cracks within the Al layer causing a reduction in its protection capability as determined by electrochemical techniques as EIS or potentiodynamic polarisation. Using $AlCl_3/[N_{3\ 3\ 3\ p}]$ [Cl] IL, a HF acid surface activation and zinc immersion treatments were done obtaining a compact intermediate layer on the Mg alloy. After that, electrodeposition of Al from $AlCl_3$-$[N_{3\ 3\ 3\ p}][Cl]$ mixed with 50 % benzene applying different currents was performed. The microstructures show spherical equiaxed grains obtained at the high current with good corrosion protection while bulk grains are obtained at the low current with low corrosion protection due to its porosity.

19.1.1.3 Steel

Steel remains the most widely used engineering material, and one of its main drawbacks is its resistance to corrosion specially for carbon steel, so different ILs were studied to improve its performance. Uerdingen et al. [6] studied the interaction of some imidazolium derivatives with different substituents and varying anions $[EtSO_4]^-$, $[C_8H_{17}SO_4]^-$, $[TOS]^-$, $[Cl]^-$ and $[Me_2PO_4]^-$ and a pair of quaternary ammonium ILs with $[MeSO_4]^-$ or $[SAC]^-$ as anions (Fig. 19.3) with carbon steel during 4 days at 90 °C and highly turbulent flow regime through rotating test. It was found that both the chemical structure of the cation and the type of anion exert dramatic effects on the corrosivity. For instance, tosylate anion increases the corrosivity, while using ethyl sulphate carbon steel was not attacked. In general, the presence of water increased corrosion rate.

The corrosion behaviour of carbon steel in contact with $[C_4mim][N(Tf)_2]$ in a range of temperature from 150 up to 350 °C [26] was studied. From 150 to 250 °C, no weight loss was detected after 48 h immersion tests and a superficial dark layer appeared in the carbon steel due to corrosion and to IL decomposition products although a detailed composition was not obtained. At higher temperatures, up to 325 °C, the dark layer is also observed but localised corrosion phenomena appeared.

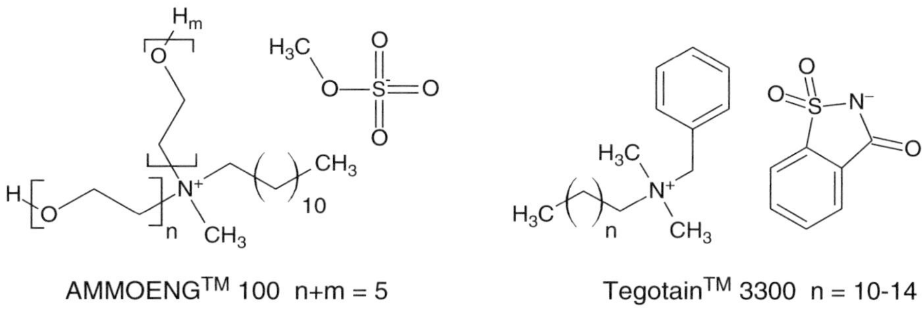

Fig. 19.3 Formulations of quaternary ammonium ILs studied in reference [6]

ILs were also used as corrosion inhibitors adding a small concentration to a corrosive media in contact with carbon steel. Many examples can be found mainly of imidazolium derivatives of the $[C_4mim]^+$ cation, with different anions such as Br^- [27], Cl^-, HSO_4^- [28], $[NO_3]^-$ or $[BF_4]^-$ [29], $[C_4dmim][BF_4]$ [29], complex imidazolium ILs as $[C_2C_4OOC_2im][Br]$, $[C_2PhC_2im][Br]$ [30], $[C_3PhC_3im][Br]$, $[C_3PhOC_3im][Br]$ [31] or pyridinium derivatives as $[C_3bPyr][Br]_2$ and $[C_4bPyr][Br]_2$ [30] in 1.0 M or 0.5 M HCl, $[dC_{18}im][Br]$ and $[C_{18}py][Br]$ [32], different vinylimidazolium cations with Br^- [33] or PF_6^- [34] anions and $[C_2mim][N(CN)_2]$ [35] in 1.0 M H_2SO_4, $[C_4mim][N(CN)_2]$ [36] in carbonated solution of monoethanolamine (MEA) and $[C_4mim][BF_4]$ [37] in alkaline chloride solution.

In most of these studies, electrochemical methods (linear polarisation, EIS, adsorption isotherm) and immersion tests are used, but spectroscopic techniques are less common resulting in incomplete characterisation of surface modification, when compared with other substrates.

In the case of HCl media, all ILs studied, mainly $[C_4mim]^+$ and other imidazole or pyridinium derivatives, regardless of the type of anion, showed the same behaviour, acting as mixed-type inhibitors. EIS data showed that inhibition takes place by strong adsorption obeying Langmuir adsorption isotherm. This adsorption has a more physical than chemical character and, based on the ΔG and ΔH values, the adsorption process is spontaneous and exothermic.

The inhibition efficiency increases with inhibitor concentration, with values from 40 up to 94 % in the case of $[C_3PhOC_3im][Br]$. In this media, Cl^- is first adsorbed onto positively charged metal surface by columbic attraction. After that, the adsorption of imidazole and pyridinium derivatives through electrostatic interactions between these positively charged molecules and the negatively charged metal surface forms a monolayer on the surface (Fig. 19.4).

In sulphuric media, all the imidazolium and pyridinium derivatives studied showed similar behaviour to that observed in HCl, as mixed-type inhibitors forming a monolayer on the surface through Langmuir isotherm, but in this case ΔG values are around $-40\ kJ{\cdot}mol^{-1}$, indicating a higher chemical adsorption character corresponding to sharing of transfer of electrons from the inhibitor molecules to the metal surface to form a coordinated type of bond. Scheme 19.1 represents the inhibition mechanism proposed by Likhanova et al. [32].

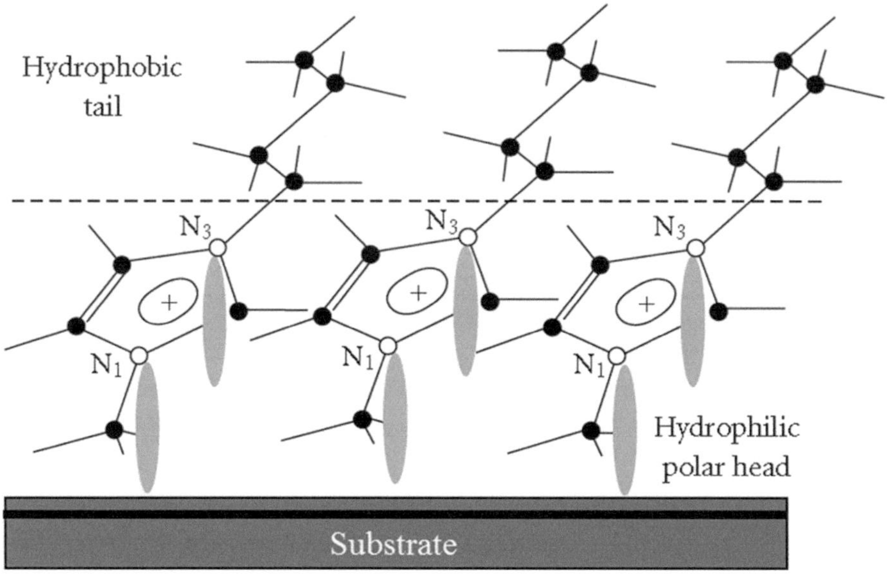

Fig. 19.4 Schematic illustration of the mechanism of corrosion protection of metallic substrate by different ILs containing $[C_4mim]^+$ cation [38]

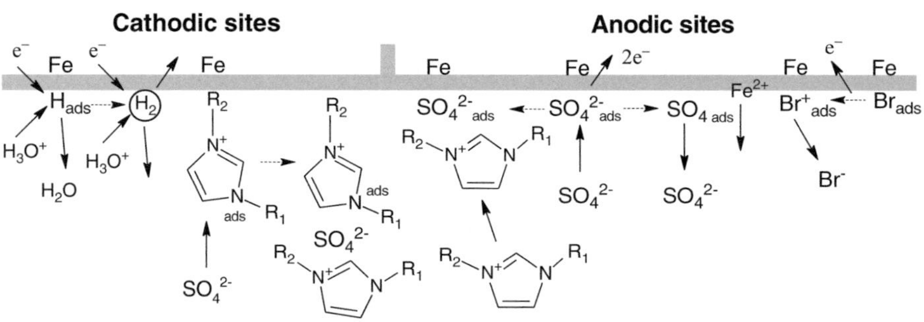

Scheme 19.1 Corrosion inhibition mechanism and adsorption model of ionic liquids on mild steel surface in aqueous sulphuric acid medium [32]

In aqueous carbonated solution of MEA, only $[C_4mim][N(CN)_2]$ was studied, showing again a mixed-type behaviour, decreasing both anodic and cathodic reactions without change in the reaction mechanism. The IL increases surface and charge transfer resistances but decreases with the increase of temperature.

In the case of alkaline chloride solution, $[C_4mim][BF_4]$ also showed a good corrosion inhibition behaviour with up to 89 % of efficiency in the case of the higher concentration used. Again, inhibition mechanism was mixed-type retarding anodic and cathodic processes through forming an adsorption film on steel surface according to Langmuir isotherm, with dominantly electrostatic interaction.

Corrosion protection using an electrodeposited thin aluminium layer was also investigated in steel. $AlCl_3$/[C_2mim][Cl] (60/ 40 mol%) mixture [39, 40] was used obtaining high quality Al layers even in conventionally pretreated mild steel, but with low adhesion. It is necessary to remove iron oxide layers using an anodic electrochemical pretreatment prior to the Al deposition. After that, re- deposition of Fe or Fe- Al alloy formation occurred before Al bulk deposition, resulting in excellent adhesion. The effect of deposition potential on the morphology and crystal size was also investigated. Nanocrystalline, bright, dense and adherent Al layers were electrodeposited at high negative potentials with an average crystallite size of 177 and 28 nm at −600 and −700 mV, respectively. Nanocrystalline Al layers electrodeposited at high potentials have higher corrosion resistance, about 1 order of magnitude higher than that of microcrystalline size due to lower corrosion current densities (I_{corr}), passivation behaviour and tendency for re- passivation.

19.1.1.4 Copper and Copper Alloys

Copper and its alloys are extensively used in different chemical equipment and processes, building construction, electronics and coinages due to their electrical, thermal, mechanical and also corrosion resistance properties, but the presence of aggressive ions like chloride, sulphates and nitrates produces localised attack. ILs have also been studied as inhibitors of copper corrosion.

Copper and brass (CuZn40) are among the various materials used by Uerdingen et al. in their work [6]. Some imidazolium derivatives with different substituents and varying anions (ethylsulphate, octylsulphate, tosylate, chloride and dimethyl-phosphate) and a pair of quaternary ammonium with methylsulphate or saccharinate as anions were used. In water-free ILs experiments, copper and brass showed to be particularly sensitive to the presence of tosylate anion, with corrosion rates between 15 and 25 mm·year^{-1} at 90 °C whereas ethylsulphate produced corrosion rates in the order of 2 mm·year^{-1}. Addition of a 10 % of water approximately doubles the corrosion rate.

The corrosion behaviour of copper and brass (Naval alloy 464) in contact with [C_4mim][Tf_2N] in a range of temperature from 150 up to 350 °C [26] was studied. For copper at 150 °C the weight loss was very high (373 g·cm^{-2}·year^{-1}) and at higher temperatures, up to 325 °C, the rupture of the sample occurred due to pitting corrosion and weight losses are not only imputable to corrosion. In the case of brass no weight loss was observed in the temperature range from 150 to 325 °C and a superficial dark layer appeared on the carbon steel surface due to corrosion and to IL decomposition products, although a detailed composition was not obtained.

Corrosion inhibition behaviour of different ILs with [C_4mim]$^+$ cation and anions as [BF_4]$^-$ [41], [Cl]$^-$, [Br]$^-$ [38] and [HSO_4]$^-$ was studied in acidic media (chloride or sulphuric). In the case of [C_4mim][BF_4], the corrosion rate of copper on dry and chloride-free IL was very low (1 μA·cm^{-2}), but corrosion rate increased with temperature and with the concentration of water and chloride to reach 23 μA·cm^{-2} in the presence of only 0.87 % of Cl^- and 1.3 % of water with pitting corrosion taking place. Chloride substantially enhanced the corrosion rate, probably due to formation of chloride complexes of Cu. Water alone also enhances corrosion but water and chloride together showed a synergistic effect, probably because aquo-chloride complexes of Cu in the ionic liquid are more stable than chloride complexes. Water promoted the presence of fluorine while chloride promoted the presence of boron in corrosion products and in degradation products of $[BF_4]^-$ anion.

Working with higher chloride concentration (1.0 M) and pH 1.0 and using $[C_4mim]$ [Cl], inhibition efficiency increased with an increase of the concentration of IL, up to a value of 95 % for 50 mM IL concentration, with similar but lower values for $[C_4mim]$ [Br].

Both ILs acted as mixed-type inhibitors affecting cathodic and anodic reactions through forming an adsorption film on copper surface according to Langmuir isotherm with dominantly electrostatic interaction (Fig. 19.4). This is the same behaviour shown by $[C_4mim]^+$, $[C_6mim]^+$ and $[C_8mim]^+$ cations with $[HSO_4]^-$ anion in sulphuric media. The best inhibition efficiency is obtained with $[C_8mim]$ $[HSO_4]$ with a value near to 93 %.

Interaction with copper of four imidazolium ILs and two protic ILs were also studied [42]. In the case of imidazolium ILs, all show reactivity toward copper with different mechanisms related to the nature of the anions. $[C_2mim][EtPO_3H]$ showed the highest mass loss due to the formation of a soluble phosphonate-copper species. In the case of $[C_2mim][C_8H_{17}SO_4]$ due to its high reactivity, precipitation of copper sulphate occurred on the copper surface, producing a very high surface roughness increase. The highest surface roughness increase was observed using $[C_6mim][BF_4]$ due to the decomposition of the $[BF_4]^-$ anion. The lowest mass and roughness changes of the imidazolium ILs were obtained for $[PF_6]^-$ with the formation of a uniform solid layer on the copper surface after reaction with $[PF_6]^-$. The $[H_2N(CH_2CH_2OH)_2]_2[Adip]$ protic IL showed the lowest corrosion rate due to formation of adsorbed layers without corrosion products corroborated by XRD, SEM- EDX, XPS and FTIR. Finally the other protic IL $[H_3N(CH_2CH_2OH)]_2[Succ]$ showed higher reactivity toward copper generating a precipitate by combination of the IL and copper oxide. Surface roughness was 75 times higher than that obtained for $[H_2N(CH_2CH_2OH)_2]_2[Adip]$ (Table 19.3).

Corrosion protection using an electrodeposited thin aluminium layer was also investigated in copper [43]. Deposition was performed from $AlCl_3$/[Amim] [Cl] mixture in different proportions. The electrodeposition of Al on copper fitted well to the theoretical model for 3- D instantaneous nucleation with diffusion-controlled growth. Different experimental parameters of deposition such as temperature, current density and molar ratio of $AlCl_3$ to [Amim][Cl] have been investigated. In the case of temperature, bright, smooth and dense deposits were obtained in the range from 313.2 to 353.2 K, but the deposits turned black and poorly adherent at higher temperature. Using different current density values, it was observed that the nucleation density on copper electrode increased with current density rising from 5 to 20 $mA{\cdot}cm^{-2}$, and the deposits displayed finer microstructures. At higher current density values, the crystallites experienced a rapid overlapping growth and deposits with larger grain size and higher roughness values

Table 19.3 Protic ammonium ILs [42]

Nomenclature and abbreviations	Cation	Anion
Di-[(2-hydroxyethy l) ammonium] succinate $[H_3N(CH_2CH_2OH)]_2[Succ]$	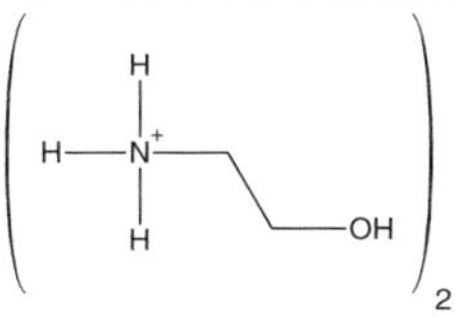	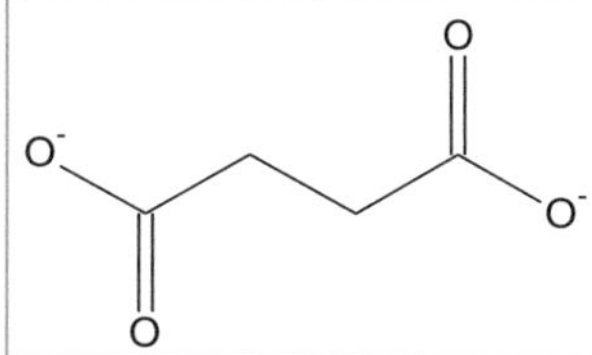
Di-[bis(2-hydroxyethyl) ammonium] adipate $[H_2N(CH_2CH_2OH)_2]_2[Adip]$	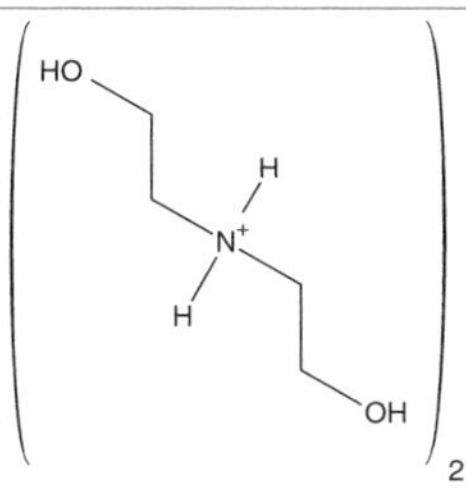	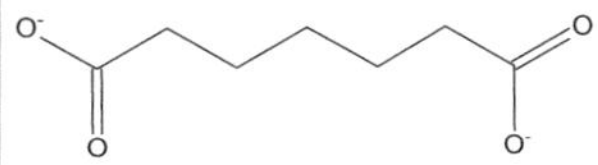

were obtained. The aluminium coatings with optimum surface morphology were obtained from 1:2.0 $AlCl_3$/[Amim][Cl] at 313.2–353.2 K and 10–20 $mA{\cdot}cm^{-2}$.

19.1.1.5 Aluminium Alloys

Aluminium is considered to be a very suitable material for the cathodic current collector in lithium-ion batteries, but it may corrode in contact with the electrolytes used in that kind of devices. The use of ionic liquids as corrosion inhibitors of aluminium and aluminium alloys was pointed out by Uerdingen et al. [6] when an Al- alloy (AlMg3) showed a very low corrosion rate (0.03 $mm{\cdot}year^{-1}$) in the presence of different imidazolium derivatives with different and varying anions (ethylsulphate, octylsulphate, tosylate and dimethylphosphate), especially in the absence of water. Adding 10 % of water, mass loss values were under 0.1 $mm{\cdot}year^{-1}$.

Immersion corrosion tests for Al2011 in different imidazolium and pyridinium derivatives with 2 to 8 chain length substituents and different anions as $[BF_4]^-$, $[CF_3SO_3]^-$, $[PF_6]^-$, tosylate and $[N(Tf)_2]^-$ during 30 days and also in aqueous solutions with 1 wt% or 5 wt% IL were performed [44]. In pure ILs Al2011 shows no mass loss was observed and only in the case of $[CF_3SO_3]^-$ some deposits were observed composed of aluminium oxide, fluorine and sulphur. In the presence of water, corrosion rate was greatly increased due to hydrolysis of the anion.

The electrodeposition of a thin aluminium layer was performed from a chloroaluminate ionic liquid ($AlCl_3$/$[C_4mim][Cl]$; molar ratio 2/ 1) on Al- Li P90 alloy [45]. A crucial aspect was the superficial treatment applied before the aluminium deposition, performing mechanical polishing inside a globe box where water and oxygen content were kept below 5 ppm. The coating obtained after 2 h at constant current density of 10 mA·cm^{-2} is quite adhering to the substrate with a thickness around 20 μm. Using EIS, the coated samples displayed a behaviour very close to that of pure aluminium increasing polarisation resistance, associated with a decrease in corrosion rate from 17 to 0.5 μA·cm^{-2}.

19.2 Ionic Liquids in Lubrication and Tribology

One of the challenges identified by Spikes in 2001 [46], in order to solve the problems of the already existing polyphenylethers and perfluoropolyethers, was the development of synthetic liquid lubricants able to withstand very high temperatures without excessive thermal or oxidative degradation or volatilisation. Another challenge was that of attaining the self-replenishing, zero wear, machine component.

In passenger cars, one- third of the fuel energy is used to overcome friction in the engine, transmission, tyres and brakes [47]. The direct frictional losses, with breaking friction excluded, are 28 % of the fuel energy.

By taking advantage of new technology for friction reduction in passenger cars, friction losses could be reduced by 18 % in the short term (5–10 years) and by 61 % in the long term (15–25 years).

The friction-related energy losses in an electric car are estimated to be only about half those of an internal combustion passenger car.

Potential actions to reduce friction in passenger cars include the use of new low-viscosity and low- shear lubricants and additives. In this sense, ILs could play a crucial role.

19.2.1 Conventional IL Lubricants

The breakthrough of ionic liquid tribology took place in 2011 with the publication [48] of the friction coefficients of a number of material contacts in the presence of imidazolium ILs with fluorine-containing anions. The first ILs described as neat lubricants were imidazolium $[BF_4]^-$ and $[PF_6]^-$ derivatives. They undergo a complex surface chemistry which depends on the chemical composition of the surfaces, the environment and the sliding conditions.

$[BF_4]^-$ and $[PF_6]^-$ anions strongly interact with metal surfaces. Recent results obtained with density functional theory (DFT) [49] found that aluminium atoms can be pulled from the surface up to 0.25 Å from the Al surface due to interaction with BF4. Li atoms show an even stronger interaction [50], being pulled up to 1 Å, while copper atoms are not pulled significantly from copper surfaces.

Since 2001 to the present moment, a growing number of research groups all over the world have developed all aspects of ionic liquid tribology. Several reviews and special issues [51–61] have compiled the information available from the point of view of materials science, IL cation and anion compositions and applications.

In their review paper, Torimoto et al. [54] considered ILs as Advanced Materials by themselves.

Since one of the review papers on IL lubricants was published in 2013 [57], the main purpose of the present chapter is to provide an overview of the most recent developments with an emphasis in the new IL formulations, properties and interactions with sliding surfaces.

The experience of the studies carried out so far, allows the definition of several criteria which should meet ILs to be considered as suitable candidates for their application as lubricants.

19.2.2 Halogen-Free IL Lubricants

From the point of view of tribological performance, the ILs should form an adsorbed layer on the sliding surfaces stable enough to withstand high contact pressures and temperatures. In case of chemical reaction between the IL and the surface, the reaction product should remain adhered to the surface without degrading it due to tribocorrosion.

In the first place, there exists a strong tendency to avoid the presence of halides, so the new so- called halogen-free ILs are attracting a growing attention [62–64].

The reasons for this shift from the originally studied perhalogenated species to the total absence of halogens are the following:

1. Corrosion of metallic surfaces by formation of metallic halides.
2. Evolution of toxic species (such as HF) from hydrolysis of the anions.

Although halogen-free ILs have been known for a long time, the first studies of halogen-free IL lubricants are fairly recent and revealed a lower friction-reducing ability than that of the halogen-containing ones.

Some imidazolium derivatives such as $[C_2mim][N(CN)_2]$ have recently shown excellent lubrication performance towards ceramic materials, including diamond-like coatings (DLC), due to the formation of nitrogen compounds on the sliding surfaces.

The halogen-free ILs exhibited low corrosion properties to steel, but provided less tribological properties in comparison with a reference $[C_4mim][N(Tf)_2]$.

Halide-free ILs were investigated as 2.5 wt% in glycerol to obtain highly biodegradable polar lubricants. The wear reduction properties were dependent on the anion, the lowest values obtained for methylsulphates, due to the formation of an iron sulphate film.

Some very recent studies on protic ammonium species solely composed of C, O, N and H are very promising in preventing wear and giving ultra-low friction even in water [42, 65, 66], although the thermal stability of these species is not so high as that of the halogen-containing imidazolium salts.

A great attention has been paid to phosphate ILs both in corrosion protection and in tribological applications, particularly in steel lubrications due to the formation of protective phosphate films.

As neat lubricants, ILs are still far from being competitive with other lubricants due to their high prices. There still remains the field of space and vacuum tribology, where the nonvolatile nature of the ILs makes them the ideal candidates [67].

19.2.3 IL Nanolubricants and Nanocoatings

Some new nanocomposite coatings from DLC/ IL [68] or even DLC/ IL/graphene [67] combinations are being developed for space applications.

In some cases, the DLC coating technology has been combined with surface texturing to improve chemical interactions with the IL [68]. The combination of surface morphology and chemical modification improves the tribological performance of DLC film. DLC- IL composite films with well-defined micro-/nanotextures were produced by the combination of ICP technology, magnetron sputtering process and dip- coating method.

DLC-IL textured film with dense dimples and further chemically modified with IL monolayer showed significantly enhanced micro-/nano-tribological performance compared with DLC- IL textured surface with sparse dimples without chemical modification.

The tribological performance of DLC- based solid-liquid lubricating dual-layer coatings on steel/steel contacts under high vacuum can be attributed to the synergy between the DLC film and the boundary IL lubrication [69].

New nanolubricants and nanocomposites are being developed by combination of ILs and carbon nanophases such as carbon nanotubes and graphene [67]. The excess graphene in the IL would tend to form irreversible agglomerates. An optimum graphene concentration of 0.075 mg/ml in IL for the composite coatings was required to exhibit the lowest friction coefficient, the highest bearing capacity and the strongest anti-irradiation in a simulated space environment. XPS spectra confirmed the formation of a fluorinated oil- containing carbon-rich tribofilm between the friction pairs further ensured the good antifriction and wear resistance performance of DLC/ IL/graphene.

For other applications, it seems more feasible to use ILs as lubricant additives and coatings, where a small amount of IL is necessary to lower friction and wear.

19.2.4 Two-Component IL Films

IL films with ordered and densely packed solid-like structure shows lower friction coefficient due to lower surface energy and the absence of meniscus effect. Camillone et al. [70] showed that with increase in the chain length, the packing

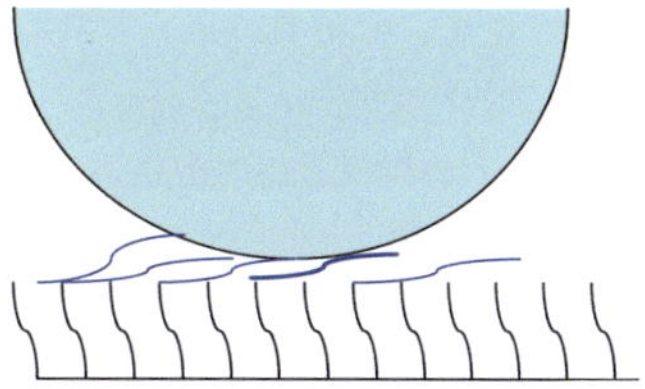

Fig. 19.5 Two-component ILs composite film formed on Si substrate: solid-like (long carbon chain) and liquid-like (short carbon chain)

density also increases. Hence, longer-chain ILs stabilised by van der Waals attraction retain molecular-scale order during shear and act as better lubricants.

Pu et al. [71] showed that an optimal proportion of liquid and solid ILs in two-component films will lead to lower adhesion and nanofriction. Two ILs with different alkyl chain length, a short four-carbon (C_4) liquid-like IL and a long 12- carbon (C_{12}) solid-like IL were blended in different proportions, showing that the two- component ultrathin film containing 80 % C_{12} gives optimal micro-/nano-tribological properties on hydroxylated Si surfaces, as compared to single-component ILs films, which is ascribed to a synergic effect between the steady solid-like IL phase as the backbone and the proper amount of flowable liquid-like IL phase with self-replenishing properties (Fig. 19.5). These researchers found that there exists a critical amount of C_4 molecules which facilitates sliding on the film surface due to low- shear strength, and helps to reduce nanofriction force; and the synergic effect of solid/liquid fractions (C_4:C_{12} (20 %:80 %)) combines high load-carrying capacity with low and stable friction coefficient. However, the inappropriate solid/ liquid ILs ratios are far from improving the tribological properties of ILs film, may even have the reverse effect.

Imidazolium-based crown-type phosphate ionic liquid ultrathin films were fabricated on silicon substrates modified by a self-assembled monolayer. The dual-layer films show enhanced micro -nano -tribological properties as compared to single-component ionic liquid [72].

The ionic moieties present in the IL molecules form adsorbed layers on metal surfaces. The thickness of these adsorbed layers would increase when long lateral chain substituents are present on the cation moiety [57]. The formation of thick stable surface layers would lead to lower friction coefficients under boundary lubrication conditions [73].

19.2.5 Quantized Friction Through IL Tribolayers

Under high pressure and/ or high temperature, ILs can react at the sliding interface giving rise to the formation of tribofilms which could be protective, having an antiwear effect. Metals such as iron or aluminium can readily react with the ILs anions to form metallic borates, fluorides, phosphates, etc. The formation of

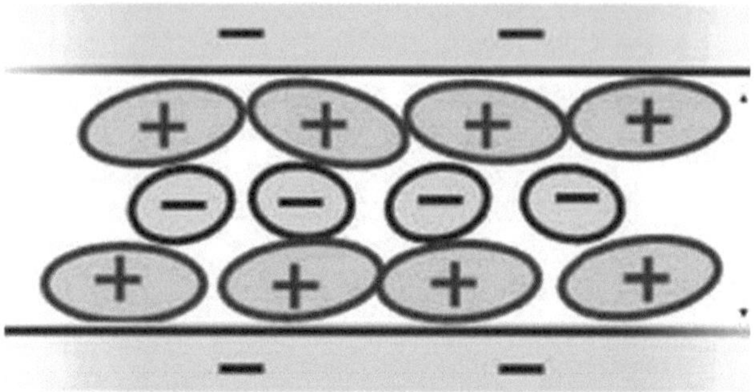

Fig. 19.6 Ordered IL ion layers at the sliding interface

tribofilms is favoured by the high reactivity of the IL anions in contrast with the conventional neutral lubricants.

We have seen that ionic liquids form layered structures at surfaces, yet it is not clear how these nanostructures relate to their lubrication properties.

Smith et al. [74] have measured the friction force between atomically smooth solid surfaces across IL films of controlled thickness in terms of the number of ion layers (Fig. 19.6). Multiple friction–load regimes emerge, each corresponding to a different number of ion layers in the film. In contrast to molecular liquids, the friction coefficients differ for each layer due to their varying composition. The high resolution measurements have revealed the following notable features for the first time in ionic liquid:

1. Discrete friction regimes for each integer number of ion layers in the film, leading to discrete multi-valued ('quantized') friction as a function of load
2. Stick–slip friction observed below a critical velocity (different for each number of ion layers)
3. Different friction coefficients measured for each film thickness. This latter point implies that the mechanism of friction across ionic liquids is qualitatively different to nonpolar liquids, where the friction coefficient is independent of film thickness. Put together, these findings point at new ways to control friction in microscopic systems and to develop ionic liquid lubricants for specialised applications

The quantized friction would be a direct result of the multiple discrete values of the adhesion-controlled contribution to the total friction, and could be a common feature for fluids able to form thin film-layered structures. The results described by Smith et al. could account for discontinuities in friction across thin films as a function of the number of layers, particularly under boundary lubrication regimes with varying liquid film thickness across the contact zone between rough surfaces.

19.2.6 Present and Future Trends in IL Tribology

19.2.6.1 Structure and Composition

Several lines of research and development can be expected to be followed in order to obtain a better understanding of IL surface interactions and lubricating performance.

In the first place, while halogens should be suppressed, boron and phosphorus should be present in order to produce protective coatings and tribolayers [75].

Hydroxyethyl groups on the imidazolium cation influence thermal stability, friction coefficient and wear, without changing significantly viscosity and corrosiveness. Increasing alkyl chain length in the imidazolium cation increased viscosity [76].

The main mechanisms which are assumed to lead to low friction values are the chemical composition of the anion and the orientation of the molecules under shear [77].

Where mobility of component ions and transfer of species to and from the interface is required (heterogeneous catalysis), multiple sterically hindered allylic functional groups could be incorporated to minimise substrate-IL interactions [78]. In situations where IL adsorption to the interface is desirable (e. g. lubrication or electrode surface restructuring), symmetric ions with localised charge centres are preferable.

A series of IL with ethylammonium cations and different anions were studied [79] by AFM force profiling to investigate the structure of adsorbed and solvation layers formed on mica surface. At least seven layers were observed for ethylammonium nitrate (EAN) at 14 °C (m. p. 13 °C), decreasing as the temperature is increased to 30 °C, due to thermal energy disrupting solvophobic forces that lead to segregation of cation alkyl tails from the charged ammonium and nitrate moieties.

The number and properties of the solvation layers can also be controlled by introducing an alcohol moiety to the alkyl chain on the cation, or by replacing the nitrate anion with formate, leading to the detection of distinct cation and anion sublayers.

Substitution of primary by secondary or tertiary ammonium cations reduces the number of solvation layers formed, and also weakens the cation layer adsorbed onto mica. The observed solvation and adsorbed layer structures are discussed in terms of the intermolecular cohesive forces.

Normal load dictates the number of interfacial ion layers and the lateral layer structure. Shear force measurements show that the lubricity of the interface changes with the number and lateral structure of the confined ion layers [80].

19.2.6.2 Influence of Water and Moisture

When $[C_4mim][CF_3SO_3]$, which is miscible with water, was used as a lubricant, using a FTIR equipped with a pin -on -disc tribometer to simultaneously measure the friction force, normal load and IR spectrum, it was seen that the water content increased with time and the ratio of water to cation remained constant even though the film thickness reduced under the lubricating condition. In contrast, when $[C_4mim][PF_6]$, which is water insoluble, was used as a lubricant, the water content was quite small and increased slightly with time, and the ratio of water to cation also remained constant. For $[C_4mim][CF_3SO_3]$, iron oxide, chromium oxide and sulphate were found on the steel pin when a CaF_2 disc was used. Iron oxide and iron phosphate were observed on the worn pin using $[C_4mim][PF_6]$ and CaF_2 disc. These results suggest that the amount of water contained in the ILs have a significant effect on the tribochemical reactions [81].

The water solubility of salts is ordinarily dictated by lattice energy and ion solvation. In ILs, lattice energy is negligible and are the differences in their hydrophobicity which can account for differences in their water solubility [82]. An equation (Eq. 19.10) to estimate water solubilities for cation-anion combinations has been developed.

$$\log x_{\mathrm{IL,\,w}}^{\mathrm{sat}} = -m \log k_{0,\mathrm{c}} + c_{\mathrm{a}} + \epsilon, \qquad (19.10)$$

where:

$x_{\mathrm{IL,\,w}}^{\mathrm{sat}}$ is the mole fraction solubility of the ionic liquid at 293 K, $\boldsymbol{k_{0,c}}$ is the cation hydrophobicity parameter. and ϵ is the random variable describing deviations from the model with unknown or unspecified causes. The slope $\boldsymbol{m}$ should ideally be unity if all model assumptions are satisfied.

Water molecules absorbed from the air are present mostly in the free (not self associated) state, bound via H- bonding with the IL anion. For concentrations of dissolved water in the range 0.2–1 mol/dm^{-3}, most of the water molecules are in the form of 1:2 H-bonded complexes [anion-(HOH)-anion] [83].

Water molecules can also form liquid-like formations with anions of strong basicity $[NO_3]^-$. The lowest H- bonding strength is found for the $[PF_6]^-$ anion.

19.2.6.3 Surface Micro-Pattern and Porosity

Experimental results obtained by Zhao et al. [84], showed that adhesive force and friction force of DLC films with microgrooves reduced effectively with increase of groove area density and incorporation of thin ILs films. The lowered adhesion and friction force were attributed to two factors including the reduced area of contact between DLC films and colloidal tip, and incorporation of thin ILs films to avoid direct contact between DLC films and colloidal tip and to facilitate sliding of colloidal tip on DLC films.

AFM analysis shows that [C_4mim][PF6] can form a wetting phase on mesoporous TiO_2 films but forms nonwetting sphere-shaped droplets (contact angle 40°) on dense TiO_2 films [85]. The weak interacting IL on dense TiO_2 helps reduce friction locally but not on the whole surface. The stable spreading film (contact angle 5°) on mesoporous TiO_2 provides low friction coefficient (0.0025).

19.2.6.4 IL-Nanophase Nanolubricants

Mixtures of SiO_2 and [C_4mim][N(Tf)$_2$] have been studied [86] as lubricants of stainless steel ball-steel surface. The tribological performance depends on the concentration of nanoparticles.

A thin composite film of graphene and IL (IL- G) was deposited on Si substrate by dip- coating [87]. The IL- G film reduces friction but increases wear with respect to IL. This is believed to be due to the discontinuous nature of the IL- G film, which is not able to prevent direct contact between surfaces.

Combinations of carbon nanotubes and ILs, where the nanotubes are modified by the IL to form a gel- like lubricant have been recently described [88].

19.2.6.5 Influence of Surface Charge and Applied Potential

The nano rheological study of two ILs confined in molecularly thin films [89] shows how surface charges drastically affect their molecular structure and flow properties.

Lateral forces vary as a function of applied surface potential and ion structure due to changes in the composition of the ion layer, from cation-enriched (at negative potentials) to mixed (at 0 V), and to anion-enriched (at positive potentials). Four ILs with different cations and the FAP (tris(pentafluoroethyl) trifluorophosphate) anion show similar nano-tribological behaviour: low friction at negative potentials and higher friction at positive potentials. At neutral potentials the behaviour is consistent with a discontinuous sliding process [90].

It has been proposed that friction forces vary with potential [91] because the composition of a confined ion layer between the two surfaces changes from cation-enriched (at negative potentials) to anion-enriched (at positive potentials).

19.2.6.6 Predictive Models

Nonequilibrium molecular dynamics for the interaction parameters between alkylammonium alkylsulfonate ILs and an iron surface have been used [92] to develop a procedure for a quantitative prediction of the friction coefficient. Changes in the frictional force are explained in terms of the specific arrangements and orientations of groups forming the ionic liquid at the surface vicinity.

Acknowledgements The authors wish to express their gratitude to the Ministerio de Economía y Competitividad (Spain) (MAT2011-23162) for financial support. T. Espinosa is grateful to the Ministerio de Educación, Cultura y Deporte (Spain) for a research grant (AP2010-3485).

References

1. Wamser CA (1948) Hydrolysis of fluoboric acid in aqueous solution. J Am Chem Soc 70(3):1209–1215. doi:10.1021/ja01183a101
2. Swatloski RP, Holbrey JD, Rogers RD (2003) Ionic liquids are not always green: hydrolysis of 1-butyl-3-methylimidazolium hexafluorophosphate. Green Chem 5(4):361–363. doi:10.1039/b304400a
3. Abbott AP, McKenzie KJ (2006) Application of ionic liquids to the electrodeposition of metals. Phys Chem Chem Phys 8(37):4265–4279. doi:10.1039/b607329h
4. Endres F, MacFarlane D, Abbott A (eds) (2008) Electrodeposition from ionic liquids. Wiley, Weinheim
5. Abbott AP, Frisch G, Ryder KS (2013) Electroplating using ionic liquids. Annu Rev Mater Res 43(1):335–358. doi:10.1146/annurev-matsci-071312-121640
6. Uerdingen M, Treber C, Balser M, Schmitt G, Werner C (2005) Corrosion behaviour of ionic liquids. Green Chem 7(5):321–325. doi:10.1039/b419320m
7. Howlett PC, MacFarlane DR, Hollenkamp AF (2004) High lithium metal cycling efficiency in a room-temperature ionic liquid. Electrochem Solid-State Lett 7(5):A97–A101. doi:10.1149/1.1664051
8. Howlett PC, Brack N, Hollenkamp AF, Forsyth M, MacFarlane DR (2006) Characterization of the lithium surface in N-methyl-N-alkylpyrrolidinium bis(trifluoromethanesulfonyl)amide room-temperature ionic liquid electrolytes. J Electrochem Soc 153(3):A595–A606. doi:10.1149/1.2164726
9. Shin J-H, Henderson WA, Passerini S (2005) PEO-based polymer electrolytes with ionic liquids and their use in lithium metal-polymer electrolyte batteries. J Electrochem Soc 152(5):A978–A983. doi:10.1149/1.1890701
10. Aurbach D, Zaban A, Ein-Eli Y, Weissman I, Chusid O, Markovsky B, Levi M, Levi E, Schechter A, Granot E (1997) Recent studies on the correlation between surface chemistry, morphology, three-dimensional structures and performance of Li and Li-C intercalation anodes in several important electrolyte systems. J Power Sources 68(1):91–98. doi:10.1016/S0378-7753(97)02575-5
11. Howlett PC, Izgorodina EI, Forsyth M, MacFarlane DR (2006) Electrochemistry at negative potentials in bis(trifluoromethanesulfonyl)amide ionic liquids. Z Phys Chem 220:1483. doi:10.1524/zpch.2006.220.10.1483
12. Randström S, Montanino M, Appetecchi GB, Lagergren C, Moreno A, Passerini S (2008) Effect of water and oxygen traces on the cathodic stability of N-alkyl-N-methylpyrrolidinium bis(trifluoromethanesulfonyl)imide. Electrochim Acta 53(22):6397–6401. doi:10.1016/j.electacta.2008.04.058
13. Zheng J, Gu M, Chen H, Meduri P, Engelhard MH, Zhang J-G, Liu J, Xiao J (2013) Ionic liquid-enhanced solid state electrolyte interface (SEI) for lithium-sulfur batteries. J Mater Chem A 1(29):8464–8470. doi:10.1039/c3ta11553d
14. Forsyth M, Howlett PC, Tan SK, MacFarlane DR, Birbilis N (2006) An ionic liquid surface treatment for corrosion protection of magnesium alloy AZ31. Electrochem Solid-State Lett 9(11):B52–B55. doi:10.1149/1.2344826
15. Howlett PC, Efthimiadis J, Hale P, van Riessen GA, MacFarlane DR, Forsyth M (2010) Characterization of the magnesium alloy AZ31 surface in the ionic liquid trihexyl

(tetradecyl)phosphonium bis(trifluoromethanesulfonyl)amide. J Electrochem Soc 157(11): C392–C398. doi:10.1149/1.3486119
16. Birbilis N, Howlett PC, MacFarlane DR, Forsyth M (2007) Exploring corrosion protection of Mg via ionic liquid pretreatment. Surf Coat Technol 201(8):4496–4504. doi:10.1016/j.surfcoat.2006.09.050
17. Howlett PC, Khoo T, Mooketsi G, Efthimiadis J, MacFarlane DR, Forsyth M (2010) The effect of potential bias on the formation of ionic liquid generated surface films on Mg alloys. Electrochim Acta 55(7):2377–2383. doi:10.1016/j.electacta.2009.11.080
18. Caporali S, Ghezzi F, Giorgetti A, Lavacchi A, Tolstogouzov A, Bardi U (2007) Interaction between an imidazolium based ionic liquid and the AZ91D magnesium alloy. Adv Eng Mater 9(3):185–190. doi:10.1002/adem.200600250
19. Shkurankov A, El Abedin SZ, Endres F (2007) AFM-assisted investigation of the corrosion behaviour of magnesium and AZ91 alloys in an ionic liquid with varying water content. Aust J Chem 60(1):35–42. doi:10.1071/ch06305
20. Sun J, Howlett PC, MacFarlane DR, Lin J, Forsyth M (2008) Synthesis and physical property characterisation of phosphonium ionic liquids based on P(O)2(OR)2− and P(O)2(R)2− anions with potential application for corrosion mitigation of magnesium alloys. Electrochim Acta 54 (2):254–260. doi:10.1016/j.electacta.2008.08.020
21. Efthimiadis J, Neil WC, Bunter A, Howlett PC, Hinton BRW, MacFarlane DR, Forsyth M (2010) Potentiostatic control of ionic liquid surface film formation on ZE41 magnesium alloy. ACS Appl Mater Interfaces 2(5):1317–1323. doi:10.1021/am900889n
22. Forsyth M, Neil WC, Howlett PC, Macfarlane DR, Hinton BRW, Rocher N, Kemp TF, Smith ME (2009) New insights into the fundamental chemical nature of ionic liquid film formation on magnesium alloy surfaces. ACS Appl Mater Interfaces 1(5):1045–1052. doi:10.1021/am900023j
23. Chang J-K, Chen S-Y, Tsai W-T, Deng M-J, Sun IW (2007) Electrodeposition of aluminum on magnesium alloy in aluminum chloride (AlCl3)-1-ethyl-3-methylimidazolium chloride (EMIC) ionic liquid and its corrosion behavior. Electrochem Commun 9(7):1602–1606. doi:10.1016/j.elecom.2007.03.009
24. Chang J-K, Chen S-Y, Tsai W-T, Deng M-J, Sun IW (2008) Improved corrosion resistance of magnesium alloy with a surface aluminum coating electrodeposited in ionic liquid. J Electrochem Soc 155(3):C112–C116. doi:10.1149/1.2828016
25. K-r L, Liu Q, Han Q, G-f T (2011) Electrodeposition of Al on AZ31 magnesium alloy in TMPAC-AlCl3 ionic liquids. Trans Nonferrous Metals Soc China 21(9):2104–2110. doi:10.1016/S1003-6326(11)60980-1
26. Perissi I, Bardi U, Caporali S, Lavacchi A (2006) High temperature corrosion properties of ionic liquids. Corros Sci 48(9):2349–2362. doi:10.1016/j.corsci.2006.06.010
27. Ashassi-Sorkhabi H, Es'haghi M (2009) Corrosion inhibition of mild steel in acidic media by [BMIm]Br ionic liquid. Mater Chem Phys 114(1):267–271. doi:10.1016/j.matchemphys.2008.09.019
28. Zhang QB, Hua YX (2009) Corrosion inhibition of mild steel by alkylimidazolium ionic liquids in hydrochloric acid. Electrochim Acta 54(6):1881–1887. doi:10.1016/j.electacta.2008.10.025
29. Shukla SK, Murulana LC, Ebenso EE (2011) Inhibitive effect of imidazolium based aprotic ionic liquids on mild steel corrosion in hydrochloric acid medium. Int J Electrochem Sci 6 (9):4286–4295
30. Ibrahim MAM, Messali M, Moussa Z, Alzahrani AY, Alamry SN, Hammouti B (2011) Corrosion inhibition of carbon steel by imidazolium and pyridinium cations ionic liquids in acidic environment. Port Electrochim Acta 29(6):375–389. doi:10.4152/pea.201106375
31. Zarrouk A, Messali M, Zarrok H, Salghi R, Ali AA, Hammouti B, Al-Deyab SS, Bentiss F (2012) Synthesis, characterization and comparative study of new functionalized imidazolium-based ionic liquids derivatives towards corrosion of C38 steel in molar hydrochloric acid. Int J Electrochem Sci 7(8):6998–7015

32. Likhanova NV, Domínguez-Aguilar MA, Olivares-Xometl O, Nava-Entzana N, Arce E, Dorantes H (2010) The effect of ionic liquids with imidazolium and pyridinium cations on the corrosion inhibition of mild steel in acidic environment. Corros Sci 52(6):2088–2097. doi:10.1016/j.corsci.2010.02.030
33. Guzmán-Lucero D, Olivares-Xometl O, Martínez-Palou R, Likhanova NV, Domínguez-Aguilar MA, Garibay-Febles V (2011) Synthesis of selected vinylimidazolium ionic liquids and their effectiveness as corrosion inhibitors for carbon steel in aqueous sulfuric acid. Ind Eng Chem Res 50(12):7129–7140. doi:10.1021/ie1024744
34. Likhanova NV, Olivares-Xometl O, Guzmán-Lucero D, Domínguez-Aguilar MA, Nava N, Corrales-Luna M, Mendoza MC (2011) Corrosion inhibition of carbon steel in acidic environment by imidazolium ionic liquids containing vinyl-hexafluorophosphate as anion. Int J Electrochem Sci 6(10):4514–4536
35. Tüken T, Demir F, Kıcır N, Sığırcık G, Erbil M (2012) Inhibition effect of 1-ethyl-3-methylimidazolium dicyanamide against steel corrosion. Corros Sci 59:110–118. doi:10.1016/j.corsci.2012.02.021
36. Barham HA, Brahim SA, Rozita Y, Mohamed KA (2011) Carbon steel corrosion behaviour in aqueous carbonated solution of MEA/[bmim] [DCA]. Int J Electrochem Sci 6(1):181–198
37. Zhou X, Yang H, Wang F (2011) [BMIM]BF4 ionic liquids as effective inhibitor for carbon steel in alkaline chloride solution. Electrochim Acta 56(11):4268–4275. doi:10.1016/j.electacta.2011.01.081
38. Scendo M, Uznanska J (2011) The effect of ionic liquids on the corrosion inhibition of copper in acidic chloride solutions. Int J Corros. doi:10.1155/2011/718626
39. Liu QX, El Abedin SZ, Endres F (2006) Electroplating of mild steel by aluminium in a first generation ionic liquid: a green alternative to commercial Al-plating in organic solvents. Surf Coat Technol 201(3–4):1352–1356. doi:10.1016/j.surfcoat.2006.01.065
40. Bakkar A, Neubert V (2013) Electrodeposition and corrosion characterisation of micro- and nano-crystalline aluminium from AlCl3/1-ethyl-3-methylimidazolium chloride ionic liquid. Electrochim Acta 103:211–218. doi:10.1016/j.electacta.2013.03.198
41. Marczewska-Boczkowska K, Kosmulski M (2009) The effect of chloride and water on the corrosion of copper in 1-butyl-3-methylimidazolium tetrafluoroborate. Mater Manuf Processes 24(10–11):1173–1179. doi:10.1080/10426910902976146
42. Espinosa T, Sanes J, Jimenez AE, Bermudez MD (2013) Surface interactions, corrosion processes and lubricating performance of protic and aprotic ionic liquids with OFHC copper. Appl Surf Sci 273:578–597. doi:10.1016/j.apsusc.2013.02.083
43. Zheng Y, Zhang S, Lü X, Wang Q, Zuo Y, Liu L (2012) Low-temperature electrodeposition of aluminium from lewis acidic 1-allyl-3-methylimidazolium chloroaluminate ionic liquids. Chin J Chem Eng 20(1):130–139. doi:10.1016/S1004-9541(12)60372-3
44. Bermudez M-D, Jimenez A-E, Martinez-Nicolas G (2007) Study of surface interactions of ionic liquids with aluminium alloys in corrosion and erosion-corrosion processes. Appl Surf Sci 253(17):7295–7302. doi:10.1016/j.apsusc.2007.03.008
45. Bardi U, Caporali S, Craig M, Giorgetti A, Perissi I, Nicholls JR (2009) Electrodeposition of aluminium film on P90 Li-Al alloy as protective coating against corrosion. Surf Coat Technol 203(10–11):1373–1378. doi:10.1016/j.surfcoat.2008.11.003
46. Spikes H (2001) Tribology research in the twenty-first century. Tribol Int 34(12):789–799. doi:10.1016/S0301-679X(01)00079-2
47. Holmberg K, Siilasto R, Laitinen T, Andersson P, Jasberg A (2013) Global energy consumption due to friction in paper machines. Tribol Int 62:58–77. doi:10.1016/j.triboint.2013.02.003
48. Ye CF, Liu WM, Chen YX, Yu LG (2001) Room-temperature ionic liquids: a novel versatile lubricant. Chem Commun 21:2244–2245. doi:10.1039/b106935g
49. Klaver TPC, Luppi M, Sluiter MHF, Kroon MC, Thijsse BJ (2011) DFT study of 1,3-dimethylimidazolium tetrafluoroborate on Al and Cu(111) surfaces. J Phys Chem C 115(30):14718–14730. doi:10.1021/jp200401h

50. Valencia H, Kohyama M, Tanaka S, Matsumoto H (2009) Ab initio study of EMIM-BF4 crystal interaction with a Li (100) surface as a model for ionic liquid/Li interfaces in Li-ion batteries. J Chem Phys 131(24):244705. doi:10.1063/1.3273087
51. Minami I (2009) Ionic liquids in tribology. Molecules 14(6):2286–2305. doi:10.3390/molecules14062286
52. Zhou F, Liang Y, Liu W (2009) Ionic liquid lubricants: designed chemistry for engineering applications. Chem Soc Rev 38(9):2590–2599. doi:10.1039/b817899m
53. Bermudez M-D, Jimenez A-E, Sanes J, Carrion F-J (2009) Ionic liquids as advanced lubricant fluids. Molecules 14(8):2888–2908. doi:10.3390/molecules14082888
54. Torimoto T, Tsuda T, K-i O, Kuwabata S (2010) New frontiers in materials science opened by ionic liquids. Adv Mater 22(11):1196–1221. doi:10.1002/adma.200902184
55. Palacio M, Bhushan B (2010) A review of ionic liquids for green molecular lubrication in nanotechnology. Tribol Lett 40(2):247–268. doi:10.1007/s11249-010-9671-8
56. Schluecker E, Wasserscheid P (2011) Ionic liquids in mechanical engineering. Chem Ing Tech 83(9):1476–1484. doi:10.1002/cite.201100110
57. Somers A, Howlett P, MacFarlane D, Forsyth M (2013) A review of ionic liquid lubricants. Lubricants 1(1):3–21. doi:10.3390/lubricants1010003
58. Angell CA, Ansari Y, Zhao ZF (2012) Ionic liquids: past, present and future. Faraday Discuss 154:9–27. doi:10.1039/c1fd00112d
59. Bermudez M-D (2010) Special issue on ionic liquids in tribology. Tribol Lett 40(2):213–284
60. Dörr N (2012) Special issue on ionic liquids as lubricants. Proceedings of the Institution of Mechanical Engineers, Part J. J Eng Tribol 226(J11):889–1006
61. Predel T, Pohrer B, Schluecker E (2010) Ionic liquids as alternative lubricants for special applications. Chem Eng Technol 33(1):132–136. doi:10.1002/ceat.200900325
62. Kondo Y, Koyama T, Tsuboi R, Nakano M, Miyake K, Sasaki S (2013) Tribological performance of halogen-free ionic liquids as lubricants of hard coatings and ceramics. Tribol Lett 51 (2):243–249. doi:10.1007/s11249-013-0159-1
63. Minami I, Inada T, Okada Y (2012) Tribological properties of halogen-free ionic liquids. Proceedings of the Institution of Mechanical Engineers, Part J. J Eng Tribol 226(J11):891–902. doi:10.1177/1350650112446276
64. Kronberger M, Pejakovic V, Gabler C, Kalin M (2012) How anion and cation species influence the tribology of a green lubricant based on ionic liquids. Proceedings of the Institution of Mechanical Engineers, Part J. J Eng Tribol 226(J11):933–951. doi:10.1177/1350650112459012
65. Espinosa T, Sanes J, Jimenez A-E, Bermudez M-D (2013) Protic ammonium carboxylate ionic liquid lubricants of OFHC copper. Wear 303(1–2):495–509. doi:10.1016/j.wear.2013.03.041
66. Espinosa T, Jimenez M, Sanes J, Jimenez A-E, Iglesias M, Bermudez M-D (2014) Ultra-low friction with a protic ionic liquid boundary film at the water-lubricated sapphire-stainless steel interface. Tribol Lett 53(1):1–9. doi:10.1007/s11249-013-0238-3
67. Liu X, Pu J, Wang L, Xue Q (2013) Novel DLC/ionic liquid/graphene nanocomposite coatings towards high-vacuum related space applications. J Mater Chem A 1(11):3797–3809. doi:10.1039/c3ta00764b
68. Zhao W, Wang Y, Zeng Z, Wu X, Chen J, Xue Q (2012) Micro/nano-texture design for improving tribological properties of DLC-IL composite films. Nanosci Nanotechnol Lett 4 (9):901–909. doi:10.1166/nnl.2012.1399
69. Liu X, Wang L, Xue Q (2011) DLC-based solid-liquid synergetic lubricating coatings for improving tribological behavior of boundary lubricated surfaces under high vacuum condition. Wear 271(5–6):889–898. doi:10.1016/j.wear.2011.03.021
70. Camillone N, Chidsey CED, Liu GY, Putvinski TM, Scoles G (1991) Surface-structure and thermal motion of normal-alkane thiols self-assembled on Au(111) studied by low-energy helium diffraction. J Chem Phys 94(12):8493–8502. doi:10.1063/1.460082
71. Pu J, Liu X, Wang L, Xue Q (2011) Formation and tribological properties of two-component ultrathin ionic liquid films on Si. Surf Interface Anal 43(10):1332–1340. doi:10.1002/sia.3718

72. Pu J, Jiang D, Mo Y, Wang L, Xue Q (2011) Micro/nano-tribological behaviors of crown-type phosphate ionic liquid ultrathin films on self-assembled monolayer modified silicon. Surf Coat Technol 205(20):4855–4863. doi:10.1016/j.surfcoat.2011.04.089
73. Xiao H, Guo D, Liu S, Pan G, Lu X (2011) Film thickness of ionic liquids under high contact pressures as a function of alkyl chain length. Tribol Lett 41(2):471–477. doi:10.1007/s11249-010-9729-7
74. Smith AM, Lovelock KRJ, Gosvami NN, Welton T, Perkin S (2013) Quantized friction across ionic liquid thin films. Phys Chem Chem Phys 15(37):15317–15320. doi:10.1039/c3cp52779d
75. Shah FU, Glavatskih S, Antzutkin ON (2013) Boron in tribology: from borates to ionic liquids. Tribol Lett 51(3):281–301. doi:10.1007/s11249-013-0181-3
76. Pejakovic V, Kronberger M, Mahrova M, Vilas M, Tojo E, Kalin M (2012) Pyrrolidinium sulfate and ammonium sulfate ionic liquids as lubricant additives for steel/steel contact lubrication. Proceedings of the Institution of Mechanical Engineers, Part J. J Eng Tribol 226 (J11):923–932. doi:10.1177/1350650112448978
77. Amann T, Dold C, Kailer A (2012) Rheological characterization of ionic liquids and ionic liquid crystals with promising tribological performance. Soft Matter 8(38):9840–9846. doi:10.1039/c2sm26030a
78. Hayes R, El Abedin SZ, Atkin R (2009) Pronounced structure in confined aprotic room-temperature ionic liquids. J Phys Chem B 113(20):7049–7052. doi:10.1021/jp902837s
79. Wakeham D, Hayes R, Warr GG, Atkin R (2009) Influence of temperature and molecular structure on ionic liquid solvation layers. J Phys Chem B 113(17):5961–5966. doi:10.1021/jp900815q
80. Elbourne A, Sweeney J, Webber GB, Wanless EJ, Warr GG, Rutland MW, Atkin R (2013) Adsorbed and near-surface structure of ionic liquids determines nanoscale friction. Chem Commun 49(60):6797–6799. doi:10.1039/c3cc42844c
81. Watanabe S, Takiwatari K, Nakano M, Miyake K, Tsuboi R, Sasaki S (2013) Molecular behavior of room-temperature ionic liquids under lubricating condition. Tribol Lett 51(2):227–234. doi:10.1007/s11249-013-0130-1
82. Ranke J, Othman A, Fan P, Mueller A (2009) Explaining ionic liquid water solubility in terms of cation and anion hydrophobicity. Int J Mol Sci 10(3):1271–1289. doi:10.3390/ijms10031271
83. Cammarata L, Kazarian SG, Salter PA, Welton T (2001) Molecular states of water in room temperature ionic liquids. Phys Chem Chem Phys 3(23):5192–5200. doi:10.1039/b106900d
84. Zhao W, Pu J, Yu Q, Zeng Z, Wu X, Xue Q (2013) A novel strategy to enhance micro/nano-tribological properties of DLC film by combining micro-pattern and thin ionic liquids film. Colloids Surf A Physicochem Eng Asp 428:70–78. doi:10.1016/j.colsurfa.2013.03.047
85. An R, Zhu Y, Wu N, Xie W, Lu J, Feng X, Lu X (2013) Wetting behavior of ionic liquid on mesoporous titanium dioxide surface by atomic force microscopy. ACS Appl Mater Interfaces 5(7):2692–2698. doi:10.1021/am400175z
86. Kheireddin BA, Lu W, Chen IC, Akbulut M (2013) Inorganic nanoparticle-based ionic liquid lubricants. Wear 303(1–2):185–190. doi:10.1016/j.wear.2013.03.004
87. Zhao W, Zeng Z, Peng S, Wu X, Xue Q, Chen J (2013) Fabrication and investigation the microtribological behaviors of ionic liquid-graphene composite films. Tribol Trans 56(3):480–487. doi:10.1080/10402004.2012.754071
88. Bou-Malham I, Bureau L (2010) Nanoconfined ionic liquids: effect of surface charges on flow and molecular layering. Soft Matter 6(17):4062–4065. doi:10.1039/c0sm00377h
89. Espejo C, Carrion F-J, Martinez D, Bermudez M-D (2013) Multi-walled carbon nanotube-imidazolium tosylate ionic liquid lubricant. Tribol Lett 50(2):127–136. doi:10.1007/s11249-012-0082-x
90. Li H, Rutland MW, Atkin R (2013) Ionic liquid lubrication: influence of ion structure, surface potential and sliding velocity. Phys Chem Chem Phys 15(35):14616–14623. doi:10.1039/c3cp52638k

91. Sweeney J, Hausen F, Hayes R, Webber GB, Endres F, Rutland MW, Bennewitz R, Atkin R (2012) Control of nanoscale friction on gold in an ionic liquid by a potential-dependent ionic lubricant layer. Phys Rev Lett 109(15), 10.1103/PhysRevLett.109.155502
92. Mendonca ACF, Padua AAH, Malfreyt P (2013) Nonequilibrium molecular simulations of new ionic lubricants at metallic surfaces: prediction of the friction. J Chem Theory Comput 9(3):1600–1610. doi:10.1021/ct3008827

Chapter 20
Industrial Applications of Ionic Liquids

Amal I. Siriwardana

20.1 Introduction

Ionic liquids (ILs) in principle are a diverse group of salts which are liquid at ambient temperatures. General definition of ionic liquid is a liquid at room temperature that consists only ions. However, the term "ionic liquid" includes an additional special definition to distinguish it from the classical definition of a molten salt (e.g., NaCl) (Table 20.1).

In molten salts, there are symmetric cations and anions making the lattice well packed and hence it requires large energy to break the lattice (Fig. 20.1). In ILs, asymmetric nature of cations of ILs makes the lattice weaker in comparison to molten slats. Therefore, ionic liquid has a low melting point ($<100\ ^\circ C$) than a molten salt. Only a liquid range below this temperature ($<100\ ^\circ C$) can enable the substitution of conventional organic solvents by ionic liquids. Room temperature ionic liquid (RTIL) is a liquid at room temperature.

During the last 10 years, various types of functionalized ionic liquids (Fig. 20.2) expressly categorized as being "task-specific" ionic liquids (TSILs) have been designed and synthesized for specific purposes such as catalysis, organic synthesis, separation of specific materials, and for the construction of nanostructured and ion conductive materials, amongst others. The salts are defined as functionalized ionic liquids when they are ionic liquids in which a functional group (FG) is covalently tethered to (a) the cation (b) the anion, or (c) a zwiterionic form of the salt. It is typically the cation that bears the reactive moiety.

Volatility of organic solvents in many applications is a huge problem in industry, especially for environment and human health. Ionic liquids are organic salts with low melting point that are being considered as green replacements for industrial

A.I. Siriwardana (✉)
TECNALIA, Parque Tecnológico de San Sebastián, Mikeletegi Pasealekua, 2, Donostia-San Sebastián, Gipuzkoa 20009, Spain
e-mail: amal.siriwardana@tecnalia.com

A.A.J. Torriero (ed.), *Electrochemistry in Ionic Liquids*,
DOI 10.1007/978-3-319-15132-8_20

Table 20.1 Comparison of molten salt and ionic liquid

Molten salt	Ionic liquid
High melting point	Liquid at low temperatures (<100 °C)
Highly viscous medium	Low viscosity

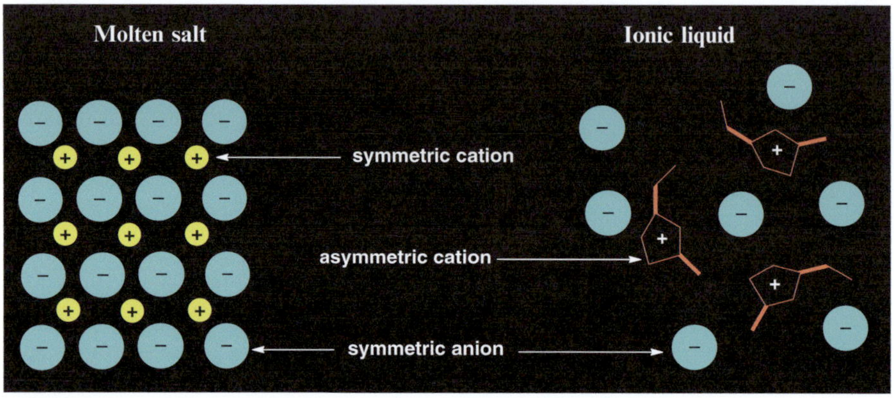

Fig. 20.1 How "ions" are packed in molten salts and ionic liquids

Fig. 20.2 Task-specific ionic liquids (TSILs)

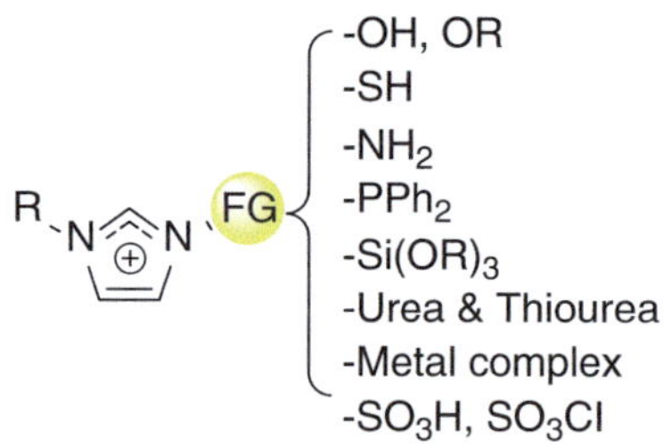

volatile organic compounds as they have practically no vapor pressure. Additionally, common problems known from volatile organic solvents such as formation of azeotropes with the reaction solvent, traces of solvent in the final product, or loss of solvent by uncontrolled evaporation can be avoided completely by using an ionic liquid as reaction medium. In this way, environmental and safety issues that arise through the use of volatile organic solvents can be solved in an efficient way.

Even though the first ionic liquids have been known since 1914, these ionic liquids have only been investigated intensively as solvents for chemical reactions in the past 15 years. In the middle of the 1980s, acidic chloroaluminate ionic liquids were successfully tested as catalysts in Friedel-Crafts reactions. The first application where an ionic liquid has been used as a catalytic solvent in biphasic catalysis was reported in 1990 by Chauvin et al. (see Sect. 20.3.2). However, the use of

Table 20.2 Issues to be addressed in ionic liquid commercialization

Scale up
Purity
Stability
Toxicity
Recycling
Disposal
Price

chloroaluminate ionic liquids has certain limitations that arise mainly from the fact that these systems are extremely reactive towards water, and very oxophillic. At the beginning of 1990s, the field of ionic liquid chemistry received a substantial boost, when the first tetrafluoroborate and hexafluorophosphate ionic liquids were described by Wilkes and Carlin. In contrast to the chloroaluminate ionic liquids, these ionic liquids with tetrafluoroborate and hexafluorophosphate ions show significantly higher stability against hydrolysis. The liquids can be handled under air, and in contact with water (at low temperatures), without immediate decomposition. The synthesis of these ILs even can be carried out in aqueous solutions.

Ionic liquids appeared as novel liquid system for the chemical industry. The scope of ionic liquids has been broadly extended into many areas as IL can reduce environmental impacts and that lead to more energy-efficient applications. With this tremendous development of ILs together with various commercially available ionic liquids in different scales, the industrial applicability of ionic liquids has begun. However, the industrialization of IL technology into different applications is rather slow. There are huge amount of patents and publications which have appeared during last decade and often describe numerous applications such as catalysis, separation, and so on. For the industrial use of ILs, some major issues must be addressed which are barriers to IL process commercialization (Table 20.2).

Ionic liquids are still in the research phase. Therefore, there are only a few industrial applications known (Fig. 20.3). However, there is a large field of potentially interesting applications (Table 20.3). Several pilots or industrial processes using ILs were publicly announced. There are few reviews which describe those applications in detail [1]. Most of the potential applications are as solvents or catalysts in many chemical reactions such as Diels-Alder, Friedel-Crafts reactions, and biocatalysis. Applications in other fields such as in separations, fluid applications, and analytical applications, are lower in numbers. There are now many companies who supply ionic liquids in gram scale to multi-ton scale. Some of the key suppliers are listed in Table 20.4. In this chapter, mainly the applications in the pilot-plant and industrial phase will be discussed. Aspects of ionic liquid stability, cost, recycling, and waste disposal will be also discussed at the end of this chapter.

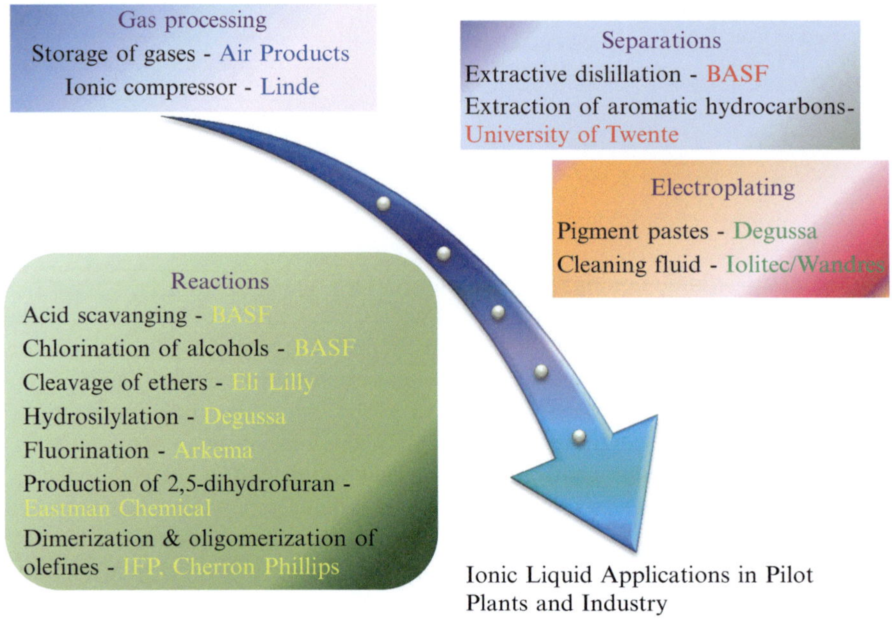

Fig. 20.3 Ionic liquids in pilot & industrial applications

Table 20.3 Summary of potential and industrial (including pilot plants) applications of ionic liquids

Applications in pilot plants and industry	Potential applications
Reactions	*Solvents for synthesis & catalysis*
Acid scavenging	Applications in reactions and catalysis
Chlorination of alcohols/Cleavage of ethers	Polymer synthesis
Hydrosilylation	Biocatalysis
Fluorination	*Electrochemical applications*
Production of 2,5-dihydrofuran	*Analytical applications*
Dimerization and oligomerization of olefins	*Separations*
Separations	Liquid separations
Extractive distillation	Gas separations
Extraction of aromatic hydrocarbons	*Fluid applications*
Electroplating	Thermal fluids
Compatibilizers for pigment pastes	Lubricants
Antistatic additives for cleaning fluids	
Gas processing	
Storage of gases	
Ionic compressor	

Table 20.4 Suppliers of ionic liquids

Sigma-Aldrich	http://www.sigmaaldrich.com/chemistry/chemistry-products.html?TablePage=16255866
Scionix	http://www.scionix.co.uk/
Cytec	http://www.cytec.com/
Merck KGaA/EMD Chemicals	http://www.emdmillipore.com/chemicals/disubstituted-imidazoliums-ionic-liquids/c_sLusHfETyAcAAAE852pCX8kx
IoLiTec	http://www.iolitec-usa.com/
Solvionic	http://en.solvionic.com/family/ionic-liquids
BASF	http://www.intermediates.basf.com/chemicals/ionic-liquids/index
SA CHEM	http://sacheminc.com/other-chemicals/tetramethylammonium-hydroxide-tmah/
DuPont	http://www2.dupont.com/FluoroIntermediates/en_US/uses_apps/ionic_liquids.html
ACROS	http://www.acros.com/_Rainbow/pdf/AO_IonicLiqui_Eur_def.pdf
Kanto Chemical Co.	http://www.kanto.co.jp/english/index.html
Solchemar	http://www.solchemar.com/
Chemada	http://www.chemada.com/index.php?page_id=46

20.2 Synthesis

20.2.1 Cation

The ionic liquid cation is generally a bulk organic structure with low symmetry. The cationic center most often involves a positively charged nitrogen or phosphorous atom. Those described so far are based on ammonium, sulfonium, phosphonium, imidazolium, pyridinium, picolinium, pyrrolidinium, thiazolium, oxazolium, and pyrazolium cations, usually completely substituted, although the more recent research has mainly focused on room temperature ionic liquids composed of asymmetric dialkylimidazolium cations associated with a variety of anions. Through modification of the cation, the properties of the liquid, notably, the melting point and liquid range [2], viscosity, [3] and miscibility with other solvents [4], can be altered. On the basis of the cation, ILs may be divided into five groups: (1) five-membered heterocyclic cations (2) six-membered and benzo-fused heterocyclic cations (3) ammonium, phosphonium, and sulfonium-based cations (4) functionalized imidazolium cations and (5) chiral cations.

20.2.1.1 Five-Membered Heterocyclic Cations

Figure 20.4 shows some five-membered cations including imidazolium, pyrazolium, triazolium, thiazolium, and oxazolium. Recently, Siriwardana et al. introduced methimazolium-based ionic liquids as a good coordinating solvent

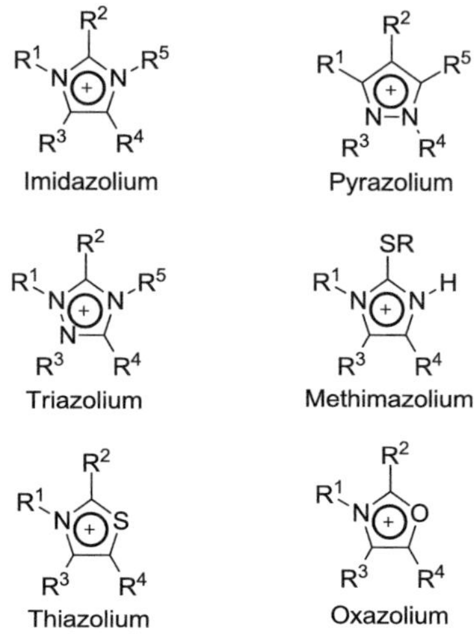

Fig. 20.4 Five-membered heterocyclic cations

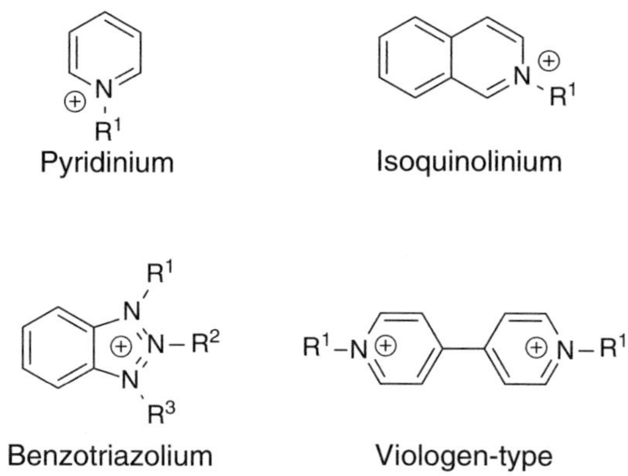

Fig. 20.5 Six-membered and benzo-fused heterocyclic cations

[5] It is generally assumed that non-symmetrical *N,N*-dialkylimidazolium cations give salts having the lowest melting points, however dibutyl, dioctyl, dinonyl, and didecylimidazolium hexafluorophosphates are also liquid at room temperature [6]. 1-Butyl-3-methylimidazolium and 1-ethyl-3-methylimidazolium cations are probably the most investigated structures of this class. As for the imidazolium species, the halide triazolium salts are solids at room temperature but metathesis to salts, such as bis(trifluoromethanesulfonyl)amide, triflate, or tetrafluoroborate, lowers the melting point to near or below room temperature.

20.2.1.2 Six-Membered and Benzo-Fused Heterocyclic Cations

Some cations which are made from heteroaromatics, have been investigated, such as pyridinium, viologen-type, benzotriazolium, and isoquinolinium (Fig. 20.5).

Moving on from the five-membered heterocyclic cations, there is a second, though less explored, class of heterocyclic room temperature ionic liquids (RTILs)-the pyridinium RTILs. These salts have been known for quite some time, but interest in them has been far less than that of the imidazolium RTILs. This is likely due to their more limited stability in the presence of nucleophiles and the toxicity of pyridine. Gorden et al. [7] have reported a series of pyridinium hexafluorophosphate salts with long alkyl chains (C12–C18) as ionic liquid crystals, some of them melt below 100 °C. A more recent area of focus in pyridinium RTILs is the viologen family of ionic liquids. While most viologens are very high melting solids, there are handfuls that do exhibit much lower melting points. Most of these materials exhibit melting points below 100 °C, but still not quite room temperature. Benzotriazolium-based ionic liquids form liquids below 100 °C depending on the length of the alkyl chain on *N*-atom [8].

20.2.1.3 Ammonium, Phosphonium, and Sulfonium-Based Cations

Tetraalkylammonium (Fig. 20.6) salts have been known for centuries. In terms of their use as RTILs, earlier studies led to the conclusion that longer alkyl chains were required to obtain room temperature melting points. These are typically prepared by alkylation of the parent amine. To obtain low melting points, typically at least two or three different alkyl groups are required to create crystal packing constraints; this usually means several steps of alkylation. Although phosphonium RTILs are certainly known and finding growing applications in organic synthesis and other areas [9], there is surprisingly little in the way of published data regarding their physical properties [10]. The hydrogen sulfate salt of the tetrabutylphosphonium cation has a melting point of 122–124 °C, while the hydrogen sulfate salt of the tributyldecylphosphonium cation is a room temperature liquid [11]. In comparison to ammonium counterparts, physical properties such as viscosity of phosphonium RTILs are generally somewhat higher than their ammonium counterparts, but decrease very rapidly with temperature. Phosphonium salts are generally high thermally stable than ammonium salts [10]. Phosphonium salts are typically made by alkylation of the parent phosphine [12]. To make the larger phosphonium cations this is straightforward, however the pyrophoric nature of the lighter alkyl phosphines makes these proves a more difficult matter. One of the least studied types of RTILs is that based on the trialkylsulfonium cation. As might be expected, the melting point and density decrease as the size of the cation increases. The viscosity,

Fig. 20.6 Ammonium, sulfonium, and phosphonium cations

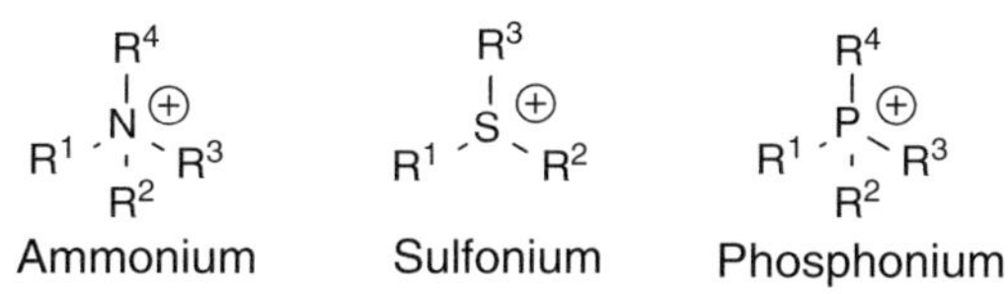

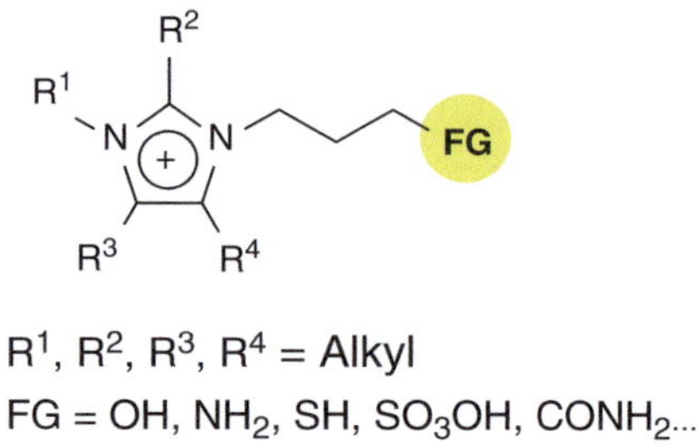

Fig. 20.7 Functionalized imidazolium cations

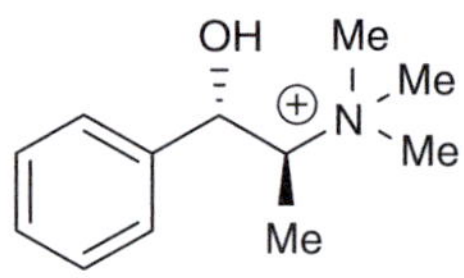

Fig. 20.8 (1S, 2R)-(+)-*N*, *N*-Dimethylephedrinium ion

however, reaches a minimum with the triethyl compound and then increases significantly for the tributyl compound, again showing that the size of the cation and anion are very important for the viscosity of RTILs.

20.2.1.4 Functionalized Imidazolium Cations

Recent advance in ionic liquids research provided another route for achieving functionalized ionic liquids (Fig. 20.7) in which a functional group is covalently tethered to the cation or anion of the ionic liquid, especially to the two *N* atoms of the imidazole ring. It was expected that these functionalized ionic liquids may further enlarge the application scope of ionic liquids in chemistry.

20.2.1.5 Chiral Cations

There are growing numbers of reports indicating that chiral ionic liquids may be useful in many areas of science and technology though synthesis and use of chiral ILs is in its infancy. For example, the use of ephedrinium-based chiral ILs (Fig. 20.8) as the gas chromatography stationary phases has been reported [13]. Chiral ILs can be prepared either from chiral starting materials [Eq. 20.1(b)] or by using asymmetric synthesis [Eq. 20.1(a)] [14].

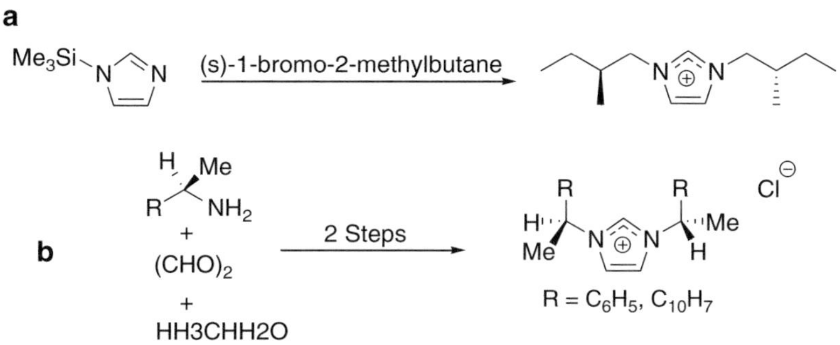

Equation 20.1 Synthesis of chiral cations

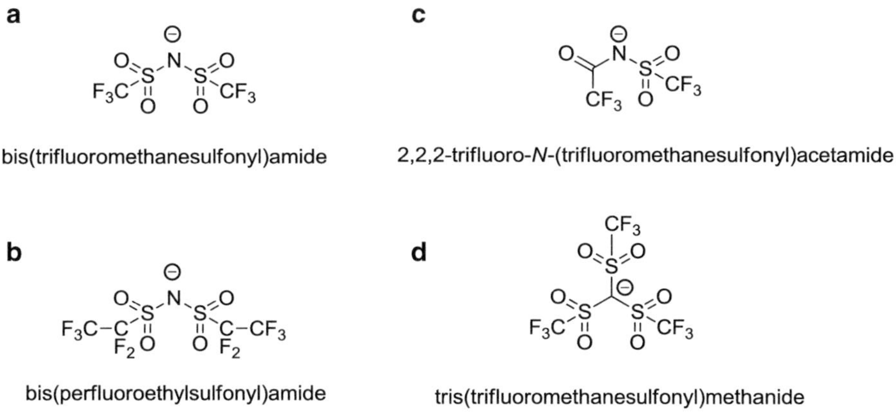

Fig. 20.9 Amide and methanide anions

20.2.2 *Anions*

Anions that form room temperature ionic liquids are usually weakly basic inorganic or organic compounds that have a diffuse or protected negative charge. On the basis of the anion, ILs may be divided into six groups: (1) ILs based on $AlCl_3$ and organic salts [15] (2) ILs based on anions like $[PF_6]^-$ [6, 16], $[BF_4]^-$ [17, 18], and $[SbF_6]^-$ [16] (3) ILs based on anions like **a** [19, 20], **b** [20, 21], **c** [8], and **d** [20] in Fig. 20.9 (4) ILs based on anions like alkylsulfates [22], alkylsulfonates [23], alkylphosphates [10], alkylphosphinates [10], and alkylphosphonates [10]

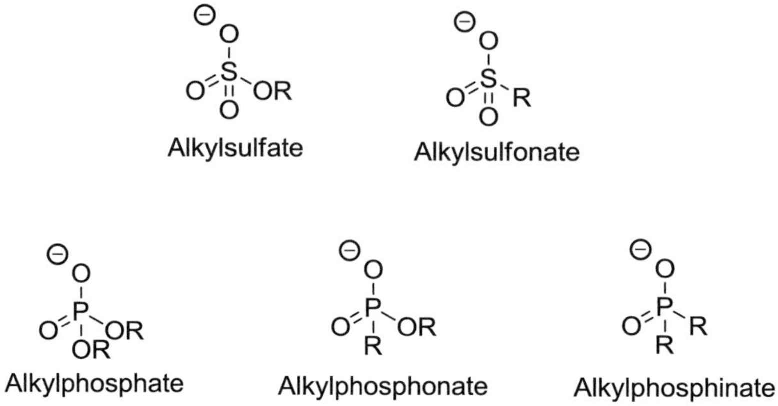

Fig. 20.10 Phosphate, phosphinate, sulfate, and sulfonate anions

Fig. 20.11 Borate and borane anions

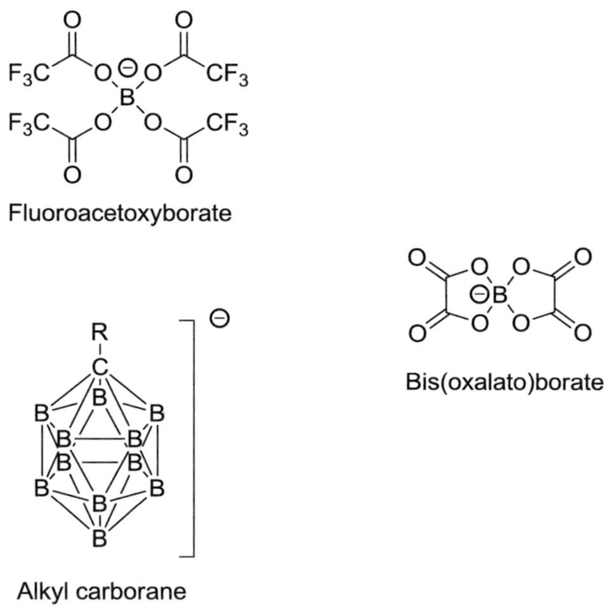

(Fig. 20.10) (5) ILs based on anions like mesylate [24, 25], tosylate ($[tos]^-$) [25], trifluoroacetate ($[CF_3CO_2]^-$) [3], acetate ($[CH_3CO_2]^-$) [8], $[SCN]^-$ [26], triflate ($[CF_3SO_3]^-$) [16, 24, 27], and dicyanamide ($[N(CN)_2]^-$) [28, 29] (6) ILs based on anions like borates [30] and carboranes (Fig. 20.11) [31].

20.2.3 Purification of Ionic Liquids

20.2.3.1 Common IL Impurities

There are four main categories of common IL impurities; water, organics (from unreacted starting materials or residual reaction solvents), halides, and metals. The most abundant impurity found in ILs is water. Many IL applications are not affected (in fact they are sometimes enhanced) by the presence of water and so excessive drying is not always necessary. However, testing in water-sensitive applications and fundamental characterization of new ILs should only be performed on very dry samples. The low volatility of most ILs allows for the removal of water by placing the wet IL under vacuum for several hours. Note however that this procedure is not appropriate for many protic ionic liquids since in those cases the conjugate acid and base species may also be volatile and loss of one of the species may occur. There can also be an issue with rigorous vacuum drying in those cases where the anion is a moderately strong base species (e.g., acetate) since small amounts of the conjugate acid can form from water and then be evaporated, leaving a hydroxide impurity. It is important to realize that vacuum drying involves mass transport of the water through the bulk of the material to the surface before it can be evaporated. Sample mass and surface area are therefore significant parameters; stirring may also be required. It is strongly recommended that reporting of procedural details includes information of sample volume and surface area in order that ensures reproducibility. It is also important to understand that during the process of evaporation of the water, or any other volatile material, the manifold pressure is often lower than the pressure at the material surface. In order to determine whether the sample has fully equilibrated with the applied vacuum one common technique involves isolating the manifold from the pump and monitoring the pressure in the manifold for approximately 20 min. When the sample is opened to the vacuum, any pressure rise greater than the leak rate of the manifold indicates continued volatilization from the sample. A typical procedure involves stirring the IL under vacuum on a standard Schlenk-line until a vacuum of approximately 0.4 mbar is maintained for 3 h. The time required for this process can be very short (a few hours) or very long (several weeks). ILs can incorporate water very strongly and in some cases vacuum conditions alone do not reduce water to a suitable level within a suitable timeframe. The sample can be heated while under vacuum, but caution should be used as heat can accelerate IL decomposition. Alternatives include vacuum azeotropic distillation with solvents such as toluene or dissolving the IL in a low boiling solvent and placing over dry molecular sieves. Chemical drying agents (e.g., potassium metal, magnesium chloride, magnesium sulfate) can be very effective for the removal of water. However, consideration should be given to the impurities these additives will likely leave behind in the IL. If none of the above methods for drying ILs are appropriate, water should be avoided altogether, e.g., the IL can be synthesized in a dry atmosphere using dried reagents and reaction solvents. Other common IL impurities are unreacted organic starting materials and residual organic reaction solvents. As with the removal of water, vacuum-induced evaporation is usually very effective. If not, the IL can be washed with a low boiling, immiscible organic solvent such as diethyl ether, which can subsequently be removed under vacuum. Suspended particles are a commonly overlooked impurity in ILs. They originate from many sources including:

- Metathesis by-products (especially silver halide salts which often precipitate as very small particles)
- Sorbents
- Molecular sieves
- Chemical drying agents

Particles have been shown to affect several fundamental IL properties [32] and therefore their presence should be carefully considered. To the knowledge of these authors no methods have been shown to completely remove particulate contamination from ILs, but filtration through a 200 nm PTFE syringe filter can reduce their levels to <10 ppm. Halide and alkali metal salts originated as metathesis by-products can remain dissolved in ILs. For hydrophobic ILs, simply washing with water until no salt is detected in the washings is very effective. With hydrophilic ILs, removal of these salts can be more complex. The IL can be dissolved in a water immiscible solvent such as dichloromethane and washed as above. However, this usually leads to a significant loss in IL yield (the loss in yield can be reduced by using very cold water). Passing the IL through silica gel has also been recommended as a method to reduce alkali metal halide salts [33]. If an IL is synthesized by precipitating an alkali metal halide salt from an organic solvent such as acetone, acetonitrile, methanol, etc. these solvents should be extensively dried. Even small traces of water can greatly increase the solubility of these salts in the IL.

Previously, Siriwardana et al. have reported the most common methods for purifying ionic liquids and also are briefly summarized in Table 20.5 [34].

Table 20.5 Purification methods of ionic liquids

Sorbents
Distillation
Zone melting
Clean synthesis
Removal of particles
Removal of water and reaction solvents (vacuum, azeotropes, Li metal)

20.2.3.2 Characterization of Impurities

Because impurities can have a profound impact on subsequent IL applications [35], all data reported involving ILs should be accompanied by an analysis report on the presence (or absence) of common IL impurities. At least one technique should be used to confirm the structure of the IL. Mass spectrometry, nuclear magnetic resonance (NMR) spectrometry, elemental analysis, and X-ray crystallography can all be useful techniques for this purpose. However, many IL impurities cannot be directly quantified using these techniques and so it should be remembered that these techniques can only be used to confirm IL structure and should not be used as evidence of IL purity. For example, in an IL contaminated with significant quantities of lithium chloride, neither lithium nor chloride ions will appear in standard mass spectrometry because of their small masses, they may not significantly alter the C, H, and N percentages in elemental analysis, and they are not detectable in standard ^{1}H and ^{13}C NMR. The presence of lithium chloride will likely prevent the growth of suitable crystals for diffraction; nonetheless, obtaining a single IL crystal suitable for analysis does not rule out the presence of lithium chloride in the bulk IL. Therefore a crystal structure should not be considered an evidence of bulk purity. Thus the researcher must consider all possible sources of contamination and systematically determine specific impurities using techniques appropriate to the trace level involved. Water content can be crudely assessed using ^{1}H NMR or infrared spectroscopy but at low levels ($<$1,000 ppm) must be accurately quantified using a Karl–Fischer titration. Organic reaction solvents or unreacted starting materials can be detected by ^{1}H NMR for quantities greater than 1 mol%. Smaller quantities can be assessed by specific detection as described in texts such as "Vogel's textbook of quantitative chemical analysis." A useful example of specific detection involves detecting unreacted 1-methylimidazole (a common IL starting material that is difficult to remove by evaporation) [36] Copper(II) chloride is added to the IL to form a colored complex with 1-methylimidazole, which can be detected at high levels visually, or more accurately quantified using atomic absorption spectroscopy. Halide impurities can be assessed using many techniques with varying degrees of accuracy. Techniques include:

- ^{19}F, ^{35}Cl NMR ($>$1 mol %)
- Titration with silver nitrate; monitored visually for qualitative analysis or using the Volhard procedure ($>$10 ppm)
- Ion selective electrodes ($>$100 ppm)
- Ion chromatography ($>$10 ppm)
- Inductively coupled plasma-mass spectrometry ($>$10 ppb)

Ion selective electrodes and ion chromatography are used for halide analysis. Metallic impurities are usually present as one of two forms; unreacted metallic starting material salts or as halide salts from metathesis by-products. Direct metal analysis is not always performed; instead the counter ions from metathesis by-products or halide salts are analyzed and if their levels are sufficiently low it

Table 20.6 Characterization of impurities, quantifying impurities, and characterization of ionic liquids

Mass spectrometry
NMR
Ion selective electrodes
Karl Fisher
Elemental analysis
Metal analysis
Ion chromatography
Single-crystal diffraction
UV/Vis for color
Chemical methods (Vollhard procedure for chloride analysis, methylimidazole $CuCl_2$ test)

is assumed the metal content is also sufficiently low. For metals present in other forms or where the counter ion is difficult to analyze, metal content can be accessed directly using atomic absorption spectroscopy (available for most metals) or ion selective electrodes (available for Na^+, K^+, Ag^+, Ca^{2+}, Mg^{2+}, Cd^{2+}, Cu^{2+}, and Pb^{2+}). It was not intended that this section exhaustively covers all possible IL impurities; instead methods for analysis of common IL impurities were suggested. Characterization of Impurities, quantifying impurities, and characterization of ILs are briefly summarized in Table 20.6.

20.3 Industrial Applications

The scope of applications of ionic liquids has been extended to many domains and now much broader than assumed. Following this tremendous development associated with the commercial availability of ILs, the industrial applicability of ILs rapidly appeared as an important aspect as demonstrated by the accelerating number of patents. However, despite of their significant benefits, their translation into viable industrial processes is far from being obvious and the industrialization of IL technologies is rather slow. For the industrial use of ILs, some major issues must be addressed such as purity, stability, toxicity, cost, waste disposal, and recycling which may be barriers to IL commercialization. Several pilot plants or industrial processes using ILs were publicly announced and some of them are considered in this section.

20.3.1 BASIL's Process

Currently, the most important application of Ionic Liquids described is their use as reaction media for chemical processes. A major breakthrough was reached with BASF's introduction of the first application (since 2002) on a commercial multi-ton scale [37–42]: the BASIL™-process, which received the "Innovation for Growth Award 2004" of ECN. BASF has developed industrial processes utilizing ionic liquids. These processes are called Basil™ processes. The abbreviation BASIL stands for "Biphasic Acid Scavenging utilizing Ionic Liquids." BASF makes use of this technology for the synthesis of alkoxyphenylphosphines (Eq. 20.2), these compounds are important raw materials for the production of BASF's Lucirin photoinitiators that are used to cure coatings and printing inks by exposure to UV radiation. These alkoxyphenylphosphines are commonly produced by reaction of chlorophenylphosphines with alcohols. Stoichiometric amounts of HCl gas is formed during the synthesis of alkoxyphenylphosphines, and this gas has to be scavenged to avoid product decomposition.

Equation 20.2 BASF's BASIL process

In the past, scavenging was done with a tertiary amine. This resulted in thick, nonstirrable slurry, which was difficult to filter off from the product (Fig. 20.12). Instead of triethylamine, 1-methylimidazole was used as an acid scavenger and immediately led to excellent results (Fig. 20.13). The Ionic Liquid formed in the reaction was methylimidazolium chloride. Methylimidazolium chloride has a

Fig. 20.12 Conventional method with no phase separation. Image copyright from BASF—The Chemical Company

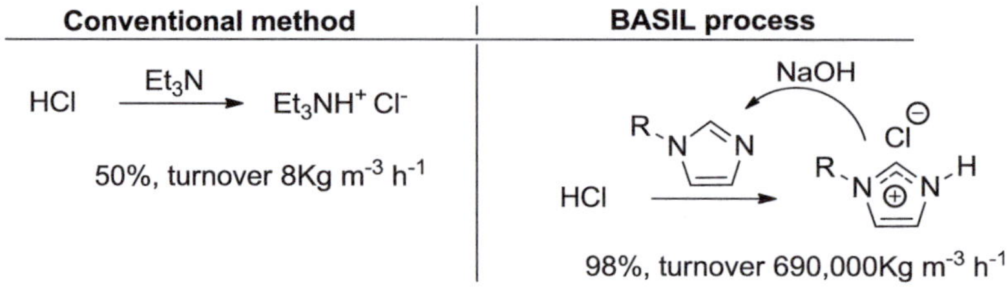

Fig. 20.13 Conventional and BASIL process for acid scavenging

Fig. 20.14 (**a**) Large reactors for conventional method. (**b**) Small reactor for BASIL's process. Image copyright from BASF—The Chemical Company

melting point of 75 °C and, therefore, is a liquid at the reaction temperature of approx. 80 °C.

After the reaction, there are two clear liquid phases that can easily be worked up by a simple phase separation. The upper phase is the pure product and no reaction solvent is needed. The lower phase is the pure Ionic Liquid and can be deprotonated with sodium hydroxide, regenerating the methylimidazole (Fig. 20.13). Comparing to the conventional method (Fig. 20.14a), BASIL process uses a much smaller reactor (Fig. 20.14b).

The ionic liquid functioned as a nucleophilic catalyst. The combined effect is a tremendous increase in the yield per unit volume time from 8 to 690,000 kg $m^{-3}\,h^{-1}$, a factor of 80,000 increased productivity (Table 20.7). This enabled BASF to carry out the reaction in a little jet reactor (Fig. 20.14b), the size of a thumb (which previously needed a 20 m^3 batch vessel). This little reactor is part of a continuously operated plant with a capacity of more than 1,000 t/annum and the reaction is now carried out at a multi-ton scale. This is a good example to show that handling of

Table 20.7 Comparison of time and yield for conventional and BASIL's process

	Conventional method	BASIL process
Space–time yield	8 kg m^{-3} h^{-1}	690,000 kg m^{-3} h^{-1}
Yield	50 %	98 %

Fig. 20.15 BASIL process with a much smaller jet reactor. Image copyright from BASF—The Chemical Company

large scale of ionic liquids in industry is practical. The plant went on stream in Q3/2004 at BASF's Ludwigshafen site (Fig. 20.15).

20.3.2 Dimersol Process

IFP [(Institut Francais du Pètrole)—Axens] was the first to operate an ionic liquid pilot plant [43]. Since the mid-1970s, IFP developed the Dimersol™ process (Dimersol-G, and -X) that upgrades light olefins by dimerization (respectively, propene and butenes) to the more valuable branched hexenes and octenes, respectively in high yield (Figs. 20.16 and 20.17) [44].

This is an important industrial process with 35 plants in operation worldwide, each plant producing between 20,000 and 90,000 t per year of dimer, with a total annual production of 3,500,000 t. The longer-chain olefins produced in the dimerization process are usually hydroformylated to alcohols (e.g., isononanols): isononanols are then converted into dialkyl phthalates, which are used as poly

Dimerization of propenes into hexenes
First industrial unit in 1977
35 units in operation worldwide
3.5 Mt products / year

Fig. 20.16 Dimerization of propenes into hexenes

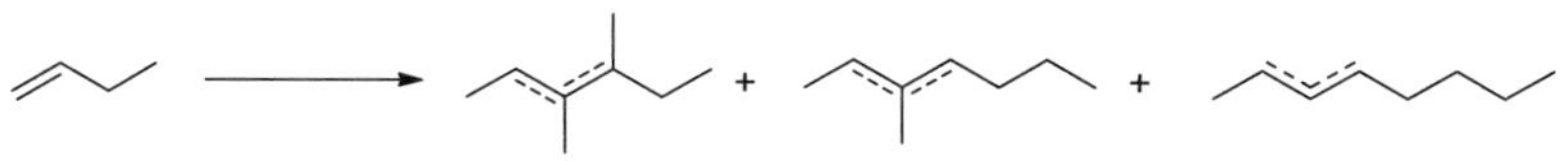

Dimerization of butenes into octenes
First industrial unit in 1980
0.4 Mt products / year

Fig. 20.17 Dimerization of butenes into octenes

(vinyl chloride) plasticizers. For this homogeneous process, the reaction is operated without a solvent in a unique liquid phase using a Ziegler-type catalyst based on nickel and activated with an alkyl-aluminum cocatalyst [45]. However, it has been found that the catalyst shows greater activity when it is dissolved in undesirable aromatic or halogenated hydrocarbons.

20.3.3 Difasol^{TM} Process

The Difasol^{TM} process [46–50] (Figs. 20.18 and 20.19) can be considered as the biphasic analogue of the Dimersol-X^{TM} process. The reaction takes place with the same nickel catalyst precursor using chloroaluminate ionic liquids as the solvent (Fig. 20.20). The activity of the catalyst is much higher than in both solvent-free and conventional solvent systems. When associated with a chloroalkylaluminium activator like $EtAlCl_2$, this mixture can react to form ionic liquids presenting mixed anions (Fig. 20.20b). The ionic liquid can then act as both solvent and cocatalyst (Fig. 20.20c). As the activity of the nickel system depends on the Lewis acidity, an accurate adjustment of the $EtAlCl_2/AlCl_3$ is required to optimize the efficiency of the catalytic system. The best results were obtained from $[C_4mim][Cl]/AlCl_3/C_2H_5AlCl_2$ (1:1.22:0.06 M ratio) mixture which is the ionic liquid phase. Because of the solubility of the Ni catalyst in the ionic liquid phase and the poor miscibility of the products, the production unit is essentially reduced to a continuously stirred-tank reactor followed by a phase separator. With this process, more efficient overall catalyst utilization and a significant increase in the yield of octenes is

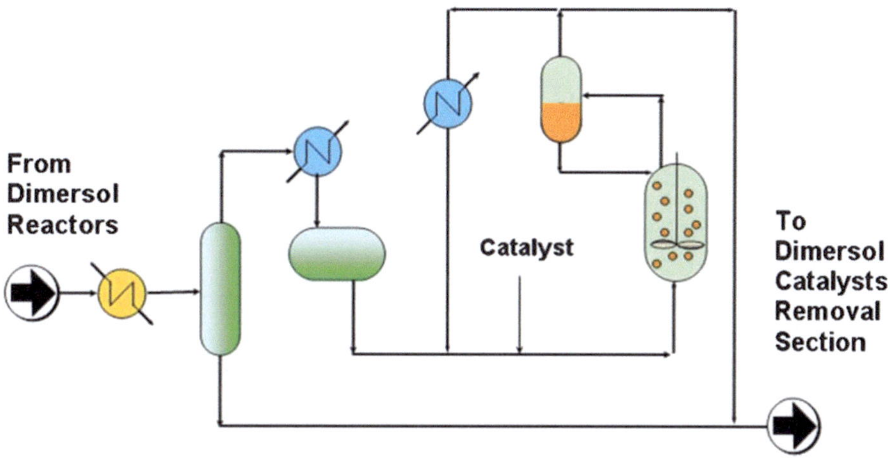

Fig. 20.18 Difasol process. Image adapted from [1c] with permission from the RSC

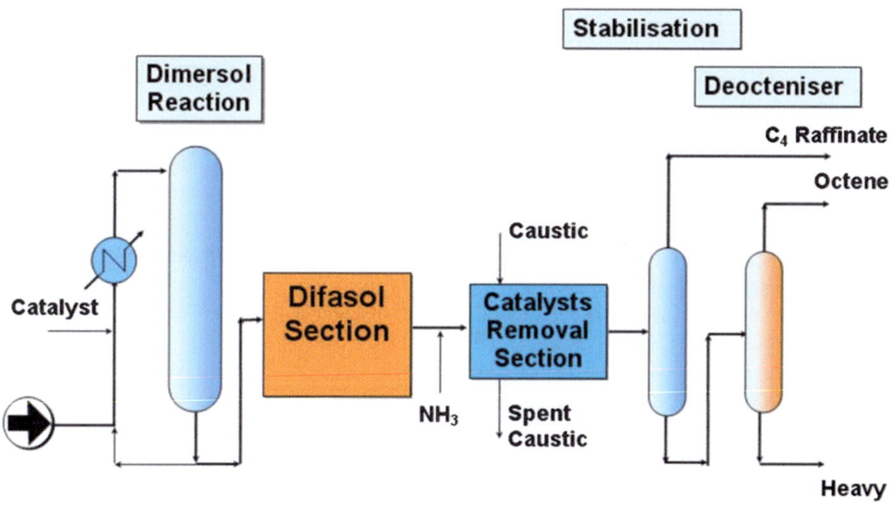

Fig. 20.19 Integrated Dimersol and Difasol processes. Image adapted from [1c] with permission from the RSC

achieved. Dimer selectivity in the range of 90–92 % with butene conversion in the range of 80–85 % can be obtained.

The advantages of the biphasic Difasol process in comparison to the homogeneous Dimersol process, are summarized in Table 20.8.

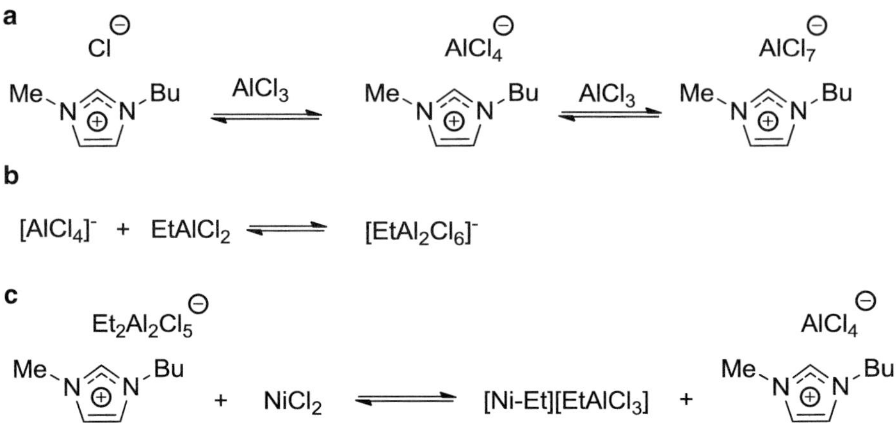

Fig. 20.20 (a) Synthesis of chloroaluminate ionic liquid. (b) Formation of the ionic liquid presenting mixed anions by using chloroalkylaluminium activator. (c) Ionic liquid acts as both solvent and cocatalyst

Table 20.8 Advantages of Difasol process in comparison to Dimersol process

Process: Difasol
Ni consumption is less. Reduced catalyst disposal and cost.
Better dimer selectivity.
No ionic liquid can be detected in products.
Much smaller reactor.
Operated in biphasic system. No co-miscibility was observed between the products and the ionic liquids. Easy product separation by settling which does not require heating and results in energy saving and reduced catalyst consumption.
Overall yield of butenes is 10 % higher and achieved in a single step even with a low concentration of alkene feed.
Maintained over a wide range of alkene concentrations with the same catalyst consumption.
Difasol feed is completely purified from the eventual impurities that could accumulate in the IL.

20.3.4 Degussa

Degussa develops ionic liquids on three distinguishable applications [hydrosilylation, paint additives, and lithium–ion batteries].

20.3.4.1 Hydrosilylation

Hydrosilylation is a widely used method for the synthesis of organomodified silanes and siloxanes. Organosilicon compounds held approximately 15 % of the entire silicon market and are accessible in a number of ways [51], including the catalyzed hydrosilylation of C=C double bond containing compounds with Si–H functional polydimethylsiloxanes (Eq. 20.3). Their aim was the synthesis of polyether-siloxanes, an important class of surface-active compounds which finds use in a broad range of industrial applications. Particularly, interesting are polyether-siloxanes, constituting an important class of surface-active compounds which find use in a broad range of industrial applications. The experimental procedure is rather simple, and can be described as a one-pot biphasic synthesis (Fig. 20.21) [52, 53].

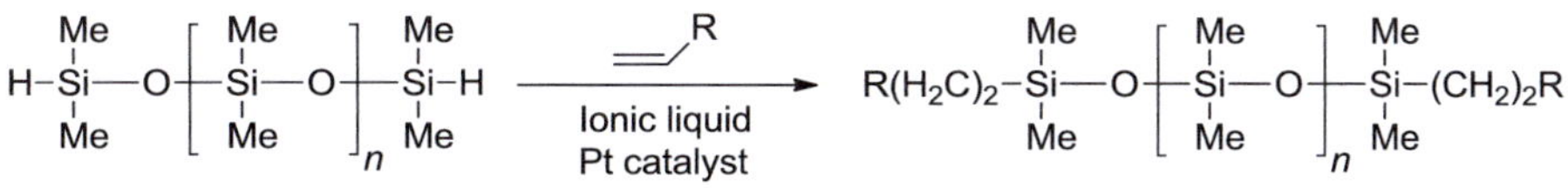

Equation 20.3 Synthesis of polydimethylsiloxanes

The hydrosilylation reaction is catalyzed by homogeneous or colloidal Pt catalysts. The major drawback of this reaction is that the catalyst cannot easily be removed from the product after completion of the reaction. To overcome this limitation, Degussa introduced the use of ionic liquid as a way of catalyst heterogenization [54, 55]. Here, ionic liquid is used as solvent and nanoparticle stabilizer.

They have studied silane-functionalized polydimethylsiloxanes with different chain lengths and functionality patterns, and polyethers and ethers with different ethylene and propylene oxide content (Fig. 20.22) [52, 54]. Several ionic $[H_2(PtCl_6)]$ and neutral $[(\mu\text{-Cl})_2\{PtCl\text{-}\ (cyclohexene)\}_2]$ platinum catalysts (Table 20.9) as well as ionic liquids [1-butyl-4-methylpyridinium tetrafluoroborate (**1a**), 1-butyl-3-methylpyridinium chloride (**1b**), 1,2,3-trimethylimidazolium methylsulfate (**2**), and TEGO1 IL K5MS (**3**)] were evaluated (Fig. 20.22).

There are several factors which are affected for the reaction which are summarized in Table 20.10. The hydrophobicity is a key factor which affects the reaction. To increase the hydrophobicity of the organosilicon products, the chain length of the polydimethylsiloxane should be increased as well as silane-functionalization of the polydimethylsiloxane should be decreased. The polyethers are more hydrophobic. If the ionic liquid layer is more hydrophobic it will separate more easily from the organosilicon products.

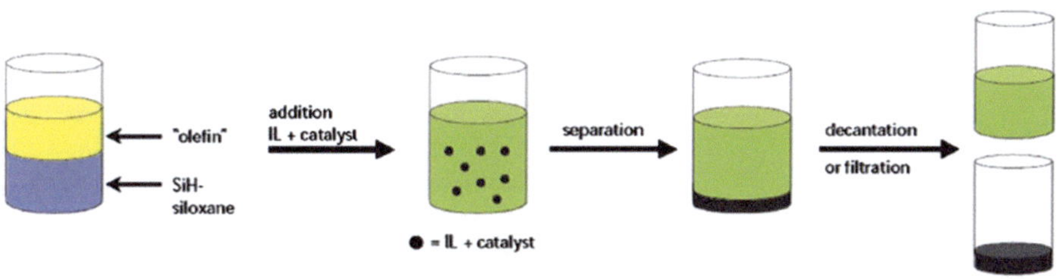

Fig. 20.21 Schematic for the hydrosilylation process (*yellow* = alkene; *blue* = silane; *green* = product; *black* = catalyst dissolved in ionic liquid phase. Image adapted from [1c] with permission from the RSC

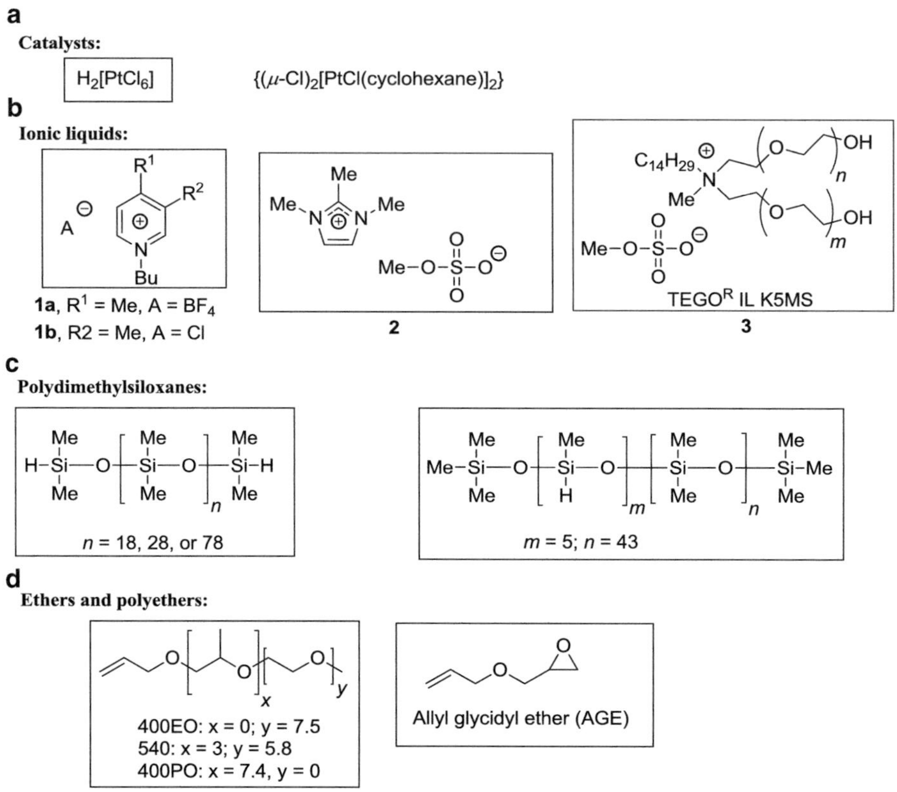

Fig. 20.22 Catalysts, ionic liquids, polydimethylsiloxanes, ethers, and polyethers which were used for hydrosilylation process

Table 20.9 Catalysts used in the process

Catalyst: $[(\mu\text{-Cl})_2\{PtCl(cyclohexene)\}_2]$	Catalyst: $H_2[PtCl_6]$
This catalyst is neutral	This catalyst is ionic
Catalyst leaches from IL	Leaching from IL is very low (below 1 ppm)

Table 20.10 Factors that affect the reaction

Hydrophilicity/hydrophobicity of the substrates
Various catalyst/ionic liquid solution combinations
Separation behavior of the ionic liquid
The nature of the catalyst

It is crucial to find an appropriate combination of a catalyst and an ionic liquid which has to be harmonized with the hydrophilicity/hydrophobicity of the product. For example, hydrosilylation reactions in the presence of ionic liquids based on 1,3-dialkylimidazolium cations did not give the desired polyethersiloxanes, as the 2-H position of the cation was too reactive. The authors assumed that formation of NHC (*N*-heterocyclic carbenes) which coordinates to the Pt center is responsible of the catalyst deactivation.

20.3.4.2 Paint Additives

Degussa has managed to develop the first industrial application of ionic liquids as performance additives to a new range of paints, for improved finish, appearance, and drying properties [53, 56, 57]. The range is marketed under the name TEGO© Dispers, and added to the Pliolite® paint range.

They have evaluated some of the non-imidazolium-based ionic liquids towards their use as performance additives during the course of their investigations as predicted in Fig. 20.23.

Without the addition of an ionic liquid to the mixture of the Pliolite® (special solvent-based coatings particularly used in the South of Europe and in the UK) coating system and one of the three colorants, the development of the color strength is miserable and poor. They used four different commercially available ionic liquids and one ionic liquid (dimethylimidazolium methyl sulfate) to evaluate their impact on the development of the color strength (Fig. 20.23). There are few advantages of using ionic liquids as dispersants which are mentioned in Table 20.11.

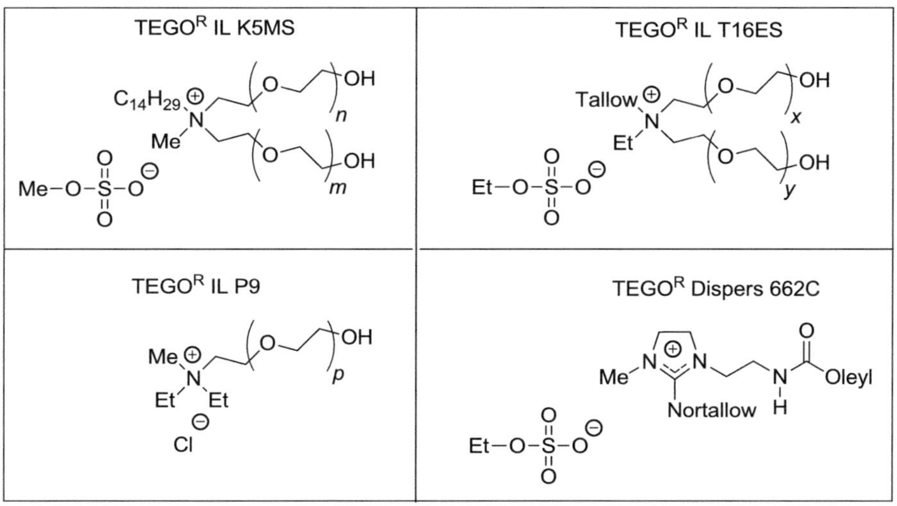

Fig. 20.23 Commercially available ionic liquids used as dispersants in paints

Table 20.11 Advantages of using ionic liquids as dispersants

Addition of only 0.4–2 % of an ionic liquid to these mixtures significantly improved the color strength
Ionic liquids are secondary dispersing agents, therefore, they do not have to be used essentially during the grinding of the pigment but can be used at any step during pigment paste manufacture
Ionic liquids provide the user and the manufacturer of commercial pigment pastes with a higher flexibility regarding the optimization of their products

Table 20.12 The ways of using ionic liquids to enhance the applicability of universal pigments

(A) White base paint + pigment paste based on universal dispersing agent + ionic liquid as third component (so-called *post addition*)-efficiency of procedure (A) is higher than (C), but the negative point of it is that it requires higher quantities of the ionic liquid
(B) White base paint already containing an ionic liquid to have the highest flexibility concerning the use of different pastes based on a variety of primary dispersant technologies
(C) White base paint + pigment paste which contains both the primary dispersant and the ionic liquid to make the colorant universal not only in the old-fashioned way but also in terms of critical solvent-based systems such as Pliolites® or isoparaffin-based alkyds

Ionic liquids can be used according to one of the three following procedures (Table 20.12) in order to enhance the applicability of universal pigment pastes, respectively to improve the color strength of the final paints, coatings, and lacquers.

By using these ionic liquids as secondary dispersing agents, universal, water-based pigment pastes can be used for all types of paints and coatings. This will reduce the use of volatile organic substances/solvents in paints and coatings in the future.

20.3.4.3 Lithium–Ion Batteries

Degussa described a precommercial study aimed at integrating ionic liquid electrolytes to substitute for volatile and flammable mixtures of organic solvents, e.g., ethylene carbonate and dialkyl carbonates, with other Degussa performance technologies into lithium–ion batteries to deliver future performance demands. While the technology for lithium–ion batteries using ionic liquids does not yet match that of current lithium batteries, there was considerable optimism that both IL-Li^+ ion and IL-polymer systems were rapidly heading in the light direction.

20.3.5 *IoLiTec*

IoLiTec (Ionic Liquids Technologies) are an interesting new company specializing in marketing ionic liquids (see Table 20.3), and developing applications. They have developed a practical and efficient technology to clean high-value and sensitive

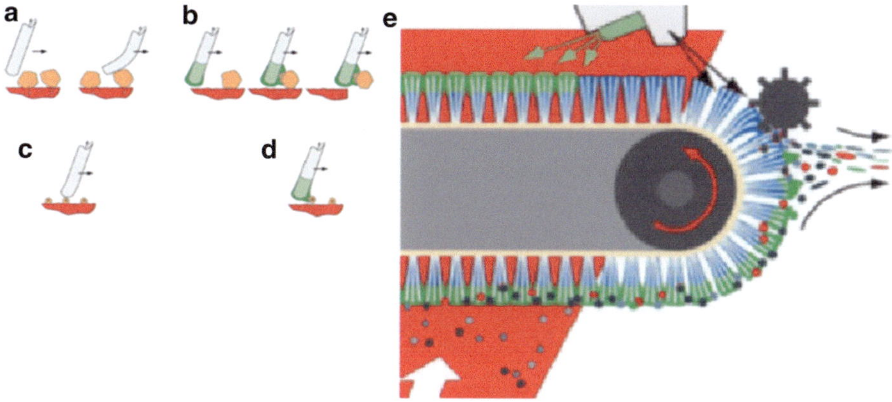

Fig. 20.24 The removal of dust particles by brush bristles (*left*) applying of conducting liquid film by a spray from a fine nozzle (*right*). Image Copyright from IoLiTec—Ionic Liquids Technologies GmbH

surfaces, using ionic liquids as antistatic cleaning agents (Fig. 20.24), The Wandres Micro-Cleaning GmbH (Buchenbach/Schwarzwald, Germany) developed a cleaning-system for high-value surfaces that uses moistened instead of dry filaments. In conventional brush cleaning, the filaments skim over small particles without removing them effectively (Fig. 20.24a, c). To keep a thin film on the filaments continuously, a very fine fog, consisting of very small droplets, is generated by using a Venturi nozzle. This film, located at the tips of the filaments, consists mainly of water. It permits even the removal of finest particles, which simply adhere to the moistened tips of the filaments (Fig. 20.24b). To avoid an electrostatic charging of the surface, the particles are neutralized by adding a supporting salt that facilitates electrical conductivity preventing recontamination from air (Fig. 20.24d). The adhering particles are removed by means of a rotor at the end of the process (Fig. 20.24e).

To reach a sufficient conductivity of the cleaning fluid, Wandres used sodium chloride as the wetting agent (supporting salt). If sodium chloride is used as a wetting agent, a solid is precipitated in the nozzle leading to encrustation and plugging of the system (Fig. 20.25, *left*). In Venturi-type nozzles, this encrustation leads to failure of the component, which in turn leads to less efficient cleaning and again a buildup of electrostatic charge. As a consequence, service technicians have to exchange the defect nozzle in comparable short service intervals. To solve the problem of encrustation, an antistatic agent is needed that is liquid at ambient conditions and leads to a conductivity of the cleaning fluid of at least 200 mS/cm. Ionic liquids can offer a unique solution to this problem, since as liquid salts they can provide electrical conductivity without precipitation of a solid, hence without formation of deposits. Ionic liquid solution to this problem was developed and established by IoLiTec in cooperation with Wandres Micro-Cleaning GmbH [58].

They have tested a set of RTILs. Especially the availability criterion led to the class of quaternary ammonium compounds, where in view of all parameters TEGO®-IL P9, a product from Degussa Industrial Specialties, Germany, was

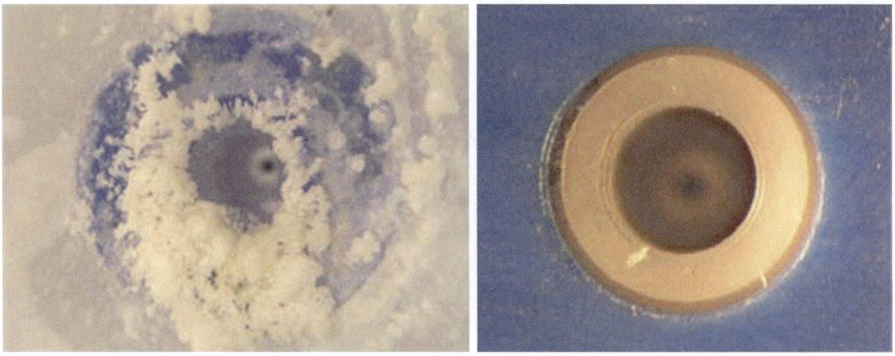

Fig. 20.25 A spray nozzle for aqueous solutions of sodium chloride after 10 h operation (*left*) and a hydrophilic ionic liquid after 35 h operation (*right*). Image Copyright from IoLiTec—Ionic Liquids Technologies GmbH

considered to be the best choice. For example, they have studied conductivity vs. concentration of aqueous solutions of Ionic Liquid TEGO®. As they expected, at higher dilutions they observed a linear correlation between concentration and conductivity. At a concentration of about 1 Mass% they observed conductivity well above 800 mS/cm, which is sufficient to prevent electrostatic charging of the cleaned surface.

The effect of replacing NaCl with ionic liquid is dramatically demonstrated in Fig. 20.25 (*right*). With these results, Wandres tested a newly designed cleaning fluid. Figure 20.25 (*left*) shows the encrustation of a sprayer-nozzle after use with a 1 % NaCl solution for 10 h. A significant deposit of NaCl is quite obvious. Figure 20.25 (*right*) shows a nozzle after use with a solution of TEGO®-IL-P9 for 35 h. Due to the reason that RTIL is not crystallized at such conditions, there is no deposit formed.

20.3.6 Central Glass Co., Ltd.

The Central Glass Company is the first firm to produce pharmaceutical intermediates (pravadoline) using ionic liquid technology. The following reaction was studied in detail by the Central Glass Co., Ltd. (Eq. 20.4).

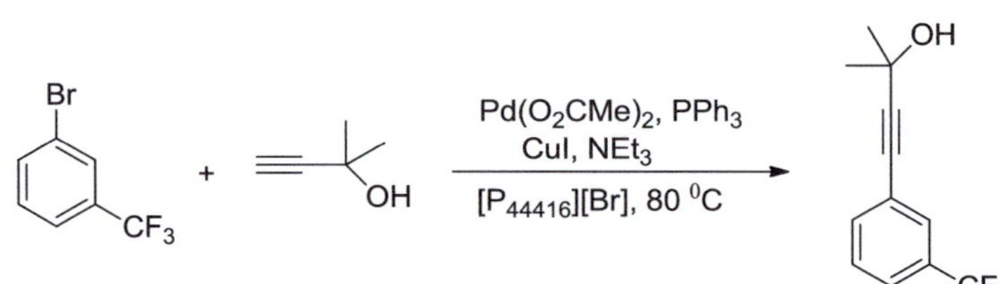

Equation 20.4 Coupling reaction studied by using ionic liquids

They have discovered that the use of 1,3-dialkylimidazolium cations led the reactions less efficient than when conventional solvents were used (it may be due to carbene formation [59]). In comparison, tetraalkylphosphonium ionic liquids led the reaction much more efficiently than organic solvents, or with neat reagents. Based on these observations, they have developed the following reaction and commercialized it (Eq. 20.5).

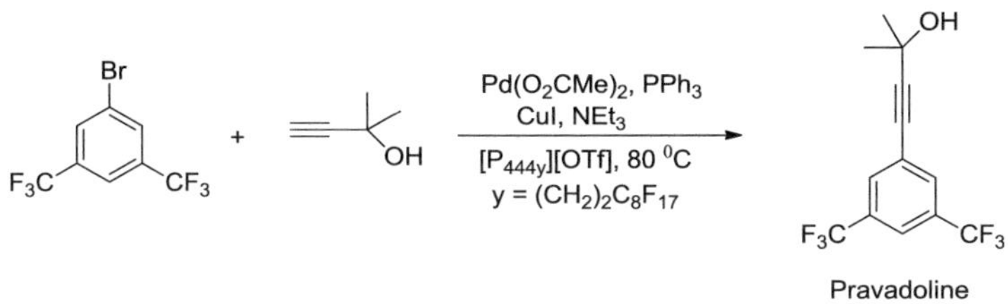

Equation 20.5 Synthesis of pravadoline using ionic liquids

20.3.7 Eastman Chemical Company

Eastman Chemical Company had been running a process for the isomerization of 3,4-epoxybut-1-ene to 2,5-dihydrofuran since December 1996 [60, 61]. This was operated by the Texas Eastman Division, with a reactor capacity of 1,400 metric tons per year, from 1996 to 2004. It actually represents the largest industrial application involving ionic liquids. Tetrahydrofuran which are synthesized by the above process, is a valuable compound useful as chemical process solvent, and as an intermediate in the preparation of polymers such as poly(tetramethyleneether) glycol. Isomerization of 3,4-epoxybut-1-ene process requires simultaneous activation by a Lewis base and by Lewis acid being an organotin compounds. The Eastman Chemical Company performed this reaction at an industrial scale using phosphonium iodide ionic liquid (Eq. 20.6) which was selected over its ammonium analogue because of its greater thermal stability (Table 20.13).

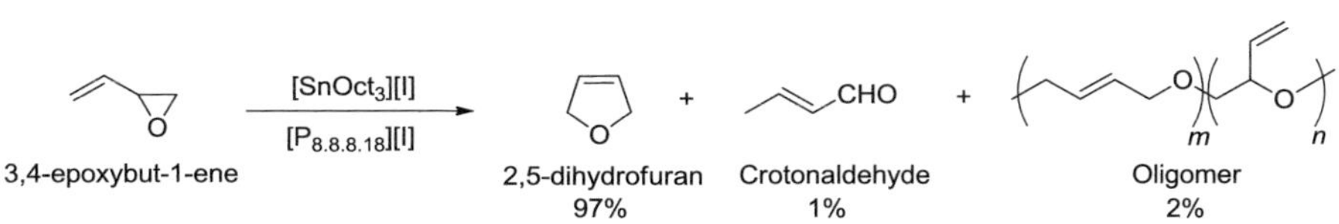

Equation 20.6 Isomerization of epoxybutene in the presence of ionic liquid

In 1986, Dr John R. Monnier of the Eastman Kodak Research Laboratories has discovered the epoxidation of butadiene to 3,4-epoxybut-1-ene [62]. In 1988, Dr Stephen N. Falling and coworkers of the Eastman Chemical Company have discovered its isomerization to 2,5-dihydrofuran [63].

Table 20.13 Phosphonium-based ionic liquids verses ammonium-based ionic liquids

Why phosphonium-based ionic liquids were selected with respect to ammonium analogues	Role of ionic liquid in this reaction	Advantages of this process
High thermal stability	Solvent	Milder reaction conditions
High hydrophobic character	Catalyst	Simplified product separation
Limited cost	By-product extractant	Ability to remove and replenish the catalyst system
Highly alkane soluble and can dissolve oligomer by-products		

The 2,5-dihydrofurane (bp = 66 °C) and the crotonaldehyde (bp = 104 °C) can then be separated from the reaction mixture by distillation due to their low boiling points, while the higher boiling oligomers remain in the IL-catalyst phase. This phase is further treated with an extractant solvent such as naphtha, to recover the oligomers and recycle the IL-catalyst. The plant is operated with three continuous, stirred-tank reactors, a wiped-film evaporator, a distillation train, and a continuous, counter-current, liquid–liquid extractor for recovery of the catalysts. The plant is now idle, because the market for the product has declined.

20.3.8 Eli Lilly

In 2004, Eli Lilly Company has described a process for demethylating of 4-methoxyphenylbutyric acid to quantitatively generate 4-hydroxyphenylbutyric acid, on both a 22 L and a 190 L pilot plant scale [64]. 4-hydroxyphenylbutyric acid is a key feedstock for the synthesis of a medicine (LY518674). Eli Lilly used melted pyridinium hydrochloride [HPyr][Cl] for demethylating of 4-Methoxyphenylbutanoic acid in 3 hours at 180 °C [64].

For preventing solidification of the high melting pyridinium hydrochloride (mp = 153 °C) during the post reaction steps, a stoichiometric amount of HCl and dilution with water were necessary. This way, a homogeneous nonviscous catalytic solution was obtained at room temperature with no indication of any thermal hazards. Following the reaction, the products are extracted from the reaction mixture by solvent extraction.

20.3.9 PetroChina

Isobutane alkylation is one of the most important processes for producing reformulated gasoline, and classical methods for alkylation of isobutene are summarized in Table 20.14.

Table 20.14 Classical methodologies for alkylation of isobutene

Alkylation of isobutane	Disadvantages	Advantages
Commercial alkylation plant: H_2SO_4 or HF	Limitations with productivity, alkylate quality, safety aspects, and operating costs	
Solid catalyst	Problems of catalysts deactivation and regeneration	
IFP process: acidic chloroaluminates [pyridine, HCl]/$AlCl_3$ (1:2 M ratio)		Provided good quality alkylate over a period of 300 h

Table 20.15 Main advantages of the gas storage technology with ionic liquids

The main advantages of this technology
Effective
Handled more easily
Deliver these reactive and hazardous gases in a safe way
It produces at least twice the performance of the main rival process, which relies on the physical adsorption of gases on solids
Reducing both the risks and hazards in the work place
The gases can be stored and transported at sub-atmospheric pressure, instead of the normal pressurized cylinders reducing risk

PetroChina have publicly announced and developed a process called "Ionikylation" for the alkylation reaction [65] employing ionic liquids [65, 66]. The ionic liquid used consists of a mixture of a conventional chloroaluminate-IL with CuCl. The purpose of addition of cuprous chloride is to enhance the acidity of the IL and when used for alkylation reaction inhibits undesirable side reactions such as isomerization and cracking. PetroChina demonstrated the high stability of this catalyst through an 8-month ageing test before a 60-day pilot scale operation. PetroChina announced in 2006 that an alkylation process has been retrofitted into an existing 65,000 t per year H_2SO_4 alkylation unit in China. Alkylation from the ionikylation process compares favorably to alkylate from H_2SO_4 (Table 20.14). Improvements in terms of alkylate yield, process unit capacity and so economics were demonstrated.

20.3.10 Air Products

Air Products has released a new technology of complexing of reactive gases in ionic liquids which is marketed as Gasguard1 Sub-Atmospheric System, Complexed Gas Technology (CGT). The main advantages of this process are summarized in the Table 20.15.

The Lewis acidity or basicity of the gases is considered in this technology. For example, Lewis-acidic gases [e.g., boron(III) fluoride] are stored in Lewis basic ionic liquids {e.g., $[C_n mim][BF_4]$}, whereas Lewis basic gases (e.g., phosphine, PH_3, or arsine, AsH_3) are stored in Lewis acid ionic liquids {e.g., $[C_n mim][Cu_2Cl_3]$ or $[C_n mim][Cu_2Br_3]$}. The gas is removed from the cylinder by the application of vacuum and can be delivered gases at sub-atmospheric pressure, instead of the normal pressurized cylinders, to the market with high purity and safety.

20.3.11 Others

Some of the other applications which are considered in pilot/commercial scales are summarized in Table 20.16, together with scale, IL and role of ionic liquid.

20.4 Challenges, Issues, and Recycling Methods of Ionic Liquids

Ionic liquids must meet a certain number of requirements to be applied in an industrial process. Changing an established industrial process for a novel technology using ionic liquids is not easy, even if it can be more profitable in the long term. The novelty and the production cost of ILs are probably one of the main barriers. The IL price must be related to the process performance and to the overall economy. Therefore, major issues with ILs in industrial applications are summarized in Table 20.17.

20.4.1 Cost of ILs

In industrial applications of ILs, the cost of the IL is a very important point. To be applied in an industrial process, ILs must meet a certain number of requirements. In general, ionic liquids are expensive in comparison to common organic solvents such as toluene, acetone, and ethanol, even if they are produced at an industrial scale. Ionic liquids are not readily accessible from cheap industrial process streams in a simple isolation or conversion step and also they are not easy to isolate and purify, due to their nonvolatile character and low melting points. The impurities of ionic liquids such as halide, water, and amine content will dramatically affect the price. Likewise, purification of ionic liquids will be a big challenge for any industrial applications (who uses ionic liquids) and manufacturer (who produce ionic liquids).

Table 20.16 Other applications with ionic liquids

Company	Scale[a]	Ionic liquid[a]	Role of ionic liquid
BASF	Pilot	[Rmim][BF_4]	IL as entrainers or separation enhancers for breaking azeotropes
BASF			Replacing phosgene: The chlorination of diols (e.g., 1,4-butanediol) using HCl as the chlorinated agent in combination with an IL
Scionix	Pilot (1,200 L)	[($HOCH_2CH_2$)N_{111}][Cl]	Electroplating of Cr: IL as electrolyte
Scionix	Pilot (50 L)		Electropolishing of stainless steel: IL as electrolyte
SASOL		[C_2dmim][NTf_2]	ILs in metathesis & olefin trimerization. Self-metathesis of oct-1-ene to yield tetradec-7-ene and ethene
BP		Franklin acidic compositions of the [$C_n$mim]Cl–$AlCl_3$ ($n=2$, 4 or 8)	Aromatic alkylation process in chloroaluminate ionic liquids
ExxonMobil			Biphasic carbonylation catalysis using conventional rhodium catalysts in Lewis-acidic ionic liquids, the electrochemical oxidation of sulfur compounds in naphtha, the use of supported ionic liquids, catalytic hydroformylation, hydrogenations, aldol condensations, and even ionic liquid synthesis
Linde	Pilot		Engineering fluids (hydraulic liquids): liquid piston
Supelco	Commercial	[1,9-di-(3-vinyl-imidazolium)nonane] [NTf_2]	Analytical chemistry: commercial GC-stationary phase
Chevron and Chevron Phillips			Refinery alkylation, emulsification and high shear mixing, oligomerization and hydrotreatment of alkenes (often to prepare lubricating oils), and removal of carbon dioxide from gas streams

[a]Blank spaces in these two columns indicate no information were found

Table 20.17 Major issues with ILs in industrial applications

IL life time: their chemical and thermal stability
Loss of chemical in the process
Recovery and recycling of ILs
Toxicity
Reliable supply—This is not a major concern now as there are companies for IL production and commercialization
Their material compatibility
Their notification requirements (REACH)

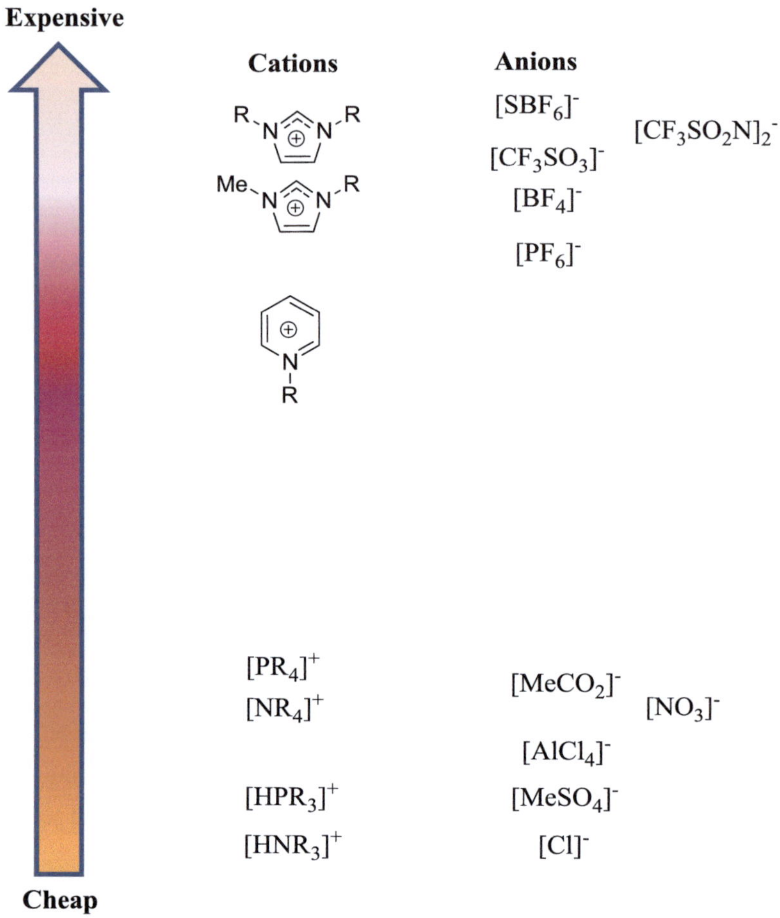

Fig. 20.26 Rough estimation of the relative prices of different cations and anions of ionic liquids

There are several companies dealing with the manufacture of ionic liquids (Table 20.3). Generally, IL with tetrafluoroborate or hexafluorophosphate anion will always be quite expensive due to the fact that the sources are not readily available for these anions. Similarly, ILs with 1-alkyl-3-methylimidazolium cations will always be more expensive than ILs with trialkylammonium cations. This is related to the fact that 1-methylimidazole is a more expensive starting material compared to a simple trialkylamine. Using simplification, it is possible to roughly estimate the relative prices of different cation and anion components of ILs (Fig. 20.26). The price of the solvent is not only the factor that decides whether a certain solvent is economically viable for a given industrial application. No one today chooses to dispose of solvents or by-products if they have commercial value. For a high priced ionic liquid which has PF_6 anion, reuse is essential to economic viability. Someone can use expensive IL for any application, but IL may be acceptable if the advantages related to their use are able to compensate for the additional cost of the solvent. The main question is if IL can be considered in the economics of an industrial production plant, as a basic investment or as a consumable. The answer to this question is closely related to the ionic liquids' chemical & thermal stability, which will decide about possible recycling and regeneration steps to reduce the ionic liquid composition in the process.

20.4.2 Stability

IL stability, lifetime and recyclability are very important factors for industrial applications. The chemical and thermal stability of ILs depend on the presence of impurities from their synthesis and their exposure to the process conditions, reactants, and products. It is very difficult to compare the stability of different ILs in a general way. All classification depends very much on the question of which stability is considered. Practical experiences show that stability against hydrolysis, redox stability, and thermal stability are very important features. But even additional aspects like the hygroscopicity of an IL may count under practical consideration as a stability problem. The uptake of water usually leads to a change in the physical and chemical behavior of the IL, which obviously will affect most industrial applications. Thus, the important question is not whether an IL is stable or unstable under general or theoretical aspects. In practice, it is much more important to know under which conditions (e.g., temperature) and in which environment (content of water in the feedstock, reactants) a given IL is stable and, most important for all industrial applications, how long it is stable. For example, chloroaluminate ILs are perfectly stable liquids if the system is absolutely water-free and the temperature is not so high. However they decompose in the presence of water, which will certainly limit their industrial application for any reaction where the feedstocks contain certain amount of water. A typical example is the decomposition of *N*,*N*-dialkylimidazolium cation to form *N*-heterocyclic carbenes under basic conditions, or the hydrolysis of $[PF_6]^-$- based ILs into HF and POF_3.

20.4.3 Waste Disposal

If ionic liquids are used in industrial applications, it is necessary to discuss as well as the problem of treatment of the IL waste. In all industrial applications, they will try to eliminate this waste by regeneration of the used IL. Indeed, the high price of the IL allows economically some treatments like extraction or stripping. But in a real application, it is very likely that a certain small amount of the IL will be lost as a waste stream. It is therefore, necessary to develop concepts for the disposal of used ILs. However, the choice of a suitable disposal concept will strongly depend on the nature of the IL. Mainly, the nature of the anion will be important. Probably, the lower the content of halogen atoms in the anion, the cheaper and easier will be a proper disposal of an IL waste. This will influence again the choice of the most suitable IL candidate for a given industrial application.

20.4.4 Recycling Methods of Ionic Liquids

There has been an explosion of interest in ILs in the last decade that has resulted in the discovery of a vast number of new ILs with a wide range of applications. However, ILs may be sensitive to contaminants and may require frequent rejuvenation or replacement. The degree of sensitivity to impurities and the rate of degradation may have a large impact on IL functionality in industrial processes. Process engineering studies are required to obtain sufficient data on ILs stability under long-term exposure to process conditions and exposure to air, moisture, heat, corrosion products, trace impurities (e.g., SO_x, NO_x, etc.) and other key industrial application components. The useful lifetime (life-cycle costs), impact of water content, sensitivity to contaminants, thermal stability, aging, IL losses, etc., over time, as well as recyclability data, are crucial in evaluating ILs viability in commercial applications. Robin Rogers (the Director of Centre for Green Manufacturing at the University of Alabama in Tuscaloosa) said: "Before we can say that ionic liquids are green, we have to look at their entire life cycle. People are calling ionic liquids green because they are not volatile, but we have to look at how they are made all the way through to recycling and disposal" [67]. Rogers added: "Researchers also need to find better ways to recycle ionic liquids." Many processes for cleaning up ionic liquids involve washing with water or volatile organic compounds (VOCs). "This needs more work. We don't want to create a secondary waste." Although ILs are invariably described as highly stable green solvents, thorough investigations to quantify their environmental impact have lagged behind the pace of other research in the area [33]. These issues are extremely important as well as the approaches for the recovery of the ILs (after their use) for further reuse (e.g., recycling). A very important challenge is to use the unique solvent properties of the ILs to develop efficient method for product separation and IL recycling [68–73]. Despite their extremely low vapor pressure that prevents IL emission to the atmosphere, they are, at least, partially miscible with water and will inevitably end up in the aqueous environment (as what happens in the application of ILs for the electrodeposition of metals). During industrial application, ILs will get mixed with other products that necessitate efficient separation and recycling of ILs for economical and ecological needs [74]. Many researchers have pointed out the major concern in the use of ILs is their relative high cost, which makes their recycling an important issue for further study [75]. Currently, ionic liquids are quite expensive in comparison with conventional molecular solvents, and large amounts of ILs are needed in the various applications, thus efficient recycling of ILs is important for their economic use, especially in large-scale applications. However, the cost of ILs is expected to decrease, due to their higher production levels, which is expected to

Table 20.18 Recycling methods of ionic liquids

IL recycling methods	Process and problems associated with the process
Multiphase process	1. Reaction products are separated from IL phase—partial miscibility of ILs in organic reaction products—dramatic issue for the overall economy of the process and product quality
	2. Use of an organic cosolvent or $ScCO_2$—more complex process and more energy consuming
Supported IL process	This process is best for gas–liquid multiphasic technology and their applicability remains limited to reactions involving low volatile reactants and products
Vacuum evaporation of volatiles	High-energy output and not suitable for all of ionic liquids and applications
Membrane separation	Higher viscosity of ILs and the treatment of membranes for IL recovery are not simple and easy
Controlled decomposition of ILs into neutral molecules	Neutral molecules can be recovered by distillation
	1. Protic ILs can be switched to the neutral amine or imidazole in the presence of a base
	2. *N*,*N*-dialkylimidazolium can be switched to alkylimidazole by heating or stable carbene can be generated in the presence of a base (Fig. 20.28)
	Problem: distillation requires high temperature under vacuum and is not realistic for large-scale IL recovery

render ILs economically competitive with organic solvents. To overcome the cost problem, some low cost and simple synthesis of the ILs are essential for their recycling and reuse. Ionic liquids are now available from a number of commercial manufacturers and from many suppliers. However, IL preparation, workup to separate/extract products and their recycling usually involve other solvents (e.g., VOCs) to extract unwanted residues, which are toxic persistent (to some extent). However, this diminishes the "green" aspect of their usage due to the cross-contamination arising from the use of organic solvents. Some ILs are certainly toxic and have been shown to have a detrimental effect on aquatic life that will tend to persist in local environment, on land/sea, etc. Thus there are worries about ILs end-of-life disposal, and materials incompatibility.

Among the published literature, it is generally recognized that ILs are easily recyclable. Undoubtedly, this is true for some specific biphasic systems that contain ILs, in particular for hydrophobic ILs, such as $[PF_6]^-$ and $[N(Tf)_2]^-$ ILs. In this case, liquid–liquid extraction is an efficient method for separation. However, there are significant issues which have to be considered (Table 20.18). Generally, *hydrophobic ILs* can be extracted with water to separate water-soluble solutes from the IL into the aqueous phase; however, this method is not suitable for *hydrophilic ILs*. Recovery of the hydrophilic ILs is more difficult in comparison to hydrophobic ILs, and the study in this field is in its infancy. To avoid cross-contamination of IL with water or vice versa, novel ways must be exploited. Supercritical CO_2 ($scCO_2$) was found to have the potential for the regeneration of hydrophilic ILs. Extraction of

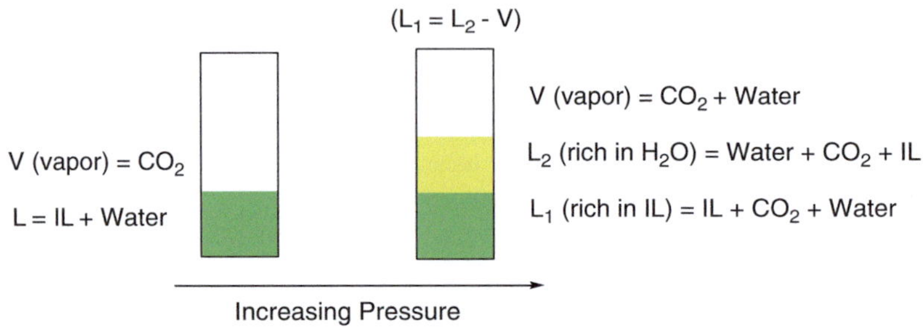

Fig. 20.27 Phase behavior of IL/water/CO_2 mixture at increasing pressure at near room temperature

solutes from ILs with VOCs obviates some of the advantages of using ILs, but $scCO_2$ extraction may offer a promise. The "green" solvent $scCO_2$ may be a good choice for the recovery of solutes and the regeneration of ILs (Fig. 20.27). While $scCO_2$ is not able to dissolve any 1,3-dialkylimidazolium ionic liquids, other supercritical fluids do [76]. The solubility of $scCO_2$ in ILs was determined to be less than 5×10^{-7} mol fraction [76]. This unique property of $scCO_2$ means that solutes can be isolated from ILs without cross-contamination of the gas phase with IL and also provides a means for recycling the ILs [76]. Recovery of the organic solute from ILs was near quantitative and achieved by simple expansion of the gas phase downstream. This efficient exchange between the two phases, coupled with the lack of mutual solubility seems to be ideally suitable for product separation and recycling of ILs. Nevertheless, this process also has drawbacks which are mentioned in the Table 20.18.

Membrane techniques have been applied to perform fine separation of undesirable constituents and these show promising results for the selective removal of volatile solutes from ILs [77]. Haerens et al. [74] investigated the use of pressure-driven membrane processes, nanofiltration, reverse osmosis, and pervaporation, as a possibility to recycle ILs from water. However, these membrane techniques for recycling ionic liquids have disadvantages, which are summarized in Table 20.18.

Distillation can be used to recover ILs from compounds with low boiling points, because of the ILs negligible vapor pressure. However, direct vacuum distillation protocol is energy consuming, particularly for nonvolatile compound/IL systems. Moreover, if the IL has a tendency to undergo hydrolytic decomposition, such as those that contain $[PF_6]^-$ ions, direct heating should be avoided or at least minimized. BASF pointed out that the recycling of ILs is easy if protonated cations (Fig. 20.28, $R^2 = H$) are used. In this case the ILs can be switched off by deprotonation (Table 20.18 and Fig. 20.28). Imidazolium cations can be deprotonated by bases to form neutral carbene molecules. The resulting *N*-heterocyclic carbenes are found surprisingly stable and can be distilled for recycling or purification purposes. It is more difficult with alkylated cations (Fig. 20.28, $R^2 = $ alkyl). Apart from purification or recycling by liquid–liquid extraction, two principal "distillation" methods have been reported. The first is the formation of distillable carbenes [78] and the second is the back alkylation of the anion (Fig. 20.28).

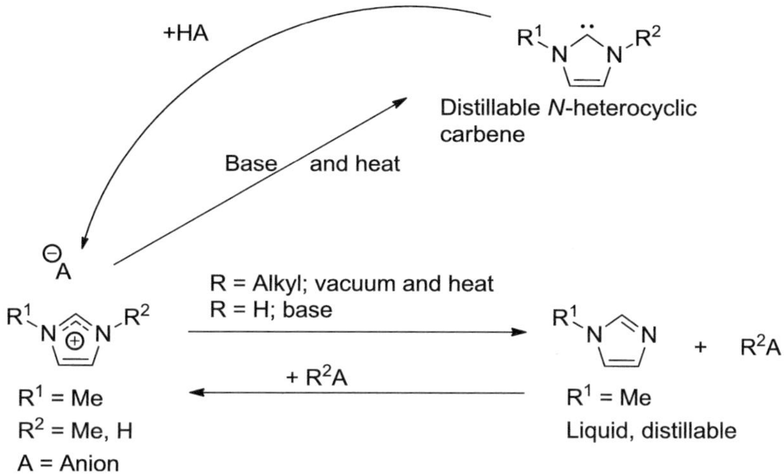

Fig. 20.28 Routes for recycling of *N,N*-dialkylimidazolium-based ionic liquids

20.5 Conclusion

ILs are interesting candidates for many solvent and non-solvent applications due to their very special physical and chemical properties. ILs are highly interesting solvents for industrial applications due to lack of vapor pressure, they eliminate vapor loss and atmospheric pollution. Therefore, they can significantly contribute to a "greener" chemistry. Moreover, ILs can enhance activity, selectivity, and stability of many industrial processes. However, the use of ionic liquids also has some disadvantages: the physical properties are not always known, their viscosity is usually higher than that of the common solvents and their toxicity is unknown. Due to the current market size and the relatively high cost of ILs, the industrial production of ionic liquids is still small or limited to lab-scale (and sometimes pilot scale) applications and hence no industrial technology is yet available for ionic liquids recycling with the objective of reuse. Most of the published works on ionic liquids recycling are made at the bench scale where the investigators are trying to extract a solute from a mixture using some suitable ionic liquid then recover the ionic liquid for further reuse. Many authors and researchers agree that we are only at the very beginning of understanding the recyclability of ILs based on the available literature in the various application fields. Understanding of ILs volatility, purity, stability, biodegradability, and toxicity is necessary for their recovery, since these determine whether an IL can be sustainably developed. In other words, there is a long way to go before large-scale implementation of ILs. Hopefully, this chapter could provide some clues to support a great deal of future research on ILs recycling and reuse.

For a successful application of ionic liquids in industrial processes, these aspects must be taken into consideration. However, the chance for a general substitution of volatile organic solvents by ILs would be greatly increased if it were possible by future research to find new ILs that overcome certain shortcomings of the existing systems. So far we do not have a single IL combining good physical properties (melting point, viscosity), high stability, low coordinating strength, low price, and easy disposal opportunities. Indeed development of such a system would boost the whole field of IL chemistry and would largely increase the potential for all industrial applications. In conclusion, it is likely that the cost and long-term stability are key properties that will decide about the industrial application of ILs. Therefore, the development of cheap and stable ILs will be of great importance for their industrial future.

References

1. (a) Oliver-Bourbigou H, Magna L, Morvan D (2010) Ionic liquids and catalysis: recent progress from knowledge to applications. 373:1–56 (b) Wytze Meindersma G, Maase M, De Haan AB (2012) Ionic liquids. Ullmann's Encyclopedia of Industrial Chemistry 548 (c) Plechkova NV, Seddon KR (2008) Applications of ionic liquids in the chemical industry. Chem Soc Rev 37:123–150
2. Seddon KR (1996) Room temperature ionic liquids: neoteric solvents for clean catalysis. Kinet Catal 37:693
3. Bonhote P, Dias AP, Papageorgiou N, Kalyanasundaram K, Gratzel M (1996) Hydrophobic, highly conductive ambient-temperature molten salts. Inorg Chem 35:1168
4. Huddleston JG, Willauer HD, Swatloski RP, Visser AE, Rogers RD (1998) Room temperature ionic liquids as novel media for "clean" liquid–liquid extraction. Chem Commun 1765
5. Siriwardana AI, Crossley IR, Torriero AAJ, Burgar IM, Dunlop NF, Bond AM, Deacon GB, MacFarlane DR (2008) Methimazole based ionic liquids. J Org Chem 73:4676
6. Dzyuba SV, Bartsch RA (2001) New room-temperature ionic liquids with C-2-symmetrical imidazolium cations. Chem Commun 1466
7. Gordon CM, Holbrey JD, Kennedy AR, Seddon KR (1998) Ionic liquid crystals: hexafluorophosphate salts. J Mater Chem 8:2627–2636
8. Forsyth SA, Pringle JM, MacFarlane DR (2004) Ionic liquids—an overview. Aust J Chem 57:113
9. Handy ST (2005) Room temperature ionic liquids: different classes and physical properties. Curr Org Chem 9:959
10. Bradaric CJ, Downard A, Kennedy C, Robertson AJ, Zhou YH (2003) Industrial preparation of phosphonium ionic liquids. Green Chem 5:143
11. De Giorgi M, Landini D, Maia A, Penso M (1987) A facile and convenient synthesis of lipophilic tetraalkylammonium or phosphonium hydrogen sulfates. Synth Commun 17:521
12. Bradaric CJ, Downard A, Kennedy C, Robertson AJ, Zhou YH (2003) Industrial preparation of phosphonium ionic liquids, in ionic liquids as green solvents: progress and prospects. In: Rogers RD, Seddon KR (eds) ACS Symposium Series. American Chemical Society, Washington, DC, p 41–56
13. Ding J, Welton T, Armstrong DW (2004) Chiral ionic liquids as stationary phases in gas chromatography. Anal Chem 76:6819
14. Ding J, Armstrong DW (2005) Chiral ionic liquids: synthesis and applications. Chirality 17:281
15. Chiappe C, Pieraccini D (2005) Ionic liquids: solvent properties and organic reactivity. J Phys Org Chem 18:275

16. Wasserscheid P, Hilgers C, Boesmann A (2002) Halogenide-free preparation of ionic fluids. EP1182196-A1
17. Holbrey JD, Seddon KR (1999) The phase behaviour of 1-alkyl-3-methylimidazolium tetrafluoroborates; ionic liquids and ionic liquid crystals. J Chem Soc Dalton Trans 2133
18. Forsyth S, Golding J, MacFarlane DR, Forsyth M (2001) N-methyl-N-alkylpyrrolidinium tetrafluoroborate salts: ionic solvents and solid electrolytes. Electrochim Acta 46:1753
19. MacFarlane DR, Meakin P, Sun J, Amini N, Forsyth M (1999) Pyrrolidinium imides: a new family of molten salts and conductive plastic crystal phases. J Phys Chem B 103:4164
20. Goldman JL, McEwen AB (1999) EMIIm and EMIBeti on aluminum—anodic stability dependence on lithium salt and propylene carbonate. Electrochem Solid-State Lett 2:501
21. Visser AE, Holbrey JD, Rogers RD (2001) Hydrophobic ionic liquids incorporating N-alkylisoquinolinium cations and their utilization in ionic-liquid separations. Chem Commun 2484
22. Holbrey JD, Reichert WM, Swatloski RP, Broker GA, Pitner WR, Seddon KR, Rogers RD (2002) Efficient, halide free synthesis of new, low cost ionic liquids: 1,3 dialkylimidazolium salts containing methyl- and ethyl-sulfate anions. Green Chem 4:407
23. Brinchi L, Germani R, Savelli G (2003) Ionic liquids as reaction media for esterification of carboxylate sodium salts with alkyl halides. Tetrahedron Lett 44:2027
24. Kitazume T, Tanaka G (2000) Preparation of fluorinated alkenes in ionic liquids. J Fluor Chem 106:211
25. Golding J, Forsyth S, MacFarlane DR, Forsyth M, Deacon GB (2002) Methanesulfonate and p-toluenesulfonate salts of the N-methyl-N-alkylpyrrolidinium and quaternary ammonium cations: novel low cost ionic liquids. Green Chem 4:223
26. Pringle JM, Golding J, Forsyth CM, Deacon GB, Forsyth M, MacFarlane DR (2002) Physical trends and structural features in organic salts of the thiocyanate anion. J Mater Chem 12:3475
27. Leveque JM, Luche JL, Petrier C, Roux R, Bonrath W (2002) An improved preparation of ionic liquids by ultrasound. Green Chem 4:357
28. MacFarlane DR, Golding J, Forsyth S, Forsyth M, Deacon GB (2001) Low viscosity ionic liquids based on organic salts of the dicyanamide anion. Chem Commun 16:1430–1431
29. MacFarlane DR, Forsyth SA, Golding J, Deacon GB (2002) Ionic liquids based on imidazolium, ammonium and pyrrolidinium salts of the dicyanamide anion. Green Chem 4:444
30. Xu W, Wang LM, Nieman RA, Angell CA (2003) Ionic liquids of chelated orthoborates as model ionic glassformers. J Phys Chem B 107:11749
31. Larsen AS, Holbrey JD, Tham FS, Reed CA (2000) Designing ionic liquids: Imidazolium melts with inert carborane anions. J Am Chem Soc 122:7264
32. Clare BR, Bayley PM, Best AS, Forsyth M, MacFarlane DR (2008) Purification or contamination? The effect of sorbents on ionic liquids. Chem Commun (Camb) 23:2689–2691
33. Scammells PJ, Scott JL, Singer RD (2005) Ionic liquids: the neglected issues. Aust J Chem 58 (3):155–169
34. Clare B, Siriwardana A, Macfarlane DR (2009) Synthesis purification and characterization of ionic liquids. In: Kirchner B (ed) Topic in current chemistry. Springer, Germany, pp 1–40
35. Seddon KR, Stark A, Torres MJ (2000) Influence of chloride, water, and organic solvents on the physical properties of ionic liquids. Pure Appl Chem 72(12):2275–2287
36. Holbrey JD, Seddon KR, Wareing R (2001) Green Chem 3:33
37. Bartsch M, Becker M, Flores M, Halbritter K, Huttenloch O, Maase M, Massonne K, Noe R, Siegel W, Less V (2003) Method for the separation of acids from chemical reaction mixtures by means of ionic fluids. WO2003062171 A3
38. Maase M (2004) Erstes technisches Verfahren mit ionischen Flussigkeiten. Chem Unserer Zeit 38:434–436
39. Rogers RD, Seddon KR (2003) Ionic liquids—solvents of the future. Science 302:792–793
40. Parkinson G (2004) Ionic liquids make an environmental splash. Chem Eng Prog 100:7–9

41. Maase M, Masonne K (2005) Biphasic acid scavenging utilizing ionic liquids: the first commercial process with ionic liquids. In: Rogers RD, Seddon KS (eds) Ionic liquids IIIB: fundamentals, progress, challenges, and opportunities. Transformations and processes, ACS Symposium Series 902, American Chemical Society, Washington, DC, p 126–132
42. Oliver H, Matthias M (2005) Method for isolating acids from chemical reaction mixtures by using 1-alkylimidazoles. WO 2005061416 A1
43. IFP (2004) Annual Report
44. Chauvin Y, Gaillard JF, Quang DV, Andrews JW (1974) The IFP dimersol process for the dimerization of C3 and C4 olefinic cuts. Chem Ind 375–378
45. Chauvin Y, Olivier H (2000) In: Cornils B, Herrmann WA (eds) Applied homogeneous catalysis with organometallic compounds. Wiley, Weinheim, pp 258–268
46. Commereuc D, Chauvin Y, Leger G, Gaillard J (1982) The dimersol process for dimerisation of olefins—a chemical approach. Revue de L'Institut Français du Petrole 37:639–649
47. Chauvin Y, Gilbert B, Guibard I (1990) Catalytic dimerization of alkenes by nickel complexes in organochloroaluminate molten salt. Chem Commun 1715–1716
48. Chauvin Y, De Souza R, Olivier H (1998) Catalytic composition for biphase catalysis, in particular using nickel complexes, and a process for the oligomerization of olefins. US patent 5723712
49. Olivier-Bourbigou H, Forestière A (2003) Ionic liquids in multiphasic reactions. In: Wasserscheid P, Welton T (eds) Ionic liquids in synthesis. Wiley, Weinheim, pp 258–280
50. Favre F, Forestière A, Hugues F, Olivier-Bourbigou H, Chodorge JA (2005) Butenes dimerisation: from monophasic dimersol(tm) to biphasic difasol(tm). Oil Gas Eur Mag 31:83–87
51. Stadtmueller S (2002) Siloxanes as additives for plastics. Polym Polym Compos 10:49
52. Weyershausen B, Hell K, Hesse U (2005) Ionic liquids IIIB: fundamentals, progress, challenges, and opportunities—transformations and processes. In: Rogers RD, Seddon KR (eds) ACS Symposium Series, American Chemical Society, Washington, DC, p 133–143
53. Hoff A, Jost C, Prodi-Schwab A, Schmidt FG, Weyershausen B (2004) Ionic liquids: new designer compounds for more efficient chemistry. Degussa Sci Newslett 9:10–15
54. Weyershausen B, Hell K, Hesse U (2005) Industrial application of ionic liquids as process aid. Green Chem 7:283
55. Hell K, Hesse U, Weyershausen B (2004) Process for the production of organically modified polysiloxanes by using ionic liquids. EP 1382630
56. Weyershausen B, Lehmann K (2005) Industrial applications of ionic liquids as performance additives. Green Chem 7:15–19
57. Lehmann K, Silber S, Weyershausen B (2005) Process for the preparation of homogeneous, storage-stable pastes, paints, lacquers by using ionic liquids as dispersing aids. EP 1566413
58. Bosmann A, Schubert T (2004) Identification of industrial applications for ionic liquids: high-performance-additives for the use in hi-tech-cleaning-solutions. Poster presented at the Green Solvents For Synthesis, Bruchsal, Germany, 3–6 October 2004. http://www.iolitec.de/en/Poster/Page-4.html
59. Stark A, Seddon KR (2007) Ionic liquids. In: Seidel A (ed) Kirk-Othmer encyclopaedia of chemical technology. Wiley, Hoboken, pp 836–920
60. Falling SN, Monnier JR, Phillips GW, Kanel JS, Godleski SA (2005) First International Congress on Ionic Liquids (COIL-1), Salzburg, Austria, 19–22 June 2005
61. Holbrey JD, Plechkova NV, Seddon KR (2006) Recalling COIL. Green Chem 8:411–414
62. Monnier JR, Muehlbauer PJ (1990) Selective monoepoxidation of olefins, US Pat., 4,897,498
63. Phillips GW, Falling SN, Godleski SA, Monnier JR (1994) Continuous process for the manufacture of 2,5-dihydrofurans from γ,δ-epoxybutenes. US 5,315,019; Falling SN, Godleski SA, McGarry L (1993) Process for the separation of oligomeric materials from a catalyst mixture. US 5,238,889

64. Schmid CR, Beck CA, Cronin JS, Staszak MA (2004) Demethylation of 4-methoxyphenylbutyric acid using molten pyridinium hydrochloride on multikilogram scale. Org Process Res Dev 8:670–673
65. Liu ZC, Zhang R, Xu CM, Xia RG (2006) Ionic liquid alkylation process produces high-quality gasoline. Oil Gas J 104:52–56
66. Liu Z, Xu C, Huang C (2004) Method for manufacturing alkylate oil with composite ionic liquid used as catalyst. US 0133056
67. Sen DJ (2006) Reflection of green chemistry. Pharm Times 38:35–39
68. Galán Sánchez LM (2008) Functionalized ionic liquids—absorption solvents for carbon dioxide and olefin separation. PhD Thesis, University of Twente, The Netherlands
69. Welton T (2004) Review ionic liquids in catalysis. Coord Chem Rev 248:2459–2477
70. Meindersma GW, Podt JG, Gutiérrez Meseguer M, de Haan AB (2005) Ionic liquids as alternative to organic solvents in liquid-liquid extraction of aromatics. In: Rogers RD, Seddon KR (eds) ACS symposium series 902, ACS, Washington, DC, p 57–71
71. Heintz A (2005) Recent developments in thermodynamics and thermophysics of non-aqueous mixtures containing ionic liquids—a review. J Chem Thermodyn 37:525–535
72. Olivier-Bourbigou H, Magna LJ (2002) Ionic liquids: perspectives for organic and catalytic reactions. J Mol Catal A: Chem 182–183:419–437
73. Holbrey JD (2004) Industrial applications of ionic liquids. Chimica Oggi 22:35–37
74. Haerens K, Van Deuren S, Matthijs E, Van der Bruggen B (2010) Challenges for recycling ionic liquids by using pressure-driven membrane processes. Green Chem 12:2182–2188
75. Verma AK, Attri P, Chopra V, Tiwari RK, Chandra R (2008) Triethylammonium acetate (TEAA): a recyclable inexpensive ionic liquid promotes the chemoselective aza- and thia-michael reactions. Monatsh Chem 139:1041–1047
76. Blanchard LA, Hancu D, Beckman EJ, Brennecke JF (1999) Green processing using ionic liquids and CO_2. Nature 399:28–29
77. Wasserscheid P, Kragl U, Kröckel J (2003) Method for separating substances from solutions containing ionic liquids by means of a membrane. WO 2003/039719
78. Earle MJ, Seddon KR (2001) Preparation of imidazole carbenes and the use thereof for the synthesis of ionic liquids. WO 01/77081

Symbols and Abbreviations

Ionic Liquid Abbreviations

Cations

$[Hmim]^+$	1-Methylimidazolium
$[C_1mim]^+$	1,3-Dimethylimidazolium
$[C_2mim]^+$	1-Ethyl-3-methylimidazolium
$[C_2dmim]^+$	1-Ethyl-2,3-dimethylimidazolium
$[C_3mim]^+$	3-Methyl-1-propylimidazolium
$[C_3dmim]^+$	2,3-Dimethyl-1-propylimidazolium
$[C_4mim]^+$	1-Butyl-3-methylimidazolium
$[C_4dmim]^+$	1-Butyl-2,3-dimethylimidazolium
$[C_4C_4im]^+$	1,3-Dibutylimidazolium
$[C_5mim]^+$	3-Methyl-1-pentylimidazolium
$[C_6mim]^+$	1-Hexyl-3-methylimidazolium
$[C_8mim]^+$	3-Methyl-1-octylimidazolium
$[C_{10}mim]^+$	1-Decyl-3-methylimidazolium
$[dC_{18}im]^+$	1,3-Dioctadecylimidazolium
$[HOCH_2CH_2\text{-}mim]^+$	1-(2-Hydroxyl)ethyl-3-methylimidazolium
$[MeOCH_2CH_2\text{-}mim]^+$	1-(2-Methoxy)ethyl 3-methylimidazolium
$[Me(OCH_2CH_2)_2\text{-}mim]^+$	1-(2-Methoxyethoxy)ethyl)-3-methylimidazolium
$[Me(OCH_2CH_2)_3\text{-}mim]^+$	1-(2-(2-Methoxyethoxy)ethoxy)ethyl)-3-methylimidazolium
$[HSO_3\text{-}Bu\text{-}mim]^+$	1-Methyl-3-(4-sulfobutyl)imidazolium
$[HSO_3\text{-}Bu\text{-}bim]^+$	1-Butyl-3-(4-sulfobutyl)imidazolium
$[HSO_3\text{-}Bu\text{-}vim]^+$	1-(4-Sulfobutyl)-3-vinylimidazolium
$[vmim]^+$	3-Methyl-1-vinylimidazolium
$[C_4vim]^+$	3-Butyl-1-vinylimidazolium
$[C_8vim]^+$	3-Octyl-1-vinylimidazolium

(continued)

A.A.J. Torriero (ed.), *Electrochemistry in Ionic Liquids*,
DOI 10.1007/978-3-319-15132-8

$[C_{12}vim]^+$	3-Dodecyl-1-vinylimidazolium
$[C_{18}vim]^+$	3-Octadecyl-1-vinylimidazolium
$[C_{22}vim]^+$	3-Docosyl-1-vinylimidazolium
$[C_2C_4OOC_2im]^+$	3-(4-Ethoxy-4-oxobutyl)-1-ethylimidazol-3-ium
$[C_2PhC_2im]^+$	1-Ethyl-3-phenethylimidazol-3-ium
$[C_3PhC_3im]^+$	3-(3-Phenylpropyl)-1-propyl-1H-imidazol-3-ium
$[C_3PhOC_3im]^+$	3-(4-Phenoxybutyl)-1-propyl-1H-imidazol-3-ium
$[FcC_1mim]^+$	1-(Ferrocenylmethyl)-3-methylimidazolium
$[Amim]^+$	1-Allyl-3-methylimidazolium
$[HPyr]^+$	Pyrrolidinium
$[C_1HPyr]^+$	1-Methylpyrrolidinium
$[C_3mPyr]^+$	1-Methy-1-propylpyrrolidinium
$[C_4mPyr]^+$	1-Butyl-1-methylpyrrolidinium
$[C_5mPyr]^+$	1-Methyl-1-pentylpyrrolidinium
$[C_3bPyr]^{2+}$	1,1′-(Propane-1,3-diyl)bis(4-methylpyridinium)
$[C_4bPyr]^{2+}$	1,1′-(Butane-1,4-diyl)bis(4-methylpyridinium)
$[C_3mpip]^+$	1-Methyl-1-propylpiperidinium
$[C_4mpip]^+$	1-Buthyl-1-methylpiperidinium
$[C_2C_6pip]^+$	1-Ethyl-1-hexylpiperidinium
$[HP_{4\ 4\ 4}]^+$	Tributylphosphonium
$[P_{1\ 4\ 4\ 4}]^+$	Tributylmethylphosphonium
$[P_{4\ 4\ 4\ 4}]^+$	Tetrabutylphosphonium
$[P_{6\ 6\ 6\ 14}]^+$	Trihexyl(tetradecyl)phosphonium
$[H_3NEt]^+$	Ethylammonium
$[HN_{1\ 1\ 1}]^+$	Trimethylammonium
$[HN_{2\ 2\ 2}]^+$	Triethylammonium
$[N_{1\ 1\ 1\ 4}]^+$	Butyltrimethylammonium
$[N_{1\ 1\ 1\ 6}]^+$	Hexyl(trimethyl)ammonium
$[N_{2\ 2\ 1\ m}]^+$	*N*,*N*-diethyl-*N*-methyl-*N*-(2-methoxyethyl)ammonium
$[N_{2\ 2\ 2\ 4}]^+$	Butyl(trimethyl)ammonium
$[N_{2\ 2\ 2\ 6}]^+$	Hexyl(trimethyl)ammonium
$[N_{3\ 3\ 3\ p}]^+$	Trimethylphenylammonium
$[N_{4\ 4\ 4\ 4}]^+$	Tetrabutylammonium
$[N_{4\ 4\ 4\ 6}]^+$	Tributyl(hexyl)ammonium
$[N_{6\ 6\ 6\ 6}]^+$	Tetrahexylammonium
$[N_{6\ 6\ 6\ 14}]^+$	Trihexyl(tetradecyl)ammonium
$[N_{1\ 8\ 8\ 8}]^+$	Methyl(trioctyl)ammonium
$[N_{8\ 8\ 8\ 8}]^+$	Tetraoctylammonium
$[N_{1\ 12\ 12\ 12}]^+$	Tridodecylmethylammonium
$[CTMA]^+$	*N*,*N*,*N*,*N*-cyanomethyl trimethyl ammonium
$[dema]^+$	Diethylmethylammoium
$[dmea]^+$	Dimethylethylamine
$[DEME]^+$	*N*-methyl-*N*-(2-methoxyethyl)ammonium
$[NHEt(i\text{-}Pr)_2]^+$	*N*,*N*-diisopropylethylamine

(continued)

$[H_3N(CH_2CH_2OH)]^+$	Ethanolammonium
$[H_2N(CH_2CH_2OH)_2]^+$	Bis(2-hydroxyethyl)ammonium
$[(HOCH_2CH_2)N_{1\ 1\ 1}]^+$	Choline
$[CH_3OCH_2CH_2\text{-Et-}Me_2N]^+$	(2-Methoxy)ethyl ethyl dimethylammonium
$[CH_3OCH_2CH_2\text{-}Et_2\text{-MeN}]^+$	(2-Methoxy)ethyl diethyl methylammonium
$[\text{Bu-}Et_2\text{-MeN}]^+$	Butyl diethyl methylammonium
$[Hpy]^+$	Pyridinium
$[C_4py]^+$	1-Butylpyridinium
$[C_4m_\beta py]^+$	1-Butyl-3-methylpyridinium
$[C_8py]^+$	1-Octylpyridinium
$[C_{18}py]^+$	1-Octadecylpyridinium
$[HmimSC_4]^+$	1-Methyl-2-butylthiolonium
$[S_{1\ 1\ 1}]^+$	Trimethylsulfonium
$[S_{2\ 2\ 1}]^+$	Diethyl(methyl)sulfonium
$[S_{2\ 2\ 2}]^+$	Triethylsulfonium
$[CDBA6]^+$	*N,N′*-(*trans*-cyclohexane-1,4-diyl)-dibenzamide
$[\text{H-betaine}]^+$	Betainium
$[PD]^{3+}$	1,3,5-Tris(3-((ferrocenylmethyl)amino)pyridiniumyl)-2,4,6-triethylbenzene
$[(N_{111})(N_{112})BH_2]^+$	(Trimethylamine)-(dimethylethylamine)dihydroborate

Anions

$[FSI]^-$	Bis(fluorosulfonyl)imide
$[BETI]^-$	Bis(perfluoroethylsulfonyl)imide
$[N(Tf)_2]^-$	Bis(trifluoromethylsulfonyl)imide
$[C_4C_4N]^-$	Bis(nonafluorobutylsulfonyl)imide
$[NTf]^-$	Brifluoromethanesulfonate
$[N(CN)_2]^-$	Bicyanamide
$[C(CN)_3]^-$	Tricyanomethanide
$[PF_6]^-$	Hexafluorophosphate
$[NO_3]^-$	Nitrate
$[B(CN)_4]^-$	Tetracyanoborate
$[BF_4]^-$	Tetrafluoroborate
$[CF_3BF_3]^-$	Trifluoromethyltrifluoroborate
$[BSB]^-$	Bis(salicylate[2-])borate
$[TFPB]^-$	Tetrakis(3,5-trifluoromethylphenyl)borate anion
$[SCN]^-$	Thiocyanate
$[HCO_2]^-$	Formate
$[CF_3CO_2]^-$	Trifluoroacetate
$[CH_3CO_2]^-$ or $[Ac]^-$	Acetate

(continued)

$[C_9H_{19}CO_2]^-$	Decanoate
$[CF_3SO_3]^-$	Trifluoromethanesulfonate
$[CH_3SO_3]^-$	Methanesulphonate
$[HSO_4]^-$	Hydrogen sulfate
$[MeSO_4]^-$	Methylsulphate
$[EtSO_4]^-$	Ethyl sulphate
$[C_8H_{17}SO_4]^-$	Octylsulfate
$[FAP]^-$	Trifluorotris(pentafluoroethyl)phosphate
$[R_2PO_4]^-$	Dialkylphosphate
$[Me_2PO_4]^-$	Dimethylphosphate
$[(EH)_2PO_4]^-$	Bis(2-ethylhexyl)phosphate
$[TPTP]^-$	Tris(pentafluoroethyl)trifluorophosphate
$[(i\text{-}C_8)PO_2]^-$	Bis(2,4,4-trimethylpentyl)phosphinate
$[DPP]^-$	Diphenylphosphate
$[DBP]^-$	Dibutylphosphate
$[EtPO_3H]^-$	Ethylphosphonate
$[(i\text{-}C_4)_2PS_2]^-$	Diisobutyldithiophosphinate
$[(OC_2)_2PS_2]^-$	O,O′-diethyldithiophosphate
$[OH]^-$	Hydroxide
$[I]^-$	Iodide
$[Cl]^-$	Chloride
$[CS]^-$	Camphorsulfonate
$[SiF_6]^{2-}$	Hexafluorosilicate
$[SbF_6]^-$	Hexafluoroantimonate
$[Cu_2Cl_3]^-$	Copper chloride
$[Cu_2Br_3]^-$	Copper bromide
$[AlCl_4]^-$	Tetrachloroaluminate
$[tos]^-$	Tosylate
$[Adip]^{2-}$	Adipate
$[Succ]^{2-}$	Succinate

Chemical Abbreviations

AA	Ascorbic acid
AC	Activated carbons
ACN	Acetronitrile
Au	Gold
γ-BL	Gamma butyrolactone
Bu_4N	Tetrabutylammonium
Cc^+	Cobaltocenium
Chi	Chitosan

(continued)

cis-Mn	*cis*-[Mn(CN)$(CO)_2$[P$(oph)_3$](DPM)] (DPM = $Ph_2PCH_2PPh_2$
cis-W	*cis*-[W$(CO)_2$$(DPE)_2$] (DPE$_5$$Ph_2PCH_2CH_2PPh_2$)
CNTs	Carbon nanotubes
Cp	Cyclopentadienyl
DEC	Diethyl carbonate
DMC	Dimethyl carbonate
DMF	Dimethyl formamide
DmFc	Decamethylferrocene
DMSO	Dimethyl sulfoxide
EAN	Ethylammonium nitrate
EC	Ethylene carbonate
EMC	Ethylmethyl carbonate
Fc	Ferrocene
Fc^+	Ferrocenium
FTO	Fluorine-doped tin oxide
HOPG	Highly ordered pyrolytic graphite
ITO	Indium tin oxide
LTO	Lithium titanium oxide
MAc	*N*-methylacetamide
MEA	Monoethanolamine
MeCN	Acetonitrile
MWCNT	Multiwall carbon nanotubes
NADH	Nicotinamide adenine dinucleotide
NHC	*N*-heterocyclic carbenes
OMCs	Ordered mesoporous carbons
PC	Propylene carbonate
	Polycarbonate
PEDOT	Poly(3,4-ethylenedioxythiophene)
PEG	Polyethylene glycol
PS	Polystyrene
Pt	Platinum
PVA	Polyvinyl alcohol
PVF	Polyvinylferrocene
RE	Rare earth
SAPG	Stress-annealed pyrolytic graphite
$scCO_2$	Supercritical carbon dioxide
SWCNT	Single wall carbon nanotubes
trans-Mn	*trans*-[Mn-(CN)$(CO)_2$\{P$(oph)_3$\}($Ph_2PCH_2PPh_2$)]

Abbreviations

AC	Alternating current
AFM	Atomic force microscopy
AIL	Aprotic ionic liquid
ANS	Analyte
ATR	Attenuated total reflectance
BDD	Boron doped diamond
bp	Boiling point
CGT	Complexed gas technology
CILEs	Carbon ionic liquid electrodes
CP	Conducting polymer
CPE	Carbon paste electrodes
CPET	Concerted proton electron transfer
CV	Cyclic voltammetry
CVD	Chemical vapor deposition
DES	Deep eutectic solvents
DLC	Diamond-like coatings
DLS	Dynamic light scattering
DRIFTS	Diffuse reflectance infrared Fourier-transform spectroscopy
DSSC	Dye-sensitized photoelectrochemical solar cells
E-ALD	Electrochemical atomic layer deposition
E-ALE	Electrochemical atomic layer epitaxy
ECC	Electrocapillary curve
ECL	Electrochemiluminescence
ECM	Electrocapillary maximum
EDL	Electrical double layer
EDLC	Electric double layer capacitors
EDR	Equivalent diffusion resistance
EDX	Energy dispersive X-ray
EE	Electrochemical mechanism with sequential electron transfer
EIS	Electrochemical impedance spectroscopy
EPT	Electron proton transfer
EQCM	Electrochemical quartz crystal microbalance
ESR	Equivalent series resistance
ESW	Electrochemical stability windows
ET	Electron transfer
EVs	Electric vehicles
EW	Electrochemical potential windows
FTIR	Fourier transform infrared spectroscopy
GC	Glassy carbon
HBD	Hydrogen bond donor
HER	Hydrogen evolution reaction

(continued)

HEVs	Hybrid electric vehicles
HFCs	Hydrogen fuel cell
HOR	Hydrogen oxidation reaction
IDE	Interdigitated electrode
IL, RTIL	Ionic liquid
ILEs	Ionic liquid-based electrolytes
IL-G	Ionic liquid-graphene composite film
IR	Infrared spectroscopy
IRRS	Internal reference redox systems/internal reference redox scale
IS	Immittance spectroscopy
JCPDS	Joint committee on powder diffraction standards
LED	Light-emitting diode
MBE	Molecular beam epitaxy
MEP	Molecular electrostatic potential
MFT	Mean field theory
mp	Melting point
NADES	Natural deep eutectic solvents
NHE	Normal hydrogen electrode
NMR	Nuclear magnetic resonance
NP	Nanoparticle
OCP	Open circuit potential
OER	Oxygen evolution reaction
OLEs	Organic solvent-based electrolytes
OPD	Overpotential deposition
ORR	Oxygen reduction reaction
OX_{ANS}	Oxidised redox component of the ANS
OX_{IRRS}	Oxidised redox component of the IRRS
QRE	Quasi-reference (or pseudo-reference) electrodes
PDDF	Pair distance distribution function
PEMFC	Polymer electrolyte membrane fuel cells
PET	Proton electron transfer
PFIL	Polyelectrolyte-functionalized ionic liquid
PILs	Protic ionic liquids
ppb	Part per billion
ppm	Part per million
PVD	Physical vapor deposition
PZC	Potential of zero charge
RDE	Rotating disk electrode
RED_{ANS}	Reduced redox component of the ANS
RED_{IRRS}	Reduced redox component of the IRRS
RFB	Redox flow batteries
rpm	Revolutions per minute
RRDE	Rotating ring-disk electrode
RTMS	Room-temperature molten salts

(continued)

SAS	Small angle scattering
SAXS	Small angle X-ray scattering
SCE	Saturated calomel electrode
SECM	Scanning electrochemical microscopy
SEM	Scanning electron microscopy
SFG	Sum frequency generation spectroscopy
SHG	Second harmonic generation
SPCE	Screen-printed carbon electrode
STM	Scanning tunneling microscope
STS	Scanning tunneling spectroscopy
ToF-SIMS	Time-of-flight secondary ion mass spectrometry
TSIL	Task-specific ionic liquids
UHV	Ultrahigh vacuum
UMEs	Ultramicroelectrodes
UPD	Underpotential deposition
UV	Ultraviolet
UV-Vis	Ultraviolet–visible
VOCs	Volatile organic compounds
VSFS	Vibrational sum frequency spectroscopy
WAXS	Wide angle X-ray scattering
XFEL	X-ray free electron laser
XPCS	X-ray photon correlation spectroscopy
XPS	X-ray photoelectron spectroscopy
XRD	X-ray diffraction
ZIL	Zwitterionic liquid

Roman Symbols

A	Electrode area
a	Microdisk electrode radius
a_i	Activity of ions
B	Porod scaling factor
C	Capacitance
	Coulomb
c	Bulk ionic concentration (also referred as C)
c_-	Concentration of anion
c_+	Concentration of cation
C_0	Capacitance at the minimum (PZC)
C_d	Double layer capacitance
C_{sp}	Specific capacitance
D	Diffusion coefficient

(continued)

D_O	Diffusion coefficient of an oxidised redox component
$D_n(R)$	Number distribution function
$D(q)$	Effective diffusion constant
D_R	Diffusion coefficient of a reduced redox component
$D(\lambda)$	Detector efficiency
d	Tip-substrate distance
	Electrolyte density
	Repeat distance
d_i	Diameter of an ion
d_m	Mass fractal dimension
d_s	Surface fractal dimension
e	Charge on the electron
E	Energy
	Potential
E'	Electric field of the visible and infrared incoming beams
$E^{\ominus\prime}$	Formal potential (also referred as $E^{0\prime}$)
$E_{1/2}$	Measured or expected half-wave potential in voltammetry
$E_{1/4}$	Potential where $I/I_d = 1/4$
$E_{3/4}$	Potential where $I/I_d = 3/4$
E_a	Activation energy
E_{end}	Ending potential in a voltammetric scan
E_m	Mid-point potential
ΔE_m	Distance between the E_m of two redox processes
E_T^N	Normalized charge transfer
ΔE_p	Peak-to-peak potential separation
E_{rev}	Reversing potential in a voltammetric scan
E_{start}	Starting potential in a voltammetric scan
F	Faraday constant-the number of coulombs per mole of electrons (F = 96485 C mol^{-1})
$F(q,..)$	Form factor
f	Scattering power of a free electron (excluding anomalous effects)
f_0	Fundamental frequency
Δf	Frequency shift
G	Guinier scaling factor
$\Delta G°$	Gibbs free energy
g	Gravitational force
$g_2(q,\tau)$	Normalized second order intensity correlation function
$H°$	Enthalpy
$H(q)$	Hydrodynamic interaction
h	Hour
I	Faradaic current (also referred as i)
I_o	Incident intensity flux
I_d	Current under diffusion control
I_e	Intensity scattered by a single electron
I_p	Peak current (also referred as i_p)

(continued)

$I(q)$	Measured intensity from a small angle scattering experiment
i_{ss}	Steady-state mass transport limited current (also referred as I_L)
i_T	Tip current
J	Raman scattering coefficient
j	Current density (also referred as J)
K	Binding constant
$k^{0'}$	Heterogeneous charge transfer rate constant (cm^2 s)
k_B	Boltzmann constant
K_{cross}	Equilibrium constant for a simple outer-sphere second-order cross-reaction
k_{cross}	Rate constant for a simple outer-sphere second-order cross-reaction
k_d	Homogeneous rate constant for "forward" reaction (the unit depends on reaction order)
L	Dimension of any inhomogeneity
M	Convolved/semiintegrated current
M_L	Mass transport limited convolved/semiintegrated current
m	Mass transfer coefficient
Δm	Mass changes
N	Number density of scatterers
n	Number of moles of electrons
	Integer
P	Pressure
	Real space vector between two inhomogeneities
Q	Charge
Q_{inv}	Scattering invariant
q	Scattering vector
R	Resistance
	Universal gas constant (8.314 J K^{-1} mol^{-1})
	Radius
R_{ct}	Charge transfer resistance
R_g	Radius of gyration
Rg_{CO}	Radius of gyration cutoff (unified equation)
R_{HS}	Hard sphere radius
R_s	Solution resistance
R_u	Uncompensated resistance
r	Real space vector between two points
	Radial distance from the center of an electrode (cm)
r_0	Electrode radius
r_g	Radius of the sheath
S	Surface area
	Siemens
$S°$	Entropy
$S(q,...)$	Structure factor
s, s_o	Unit vector of scattered and incoming X-rays, respectively
T	Temperature
T_g	Glass transition temperature
T_m	Melting point

(continued)

Tr	Transmission
t_s	Material thickness
V	Voltage (also referred as U)
ν	Particle volume fraction
ν_{p}	Volume occupied by a single point
W	Warburg element
w	Mass fraction
x	Distance from a planar electrode
x_2	Distance between the electrode surface and the inner Helmholtz plane

Greek Symbols

α	Electron transfer coefficient
$\beta(q)$	Optical transfer function
γ	Decay exponent
γ_{s}	Interfacial tension
$\gamma_{\mathrm{o}}(r)$	Normalized auto-correlation function
ε	Dielectric constant
	Permittivity
ε_{r}	Relative permittivity
η	Efficiency of the process
	Viscosity
η^{-1}	Fluidity
$\overline{\eta^2}$	Square of the volume average contrast
Λ	Molar conductivity
λ	Wavelength of light (nm)
μ_q	Quartz shear modulus,
μ/ρ	Mass absorption coefficient
υ	Scan rate
ρ	Scattering length density
ρ_{q}	Quartz density
ρ_s	Mass density of a material
$\overline{\rho}^2$	Auto-correlation of scattering length density function
$\rho^2(\lambda)$	Scattering power (includes absorption)
σ	Ionic conductivity
	Standard deviation
τ	Cell time constant
	Time delay between two measurements
ϕ	Phase
ω	Electrode angular velocity

Index for First Volume

A.A.J. Torriero (ed.), *Electrochemistry in Ionic Liquids*,
DOI 10.1007/978-3-319-15132-8

Index for Second Volume

A.A.J. Torriero (ed.), *Electrochemistry in Ionic Liquids*,
DOI 10.1007/978-3-319-15132-8

MIX
Papier aus verantwortungsvollen Quellen
Paper from responsible sources
FSC® C105338

If you have any concerns about our products,
you can contact us on
ProductSafety@springernature.com

In case Publisher is established outside the EU,
the EU authorized representative is:
Springer Nature Customer Service Center GmbH
Europaplatz 3, 69115 Heidelberg, Germany

Printed by Libri Plureos GmbH
in Hamburg, Germany